JOINING TECHNOLOGIES

Edited by **Mahadzir Ishak**

Joining Technologies
http://dx.doi.org/10.5772/61698
Edited by Mahadzir Ishak

Contributors

Marcello Cabibbo, Archimede Forcellese, Michela Simoncini, Thomas Gietzelt, Volker Toth, Andreas Huell, Simona Jevšnik, Savvas G. Vassiliadis, Dragana Grujić, Zoran Stjepanovič, Senem Kurşun Bahadır, Kirk Fraser, Laszlo Kiss, Lyne St-Georges, Raoelison Rija, Thaneshan Sapanathan, Nicolas Buiron, Mohamed Rachik, Ruham Pablo Reis, Thiago Larquer, Jerzy Winczek, Junling Hu, Zhenghua Rao, Hai-Lung Tsai, Gleb Andreevich Turichin, Uday Basheer, Alias Mohd Noor, Srithar Rajoo, Ahmad-Fauzi M. N., Mahadzir Ishak

Published by InTech

Janeza Trdine 9, 51000 Rijeka, Croatia

Publishing Process Manager Andrea Koric

Technical Editor SPi Global

Cover InTech Design team

First published September, 2016

Printed in Croatia

Additional hard copies can be obtained from orders@intechopen.com

Joining Technologies, Edited by Mahadzir Ishak

p. cm.

Print ISBN 978-953-51-2596-9
Online ISBN 978-953-51-2597-6

For EU product safety concerns:
IN TECH d.o.o., Prolaz Marije Krucifikse Kozulić 3, 51000 Rijeka, Croatia,
info@intechopen.com or visit our website at intechopen.com.

Meet the editor

Associate Professor Dr. Mahadzir Ishak received his undergraduate degree at the Department of Mechanical Engineering,Ehime University, Japan, in 2000. In 2002, he obtained his master's degree from Tokushima University, Japan. He worked as an engineer at welding shop in automotive-related company before becoming academician at Universiti Malaysia Pahang (UMP). He obtained his PhD in Mechanical Engineering from Ibaraki University, Japan, in 2010. Since 2007, he has been involved in welding research such as laser welding on light alloys, arc welding on dissimilar metals, friction-stir welding, as well as other joining technologies. He teaches joining and welding, manufacturing processes, and material engineering courses for undergraduate students at UMP. He has authored three book chapters and has published more than 38 papers in scientific journals. He held more than 40 presentations at international conferences. Currently, he supervises four PhDs and eleven masters by research students related to joining and welding research.

Contents

Preface

Joining is one of the most important processes in manufacturing. This technology has vastly improved and is now extensively used in numerous industries. It is important because many products are impossible to be produced in a single piece and the cost of a product is cheaper when they are fabricated as many individual components prior to final assembly.

One of the advantages of joining is that the product can be disassembled for repairs. Joining technology can be classified into three main processes such as metallurgical process (welding, soldering, brazing, heat application), mechanical processes (riveting, bolt and nut, screw, key), and chemical process (adhesion). This book, however, focuses on joining technology involved with metallurgical process especially welding process and application of joining technique for textile materials.

This book is intended to share recent research and knowledge related to joining technology especially welding process. This book contains 11 chapters. Chapter 1 is an introduction to joining technology, it briefly explains the welding process and gives the overview of all topics covered in this book. Chapter 2 describes new approaches in joining of weld sheet aluminum alloys using friction-stir welding (FSW), and Chapter 3 introduces a new approach to simulate friction-stir welding (FSW) process by smoothed-particle hydrodynamics (SPH). Chapters 4 and 5 are related to a method of controlling arc motion using the magnetic oscillation method of GTAW and development of a comprehensive two-dimensional model for GMAW using volume of fluid approaches, respectively.

Chapter 6 describes model approaches for temperature field and phase transformation analysis of butt weld using arc welding. Chapter 7 discusses the uniqueness of laser and hybrid laser-arc welding on metals and its potential application in the industry. Chapter 8 explains various studies that have been conducted on joining between ceramic and metals as well as its challenges. Chapter 9 deals with the technology of diffusion for welding application.

Chapter 10 describes the potential uses of magnetic pulse welding in industries. It describes the magnetic pulse welding process, its potential application, its weldability on joining metal as well as numerical simulation works on its interface behavior and multi-physics. Finally, Chapter 11 explores methods such as fusion, hot air, and hot wedge welding for joining textile product, basic knowledge and working principle of these technologies, as well as application opportunities.

We hope that the knowledge and updates of recent research achievements on joining technology that we have shared in this book will be useful for researchers, engineers, students, and others that work in the joining-related area.

I would like to express my thanks to all authors who contribute their findings related to joining technology in this book. Without their expertise and contributions, this book could not have been compiled.

Dr. Mahadzir Ishak
Faculty of Mechanical Engineering,
Universiti Malaysia Pahang,
Pahang, Malaysia

Chapter 1

Introductory Chapter: A Brief Introduction to Joining and Welding

Mahadzir Ishak

Additional information is available at the end of the chapter

http://dx.doi.org/10.5772/64726

1. Introduction

Joining is an important process in a number of industries, such as aerospace, automotive, oil, and gas. Many products cannot be fabricated as a single piece, so components are fabricated first and assembled later. Joining technology can be classified as a liquid-solid-state process and mechanical means. Liquid-solid-state joining includes welding, brazing, soldering, and adhesive bonding. Mechanical joining includes fasteners, bolts, nuts, and screws.

Metal joining is a process that uses heat to melt or heat metal just below the melting temperature. Joining metal by fusion is known as fusion welding. Without fusion, the process is known as solid-state welding. Fusion welding includes arc welding and laser welding, Whereas solid-state welding such as friction stir welding (FSW) where process occurred below the melting temperature.

2. Fusion welding (arc welding)

Fusion welding is known as non-pressure welding, in which edge samples to be joined with the filler metal are heated above the melting points to create a weld pool and allow solidification. Gas tungsten arc welding (GTAW) and gas metal arc welding (GMAW) are categorized under fusion welding. GTAW and GMAW are mostly used by the welder to weld both ferrous and non-ferrous metals. In fusion welding, inert gases, such as argon (Ar), helium (He), and carbon dioxide (CO_2), are used for surrounding the electrode and molten metal from the welded metal. These inert gases will eliminate the formation of metal oxides and nitrides, which can lower the ductility and toughness of the welded metal.

3. GMAW

GMAW is also known as metal inert gas (MIG) welding in which an external gas, such as argon, helium, carbon dioxide, argon + oxygen, and other gas mixtures, is used as a shielding gas [1]. Consumable electrode wire, having the same or approximate chemical composition to that of parent metal, is continuously fed from a spool to the arc zone. The arc from the welding parameters (voltage and current) heats and melts the samples' edges and the filler wire. The fused filler metal is supplied to the surface of the workpiece, fills the weld pools, and forms the joint between the workpieces similarly or dissimilarly. The overall process in GMAW is described as a semi-automatic method because of the automatic feeding of the filler rod while the welder controls only the position and speed of the torch. GMAW can weld almost all metals and alloys, aluminum alloys, and stainless steel [2].

4. GTAW

GTAW is also known as tungsten inert gas (TIG) welding, in which heat from an electric arc is used. The arc sparks between a tungsten non-consumable electrode and the workpiece [1]. The molten pool is shielded by an inert gas such as argon, helium, and nitrogen. The shielding gas prevents the molten pool from atmospheric contamination. The heat produced by the arc melts the samples' edges. The filler rod can be used if required, especially in welding aluminum. GTAW produces a high quality weld of most metals because it does not use flux. An externally supplied shielding gas is necessary because of high temperatures involved to prevent metal from oxidation. Direct current is typically used, and its polarity is important as this welding method still uses current and voltage as critical parameters. Given that the tungsten electrode is not consumed during welding, a stable and constant arc is preserved at a constant current level. The filler metals used are usually similar to the parent metals to be welded, without using flux. The shielding gas used is normally argon or helium (or a mixture of gases). GTAW is used for a wide variety of metals and applications. Metals that usually can be welded by GTAW are aluminum, magnesium, titanium, and copper and its alloy. The tungsten electrode is usually in contact with a water-cooled copper tube (contact tube), which is connected to the welding cable from the terminals. Both the weld current and electrode must be cooled to avoid overheating during welding.

5. Laser welding

Laser welding has shown remarkable progress as a high-efficiency welding technique through the years. The process of Laser welding for metal is based on melting metal under a highly concentrated beam of radiation that is focused on the surface metal to join two parts. Radiation is partially absorbed by the upper layer of the metal, causing it to heat to the melting point. The important processing parameters involved in laser welding include laser properties (average and peak power, beam quality, beam diameter, wavelength, and focal length), weld setting (focus position toward the material surface, weld type, and shielding gas), and physical properties of the parent metal. There are two types of welding area, namely conduction or keyhole mode. The obvious width and depth difference in this welding area is due to the energy *E* and peak power density *PPD* applied.

Laser welding has many advantages over the conventional joining method, such as deep penetration, low heat input, small heat-affected zone (HAZ), and high speed. In terms of production, some of the advantages of laser welding are high speed, high process productivity, flexibility in control, and automation. Three common types of laser machines, namely CO_2, Nd:YAG, and fiber lasers, are widely used in the industry for welding purposes. CO_2 is known as a gas laser with a wider wavelength compared with solid-state lasers Nd:YAG and fiber lasers. Unlike solid-state lasers, the wide wavelength of the CO_2 laser results in poor absorption by a wide range of materials. Meanwhile, the fiber laser presents several advantages over the Nd:YAG laser because of the former's compact design, good beam quality, and low cost of ownership and maintenance.

6. FSW

FSW is a welding process that involves solid-state joining; this process has expanded rapidly since its development in 1991 by The Welding Institute, UK [3–6]. FSW is a solid-state welding technique that does not involve melting and occurs below the melting point. It uses a rotating tool to generate necessary heat for welding. This tool consists of three parts: the shank, shoulder, and pin. The shank is the part where the tool is attached to the FSW machine, whereas the shoulder and pin are attached to the workpiece. The shoulder and pin provide additional frictional treatment and prevent the plasticized material from escaping from the weld region. During FSW, the rotating tool moves along the joint of two plates that generate heat. This tool then recirculates flow of the plasticized material near the tool surface. The size of the tool shoulder is larger than that of the pin tool. The FSW tool serves two main functions, namely workpiece heating and material movement to produce a joint [4]. Heating is produced by friction of the pin and workpiece and plastic deformation of the workpiece. The heat that is produced will soften the material around the pin, and tool rotation will move the material from the front of the pin to the back of the pin. The result of this process is a joint produced in solid state.

FSW can be utilized in a wide variety of industries, such as automotive, aerospace, maritime, and railway [3, 4, 7–11]. FSW has been considered the most substantial joining process in the past decade because it offers many advantages such as energy efficiency, environmental friendliness, and versatility [4]. Compared with arc welding, FSW uses less energy and does not require a shielding gas and flux, thereby making this process an eco-friendly one. This joining process does not need any filler, so it is suitable to join many types of dissimilar metals. FSW is a technique that can avoid drawbacks from common fusion welding because FSW can be conducted under solid state. Several problems (e.g., spatter, hot cracking, and distortion) in other types of welding are eliminated by using FSW [12]. Defects such as voids, lack of penetration, and broken surface can be minimized by using this welding technique.

7. Overview of the chapters

Chapter 2: “New approaches to the Friction Stir Welding of aluminum alloys” written by Marcello Cabibbo, Archimede Forcellese, and Michela Simoncini. The main contribution of this chapter is the presentation of two new methods to weld aluminum alloys sheets by using FSW.

Chapter 3: “A mesh-free solid mechanics approach for simulating the friction stir welding process” written by K. Fraser, L. St-Georges, and L. I. Kiss. The main contribution of this chapter is the introduction of a new approach to simulate FSW by smoothed particle hydrodynamics (SPH). This approach can determine elastic and plastic deformation, residual stresses, temperature, and material flow in the same model.

Chapter 4: “Gas Tungsten Arc Welding with Synchronized Magnetic Oscillation” written by Thiago Resende Larquer and Ruham Pablo Reis. This chapter describes a method of controlling arc motion using the magnetic oscillation method of GTAW. The good coordination of magnetic oscillation and the welding process can influence the delivery of arc energy to the welded metal, thereby controlling the formation of weld bead.

Chapter 5: “A Comprehensive mode of the transport phenomena in Gas Metal Arc Welding” written by J. Hu, Z.H. Rao, and H.L. Tsai. This chapter explains the development of a comprehensive two-dimensional GMAW model, which considers the effect of arc plasma, electrode condition, droplet formation, detachment transfer, impingement onto the workpiece and the weld pool, and weld formation. This model uses volume of fluid approaches to track the free surface, which can eliminate the requirement of boundary condition at the interface.

Chapter 6: “The analysis of temporary temperature field and phase transformation in one-side butt welded of steels flats” written by Jerzy Winczek. This chapter describes model approaches for temperature field and phase transformation analysis of butt welding. This model is verified by metallographic observation of the butt weld workpiece, which was welded using an arc welding machine.

Chapter 7: “Laser and Hybrid laser-arc welding” written by G.A. Turichin. This chapter explains the technology of laser and hybrid laser-arc welding. It discusses the uniqueness of laser and hybrid laser-arc welding on metals and its potential application in the industry.

Chapter 8: “Current issues and problems in the joining of ceramic to metal” written by Uday M.B., Alias Mohd Noor, and Srithar Rajoo. This chapter describes the challenges of joining between ceramics and metals. These two metals have significantly different properties, so joining of these materials is difficult. This chapter explains various studies that have been conducted on joining between ceramic and metals.

Chapter 9: “Diffusion Bonding: Influence of Process Parameters and Material Microstructure” written by Thomas Gietzelt, Volker Toth, and Andreas Huell. This chapter deals with the technology of diffusion for welding application. The parameters that influence mechanical properties and the microstructure, as well as those involved in diffusion welding, are discussed.

Chapter 10: "Applying Heat for Joining Textile Materials" written by Simona Jevšnik, Senem Kurson Bahadir, Dragana Grujić, and Zoran Stjepanovič. This chapter explains the application of joining technology in the textile industry. It explores methods such as fusion, hot air, and hot wedge welding for joining textile. The basic knowledge and working principle of these technologies, as well as application opportunities, are presented in this chapter.

Chapter 11: "Magnetic Pulse Welding: An Innovative Joining Technology for Similar and Dissimilar Metal Pairs" written by T. Sapanathan, R.N. Raoelison, N. Buiron, and M. Rachik. This chapter focuses on magnetic pulse welding process, its potential requirements, interfacial kinematics of the welding, weld features as well as interfacial behaviors, and multi-physics numerical simulations. The magnetic pulse welding is recognized as one of the promising joining method to weld similar and dissimilar metal, which provides many attractive advantages.

Author details

Mahadzir Ishak

Address all correspondence to: mahadzir@ump.edu.my

Faculty of Mechanical Engineering, Universiti Malaysia Pahang, Pahang, Malaysia

References

[1] Please check and confirm the inserted publisher location for Refs. [1 and 2].S. Kalpakjian and S. R. Schmid, Manufacturing, Engineering & Technology, sixth edition in SI Units, pp. 869–875, Upper Saddle River: Prentice Hall, Pearson, 2010.

[2] S. Kou, Welding Metallurgy, second edition, pp. 3–26, New York: Wiley, 2002 (A Wiley-Interscience Publication).

[3] B. T. Gibson, D. H. Lammlein, T. J. Prater, W. R. Longhurst, C. D. Cox, M. C. Ballun, K. J. Dharmaraj, G. E. Cook, and A. M. Strauss, "Friction stir welding: process, automation, and control," J. Manuf. Process.vol 16, pp. 1"18, Jun. 2013

[4] R. S. Mishra and Z. Y. Ma, "Friction stir welding and processing," Mater. Sci. Eng. R Rep., vol. 50, no. 1–2, pp. 1–78, Aug. 2005.

[5] D. Burford, C. Widener, and B. Tweedy, "Advances in friction stir welding for aerospace applications," 6th AIAA Aviation Technology, Integration and Operations Conference (ATIO), vol. 6 no. 14, pp. 3–7, 2006.

[6] W. J. Arbegast, "Friction stir joining: characteristic defects," Advanced Materials Processing Center MET,vol. 6,pp. 1-30October, 2003.

[7] X. Xu, X. Yang, G. Zhou, and J. Tong, "Microstructures and fatigue properties of friction stir lap welds in aluminum alloy AA6061-T6," Mater. Des., vol. 35, pp. 175–183, Mar. 2012.

[8] N. Shanmuga Sundaram and N. Murugan, "Tensile behavior of dissimilar friction stir welded joints of aluminium alloys," Mater. Des., vol. 31, no. 9, pp. 4184–4193, Oct. 2010.

[9] P. Pourahmad and M. Abbasi, "Materials flow and phase transformation in friction stir welding of Al 6013/Mg," Trans. Nonferrous Met. Soc. China, vol. 23, no. 5, pp. 1253–1261, May 2013.

[10] R. S. Coelho, A. Kostka, J. F. dos Santos, and A. Kaysser-Pyzalla, "Friction-stir dissimilar welding of aluminium alloy to high strength steels mechanical properties and their relation to microstructure," Mater. Sci. Eng. A, vol. 556, pp. 175–183, Oct. 2012.

[11] M. Assidi, L. Fourment, S. Guerdoux, and T. Nelson, "International Journal of Machine Tools & Manufacture Friction model for friction stir welding process simulation: calibrations from welding experiments," Int. J. Mach. Tools Manuf., vol. 50, no. 2, pp. 143–155, 2010.

[12] B. Fu, G. Qin, F. Li, X. Meng, J. Zhang, and C. Wu, "Friction stir welding process of dissimilar metals of 6061-T6 aluminum alloy to AZ31B magnesium alloy," J. Mater. Process. Tech. vol. 218, pp 38-47 , 2014.

New Approaches to the Friction Stir Welding of Aluminum Alloys

Marcello Cabibbo, Archimede Forcellese and
Michela Simoncini

Additional information is available at the end of the chapter

http://dx.doi.org/10.5772/64523

Abstract

Friction stir welding (FSW) is a technique able to guarantee welding advantages such as the easy control of tool design, rotation speed, and translation speed. This is also a reason for a continuous research activity to optimize the effect of the different welding parameters and tool-metal setups. In this contribution, two innovative welding methodologies are presented and discussed. A first new FSW configuration was defined as double-side friction stir welding (DS-FSW). In the DS-FSW, the welding is performed on both sheet surfaces, that is, the first welding is followed by a second one performed on the opposite sheet surface. In this chapter, the effect of the welding parameters, tool configuration and sheet positioning on the yield, ultimate strength, and ductility of an aluminum plate, its microstructure and its post-welding formability are discussed. A second new FSW configuration consists of a pin rotation around its centerline welding direction by 0.5 and 1.0 mm. This was defined by authors as RT-type configuration and it is characterized by a welding motion of the pin tool obtained by the combination of two different movements occurring simultaneously.

Keywords: FSW, tensile strength, ductility, LDH, FLC, hardness

1. Introduction

The continuing scientific and technological attention to reduce vehicle weight and emissions resulted in a diffuse use of Al alloys as a substituting metal with respect to steel, in many applications that were formerly dominated by steel. This change of technology metallic material has strongly promoted tailor-welded blanks new solutions especially in the vehicle applications [1, 2]. Laser-beam welding and friction-stir welding (FSW) are currently considered to be the most prospective welding processes.

Anyhow, the common difficulties involved in laser welding of aluminum alloys include porosity, hot cracking, poor coupling (due to the high reflectivity of the metal), and degradation of the material properties in the heat-affected zone (HAZ) [3–5]. Despite several advantages offered by laser beam welding applied to aluminum alloys, this welding technology usually suffers from seam imperfections such as notches, which reduce the mechanical properties of the joint [5]. In order to overcome such drawbacks, the friction stir welding (FSW) has strong potentials against the laser beam welding, as it is a solid-state welding technology [6, 7]. In this sense, FSW is surely considered to be the most significant development in the metal joining techniques over the past two-to-three decades. The nonwelded nugget zone (NZ) makes this welding technology an energy-effective one. It is also an environmentally friend and a versatility welding technique often considered as a "green" technology. In fact, compared to the fusion welding processes, FSW consumes less energy with very low fraction of wasted material and a drastic reduction of dangerous fumes production [3–7].

Moreover, FSW produces a high-quality joint, compared to other conventional fusion welding processes. It is also a welding process particularly suited for joining nonmetal materials to metals, especially in those cases where it is not possible by using conventional fusion methods [8, 9]. Its key factors and main properties consist of the welding nature of the FSW metals. The weld zone undergoes a solid-state process promoted by the frictional heat between a rotating tool and the welding metal. The plasticized zone, induced in the material by the rotating tool, is further extruded from the leading side (advancing side, AS) to the trailing side (retreating side, RS) of the tool during its steady translation along the joint line [10]. Neither filler material nor shielding gas is required. The temperature involved is typically some 50–100°C below the metal melting point, and thus there is no volume change during joining. Moreover, it is generally agreed that FSW, compared to the fusion welding techniques, induces rather low residual stresses after welding. This also implies process-reduced manufacturing costs [11].

As for the welded alloy mechanical properties acquired after the welding process, the FSW generally guarantees better tensile, bend, and fatigue properties than fusion welds. Taking advantages of these positive factors, this process has already been applied to a great variety of aluminum alloys, other than many other metallic materials. In the case of the aluminum alloys, the FSW technique has found many applications, such as external fuel tank of rockets, stock of railways, bridges [12, 13], to cite but few. Other interesting applications of FSW in the aerospace industry include fuselage, structural parts, cryogenic tanks, etc. [10]. Other interesting applications also include the marine applications (like offshore industry) [10, 14].

The microstructure modifications occurring at the central FSW zone (i.e., NZ) most usually consists of dynamic recrystallization resulting in the formation of fine equiaxed grains [8, 16]. This recrystallized zone can slightly reduce the welded alloy mechanical properties. For this reason, an accurate choice of the process parameters (rotational speed, welding speed, tilt angle, and sinking) and of the tool geometry (pin and shoulder geometry and size) is required. In fact, by increasing the pin rotational speed or by decreasing the welding line progression, the alloy mechanical properties can usually be optimized [15, 16].

One of the main welding defects, from which FSW is likely to be affected, is the oxide layer formation on the butt surface ("kissing-bond" phenomenon). The kissing-bond generally means a partial remnant of the unwelded butt surface below the stir zone. Its formation is mainly attributed to insufficient plunging of the welding tool during FSW [17], and it is usually responsible of the formation of small geometric discontinuities into the NZ [18].

2. Novel approach to the FSW process

In this context, the present contribution shows the effect of the process parameters, tool geometry, and size on macromechanical and micromechanical properties of FSWed joints by using a conventional pin and a nonconventional pinless tool configuration. The potential advantages offered by the pinless tool configuration can be fully exploited only as thin sheets are welded since, as the thickness increases, the shoulder influence becomes ever more localized to the top sheet surface.

A new FSW approach is here presented. This was developed to promote a better joint formability and it consists of carrying out the FSW process on both the sheet surfaces. In this process, the first welding operation is followed by a second welding performed at the plate opposite surface. Such an innovative methodology has been defined by these authors as double-side friction stir welding (DS-FSW) [19, 20]. This new FSW methodology has proven to be able to seal the geometric discontinuities, possibly produced by the first welding process, by means of the second welding operation performed at the opposite surface at the same experimental conditions. In addition, this new approach allows more uniform hardness values across the NZ. Moreover, the recrystallized grain size across the NZ is more homogeneous with respect to the surrounding FSW zones, compared to the conventional FSW, as shown by Cabibbo et al. [20]. Such improvement in the joint quality is very attractive, especially in those cases where the joint materials are meant to be subjected to post-welding forming operations. The hardness and local Young's modulus, determined by nanoindentation, were used to probe the overall weld joint strength. Nanoindentation profiles are also used to correlate the sub-micrometer hardness values to the corresponding FSW microstructure, and finally to properly correlate the welded plate formability with the welded sheet microstructure and micromechanical response.

A further novel approach to the FSW process (defined by authors as RT-type [21]) is also reported. This new configuration consists of a combination of different plate-to-pin motions. In one configuration, the pin axial spin rotation is set perpendicularly to the sheet blanks travelling along the welding line, with a lateral rotation radius $R = 0$, 0.5 and 1 mm. In a second configuration, the pin translation along the welding plate is set parallel to the welding line. Both these new welding approaches were compared with the conventional FSW practice, in which the welding motion occurs linearly along the welding line (and this conventional configuration is here defined by authors as T-type). With this respect, the two here proposed new configurations were also characterized using tools with different pin heights. These involved different sinking values during FSW. The study of the new setup also includes plate heat treatments, such annealing, prior and after the FSW. The effect of the radius R, pin height, and annealing treatment on microstructure, micromechanical and macromechanical properties is here discussed in order to define the process condition and the temper state that allows to obtain defect-free joints, without the occurrence of the oxide defects of kissing-bonds, and faint zigzag line pattern in the NZ.

3. Description of the new FSW process setups

FSW experiments were carried out using a computer numerical control (CNC) machining center.

3.1. DS-FSW method

As for the DS-FSW method, a conical pin tool geometry (H13 steel of HRC = 52), with a shoulder diameter equal to 12 mm and cone base diameter and height of the pin of 3.5 and 1.7 mm, respectively, with a pin angle of 30°. A 19-mm-diameter rotating tool was used. All the welding experiments were carried out with a nutting angle equal to 2°.

In DS-FSW, the first welding is followed by a second one, performed at the opposite surface, with respect to the first welding operation. Two different sheet positions, with respect to the welding tool, were investigated and are here presented:

(1) AS-AS, in which the sheet is placed in the AS, at the first FSW operation, and it is maintained in the same side also at the second FSW passage at the opposite surface;

(2) AS-RS, in which the sheet, placed in the AS at the first FSW, to be reversed, in the RS, at the second FSW passage at the opposite surface.

In both configurations, the effect of tool configurations on the quality of the DS-FSW joints was also analyzed.

In **Table 1** are reported the different tool configurations and sheet positions used in the DS-FSW. The used blanks were 180 mm in length, 85 mm in width, and 2 mm in thickness. The FSW was performed by fixing the welding line perpendicular to the rolling direction.

Sheet position	Tool configuration for the first pass – and second pass
AS-AS	Pin-pin
AS-RS	Pin-pin
AS-AS	Pin-pinless
AS-RS	Pin-pinless
AS-AS	Pinless-pinless
AS-RS	Pinless-pinless

Table 1. DS-FSW configurations (in terms of tool used and sheet arrangement).

http://dx.doi.org/10.5772/64523

The effect of the process parameters on the conventional and the DS-FSW was inferred using homologous rotational speed values (ω), which ranged 1200–2500 rpm, and same welding speed (v), equals to 60 and 100 mm/min. The conventional FSW was carried out using a tool sinking of 0.2 mm, while the DS-FSW was performed with a sinking of 0.15 mm in the first pass and 0.05 mm in the opposite surface. These welding parameters were set on the basis of the results obtained by preliminary tests, carried out using different tool sinking values, showed the need to perform the second pass with a sinking lower than that of the first one in order to prevent the occurrence of fracture.

In **Table 1** DS-FSW AS-AS pin-pin consists of maintaining fixed the AS and RS for both welding procedures; AS-RS pin-pin consists of reversing the AS into RS, from the first to the second welding procedure.

The third and fourth configuration differs from the first two only in the absence of the pin during the second welding process. In the last two (AS-AS, and AS-RS pinless-pinless), the welding process was performed with no pin in both processes. **Figure 1** shows a schematic representation of the three DS-FSW configurations used here.

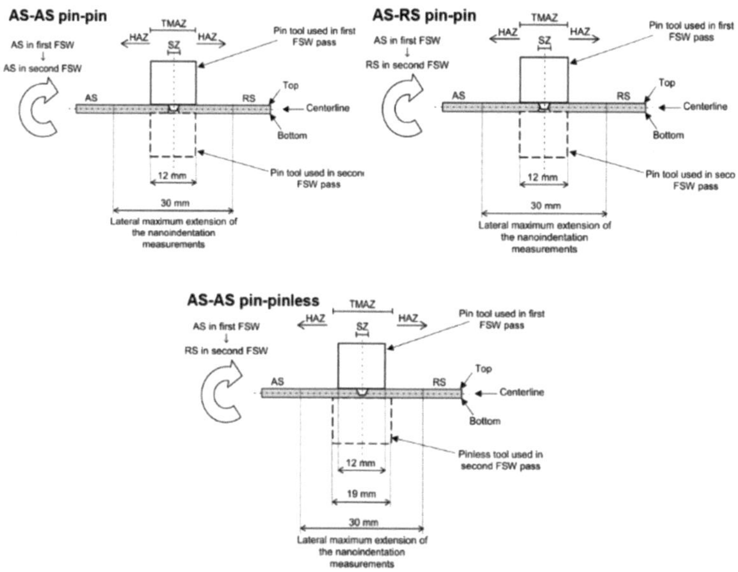

Figure 1. Representation of the three DS-FSW configurations: AS-AS pin-pin (left side); AS-RS pin-pin (center); AS-AS pin-pinless (right side).

3.2. Pin rotation deviation from centerline (RT-FSW) method

As for the pin rotation configuration method, the innovative approach to the FSW process was defined by authors as RT-type. For this purpose, a conical pin tools in H13 steel (HRC = 52) with a 2.3 mm pin height, 3.9 mm in diameter at the shoulder, a 30° pin angle, and a shoulder diameter of 15 mm (applying a vertical force of 1.7 kN) was used (**Figure 2**).

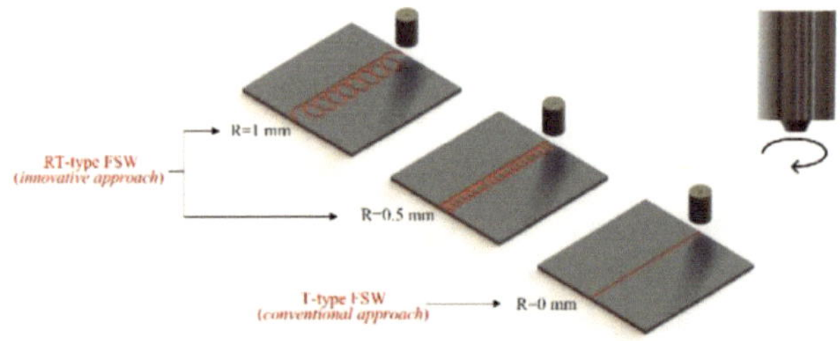

Figure 2. Comparison between conventional (R-type) and T-type FSW configurations.

The welding motion combines two different plate-to-pin mutual motion setups:

(1) a pin axial spin rotation sets perpendicular to the sheet blanks, changing the rotation along the plate centerline by a radius equal to R (=0, corresponding to the conventional FSW, 0.5, and 1 mm);

(2) a pin translation along a direction parallel to the welding centerline line.

The RT-type FSW innovative approach was compared with the conventional T-type (linear welding motion, i.e., for $R = 0$). In both the RT-type and T-type FSW processes, the stirring action was exerted by the pin tool rotation around its axis; the pin tilt angle was set at 2°, with respect to the normal direction to the plate surface. The RT-type and T-type FSW were performed using a pin rotational speed, $\omega = 2000$ rpm, and a transverse speed, $v = 30$ mm/min. All experiments were carried out with a tool plunging speed of 1.5 mm/min. The above reported setting parameters were chosen by an optimization FSW processing study reported in [22], where the effect of the welding parameters and tool configuration on micromechanical and macromechanical properties of FSW joints in AA5754 sheets were investigated. The AA5754 was subjected to an annealing treatment at 415°C/3 h, both prior (AA5754-O), and after (post-weld annealing, PWA), followed by furnace cooling.

4. The material

In both the cases, aluminum alloys were tested. In the first methodology, a heat-treatable AA6000-series alloy (AA6082) was used; whereas in the second methodology, a nonheat-treatable AA5000-series alloy (AA5754) was welded.

The AA5000-series alloys, such as the AA5754, are widely used in automotive, aerospace, marine, and military applications. They are characterized by a good strength-to-weight ratio, an appropriate weldability, and a good corrosion resistance. This class of aluminum alloy is difficult to join by conventional fusion welding techniques. This is mainly due to a dendritic structure, which typically forms in the melted zone and it seriously weakens the mechanical properties of the joint AA5000 series alloys. In this context, FSW has emerged as a promising solid-state process with the successfully overcome the fusion welding problems, making the welding process of AA5000 series alloys a sound one [23]. FSW of AA5754 is thus a promising technique to obtain sound welded joints, either in similar [24, 25] and dissimilar [26, 27] welding combinations (using AA5757 or AA5083).

In the present case, the AA5754 sheets were produced by twin roll continuous casting followed by cold rolling to give a H111 (EN485) metallurgical initial status and a thickness of 2.5 mm.

The AA6000-series are widely used because of their good weldability, corrosion resistance, and immunity to stress-corrosion cracking. These are known to be among the most used aluminum alloys for extruded components [28]. In fact, AA6082 (Al-Mg-Si) typical application include aeronautics, automotive, and recreation industries. In the present case, cold-rolled sheets of AA6082 were used to show the soundness of the DS-FSW.

5. Experimental findings and evidence for sound and better FSW joints

5.1. The DS-FSW method

There is a strong need for an improvement in ductility and formability of FSWed joints. Some previous studies reported significant mechanical improvements by carrying out multipass [29], double lap [30], reverse dual rotation [31] FSW, and FS spot welding [32]. With this respect, the DS-FSW showed better strength, elongation, and formability of FSWed aluminum joints. The DS-FSW was proven to induce the serration of the geometric discontinuities, thus promoting a significant microstructure homogeneity at the NZ.

5.1.1. Mechanical properties

Figure 3 shows typical nominal stress versus nominal strain curves of FSWed joints in AA6082 obtained under different values of the rotational speed and welding speed. The joints ductility is shown to be lower in the NZ, with respect to the base metal (BM), irrespective of the welding parameters and process methodology [22]. In general, in terms of both the ultimate values of tensile strength and elongation, the conventional FSWed joints show a tensile behavior better than the one exhibited by the DS-FSWed joints. Actually, the conventional FSW process requires a high sinking value in order to generate the frictional heating allowing the material flow necessary to obtain sound joints, according to Mishra and Ma [7]. Thus, in the first pass, by using the same tool sinking as of conventional FSW produces a step in the blank surface that acts as a notch during the second pass. Therefore, the tool sinking value imposed in the second pass had to be further decreased in order to reduce the formation of surface defects. The pinless-pinless configuration has provided the worst tensile properties. In particular, the AS-AS configuration showed low mechanical properties of the joint, while the AS-RS configuration did not reach a sound weldment.

The mechanical behavior is strongly improved when welding is performed using the pin-pinless configuration. In this case, ductility levels similar to the ones showed by the conventional FSWed samples were obtained. In this case, the tensile fracture occurred at the HAZ, in the RS zone. The tensile properties of the joints are slightly affected by the rotational and welding speeds. This is not the case in the DS-FSW pinless-pinless tool configuration, which exhibits ultimate tensile strength (UTS) and UE values strongly dependent on the process parameters (**Figure 3**). As a matter of fact, as the thickness increases, the shoulder influence becomes ever more localized near the top surface of the sheet and, consequently, the stirring action becomes less and less effective. With this respect, in published work by these authors, the FSW capability to obtain sound joints in 1- and 1.5-mm-thick sheets using a pinless tool was widely documented [20, 22, 33].

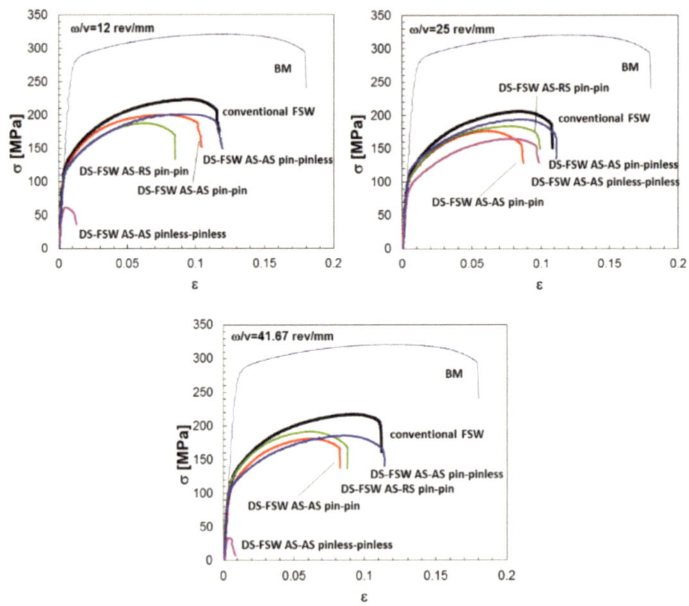

Figure 3. Tensile stress-strain curves of the DS-FSW, with different welding parameters and tool configurations.

5.1.2. Post-welding formability

In comparison with conventional fusion welding techniques, one of the most important advantages offered by FSW is the relatively high post-welding formability. In this sense, the conventional FSW and the DS-FSW formability, obtained under the same process conditions, was detected. For this purpose, limit dome height (LDH) analyses were carried out. These values represent the punch stroke at the peak of the load versus the stroke curves. It actually represents the dome height of the deformed samples at the onset of necking, and the results are reported in the plot of **Figure 4**. These data agree with the joint ductility obtained by tensile tests (**Figure 3**). The LDH values were lower than those obtained on the BM, no matter what welding methodology was used. Such results reveal that a noticeable formability reduction along the welding zone [18, 20, 22, 33–35]. More specifically, the B arrangement leads to a LDH value lower than the T arrangement (as reported by the letter B and T in **Figure 4**, and according to the configuration reported in **Figure 5**).

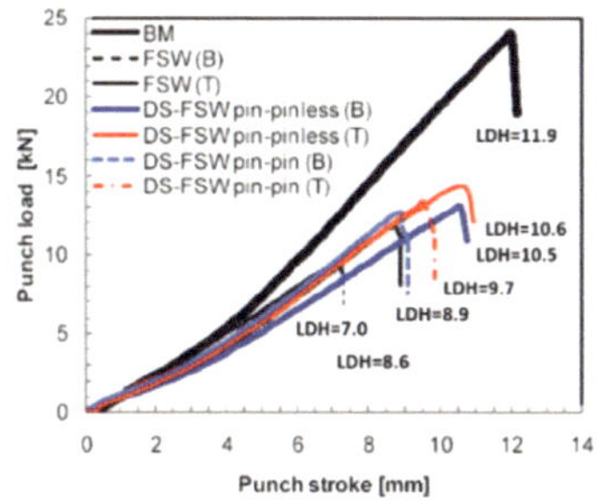

Figure 4. Hemispherical punch test for different testing conditions (LDH is the Limit Dome Height).

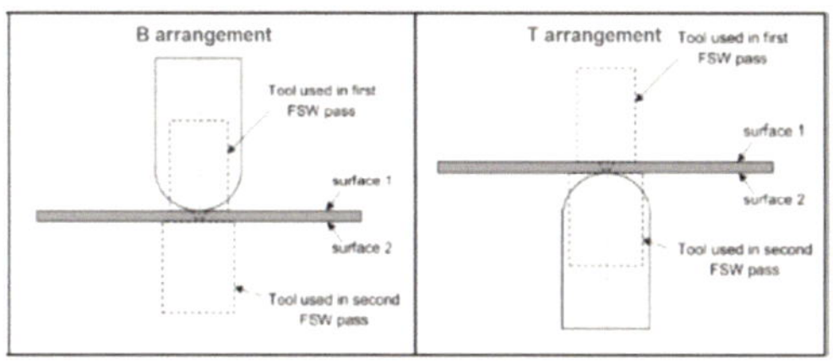

Figure 5. Hemispherical punch test configurations. In the B arrangement, surface 1 is in contact with the punch; in the T arrangement surface 1 is opposite.

In the B arrangement, the local stress field intensity rise, caused by the notch, is responsible of the FSW sample failure at the geometric discontinuity. In the T arrangement, that is, FSWed blank having the notch in contact with the punch, the failure of the deformed joint occurs at the step produced by the sinking action applied by the shoulder [19]. This is mainly due to the biaxial tensile stress state to which the notch is subjected. This appeared to be less severe in the T arrangements, with respect to that in the B arrangement.

The DS-FSW joints showed LDH values higher than those measured on the conventional FSWed joints. This is likely to be attributed to the beneficial effect of the second pass of the DS-FSW. This second welding induces a dual beneficial effect: it allows both the closure of the geometric discontinuity, and the reduction in the height of the step produced by the first welding on the opposite plate surface. Furthermore, the DS-FSW is characterized by more uniform recrystallized grains across the NZ, and also partially across the thermo-mechanical affected zone (TMAZ), than in the case of the conventional FSW [19]. Finally, the joints obtained using the pin-pinless tool configuration lead to LDH values higher than the ones obtained by using the pin-pin configuration, irrespective of the sheets arrangement. Thus, in the pin-pinless configuration, the LDH value was only ~12% lower than that of the BM. This result appears to be virtually independent of the sheet arrangement. On the other hand, in the pin-pin tool configuration, the LDH reaches values of ~19 and 25% lower than that of BM in the T and B arrangements, respectively (**Figure 4**).

A more accurate evaluation of formability is obtained by means of the forming limit curves (FLCs). This is obtained by plotting the major strain versus minor strain data (**Figure 6**). It resulted that the formability of the BM is always higher than that of the welded joints. In the stretching side of the FLD, for a given minor strain, the major strain measured on the DS-FSWed joints appeared systematically higher than that provided by the conventional FSWed ones. This result agrees with the behavior exhibited by the LDH (**Figure 4**). The higher vertical position of the FLCs confirms that formability is strongly improved when the DS-FSW technique is used. The comparison among the different FLCs obtained in the DS-FSW, using both the pin-pinless and the pin-pin configurations, shows that the FLCs are scarcely affected by the tool configuration used in the second pass. Finally, the B arrangement is characterized by the lowest major strain values, and this agrees with the LDH results shown in **Figure 4**. This differentiation tends to vanish when DS-FSW is used. It is noteworthy to observe that the process methodology and sample arrangement also affect the extension of the FLCs. In fact, **Figure 6** shows FLCs smaller extension in the welded joints, compared to the BM. Such behavior is almost negligible in the drawing zone of the FLD. This becomes significant in the stretching region, as confirmed by LDH values shown in **Figure 6**.

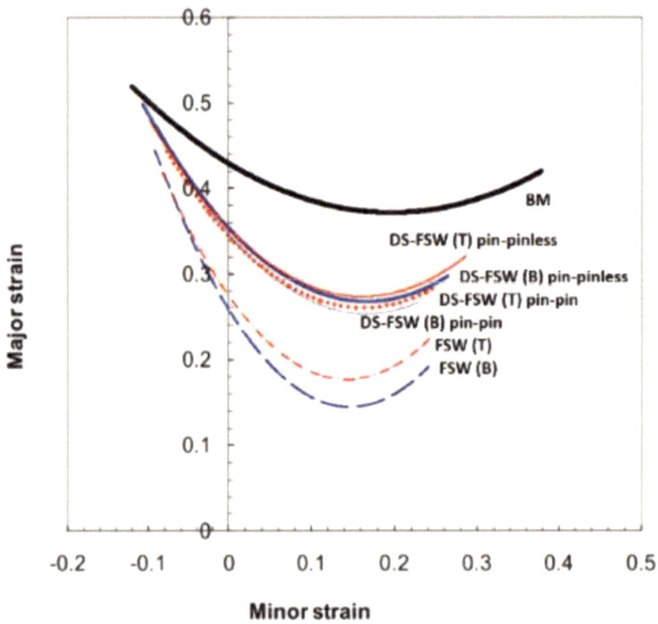

Figure 6. Forming limit curves (FLC) obtained using different welding processes and sample arrangements.

Figures 7 and **8** show the nanoindentation experimental results, in terms of harness and Young's modulus. The main difference between the pin and the pinless FSW consists of the low hardness (*H*) and elastic modulus (E_r) values obtained at the TMAZ. In the conventional pin FSW, low *H* and E_r are in the retreating TMAZ, whereas in the pinless FSW, both the advancing and the retreating TMAZ experienced such low *H* and E_r values. In both the cases, these lower values accounted for a drastic hardness reduction, being three times less, and an elastic modulus reduction of almost ten times, with respect to the BM values.

http://dx.doi.org/10.5772/64523

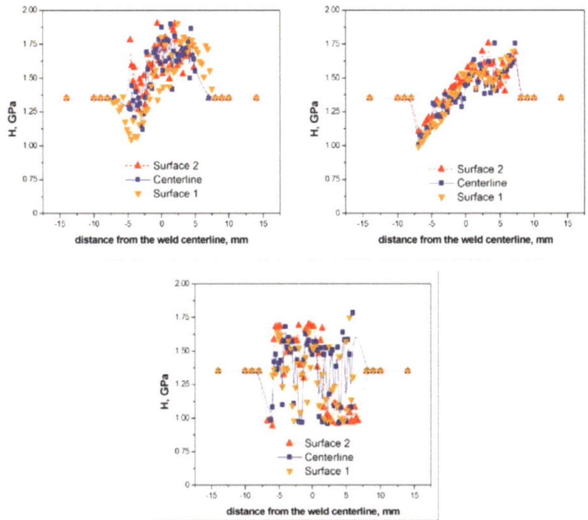

Figure 7. Nanoindentation hardness profiles for the different configurations used.

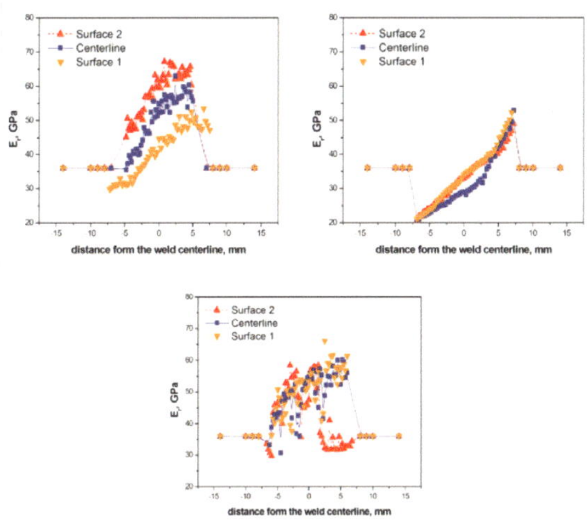

Figure 8. Nanoindentation Young's modulus profiles for the different configurations used.

Moreover, in both the pin and pinless conventional FSW, the lower H and E_r values, recorded at the surface in contact with the shoulder, also pertain to the outer part of the SZ. On the contrary, in both pin and pinless FSW, the profile along the centerline of the sheet section did not show any reduction of H and E_r across the NZ, TMAZ, and HAZ. The hardness, at the section centerline of the NZ, is constantly higher than the BM, with values that peak at 2.05 GPa. In particular, the two AS-TAMZ and RS-TMAZ showed values ranging 1.40–1.55 GPa, while the hardness in the NZ ranged from 1.65 to 2.05. Along the centerline of the pinless FSWed section, the hardness profile appeared considerably more uniform than the one obtained in the pin FSW. In the former case, the hardness ranged from 1.75 to 2.10 GPa, across the characteristic regions of the FSW joint.

Uniform *H* and E_r values were obtained across the welded zone in all the three DS-FSW configurations. This was not the case in the two conventional, where hardness appeared far from being uniform across the FSW joint regions. In particular, in the pin-pin AS-AS DS-FSW, the hardness decreased in the TMAZ AS, while the rest of the welded zone (i.e., NZ and RS-TMAZ) showed hardness values significantly higher than the ones of the BM. This trend was common to all the three profiles (upper surface, centerline, lower surface). The elastic modulus increased up to 50%, in the NZ. The top values were reached in the surface where the second FSW took place (Surface 2). Quite similar hardness trends were found in the pin-pin AS-RS DS-FSW. In particular, in the TMAZ RS (at the second welding), the elastic modulus steadily increased from values of almost half respect the BM, to reach values of some 30–35% higher than those of BM, in the TMAZ AS. Finally, in the pin-pinless AS-AS DS-FSW, *H* and E_r profiles, taken along the upper-surface, centerline, and lower-surface, were essentially similar to those observed in the pin-pin AS-AS DS-FSW. The only significant difference was the rather fuzzy and wavy hardness trend obtained in this case, at the AS-TMAZ, SZ, and RS-TMAZ.

The better formability of the DS-FSW, with respect to the conventional FSW, is most likely related to the local elastic modulus uniformity (i.e., the reduced Young's modulus) across the weld, and to the less dramatic hardness variation, from top to bottom of the sheet section.

5.1.3. Microstructure of joints

Table 2 reports the mean grain size as a function of welding methodology and tool configuration, measured at different zones of the welded joints. The considerably small grain size in the SZ, combined with the equiaxed grain shape, implies the occurrence of dynamic recrystallization because of the very high levels of deformation and temperature reached in such zone during FSW [35].

FSW configuration	Mean grain size, μm				
	AS/TMAZ	Surface 1/SZ	Middle/SZ	Surface 2/SZ	RS/TMAZ
Conventional	14 ± 2	7.8 ± 0.3	7.6 ± 0.3	6.9 ± 0.3	14 ± 2
DS-FSW pin-pin	12 ± 2	6.3 ± 0.2	6.2 ± 0.2	6.4 ± 0.2	13 ± 2
DS-FSW pin-pinless	12 ± 2	5.7 ± 0.2	5.7 ± 0.2	5.8 ± 0.2	13 ± 2

Table 2. AA6082 DS-FSW mean grain size; the BM SZ had a mean grain size of 20 ± 2 μm.

In the conventional FSW, the grain size within the SZ tended to increase near the top of the weld zone, and this is chiefly due to the temperature variation within the weld zone [36]. In all the three DS-FSW configurations described here, the observed grain size uniformity and morphology across the SZ, from surface to surface, greatly favored the soundness and better post-welding response of the welded Al-sheets.

5.2. The friction stir welding method by pin rotation deviation from centerline: RT-FSW

Rotation of the pin from its centerline progression, during the FSW process, is intended as a further possible improvement in the welded soundness of aluminum plates.

5.2.1. Mechanical properties

Figure 9 shows typical stress-strain tensile curves of both as-received and FSW AA5754 sheets. The ultimate tensile strength (UTS) and plastic elongation (El) values are reported in **Figure 10**. The mechanical response is quite different as the pin moves in RT-type configuration with R = 0.5 mm. In this case, the UTS reduction respect to the unwelded sheet was 25%, that is double that for R = 0 mm (T-type FSW). At R = 0.5 mm, ductility (El) was 78%, respect to the unwelded sheet, and thus it was three times lower with respect to the ductility obtained at R = 0 mm (in the T-type FSW). UTS reduction, compared to the unwelded sheet, accounted for 20 and 22%, respectively; the ductility reduction was 48 and 74%, accordingly.

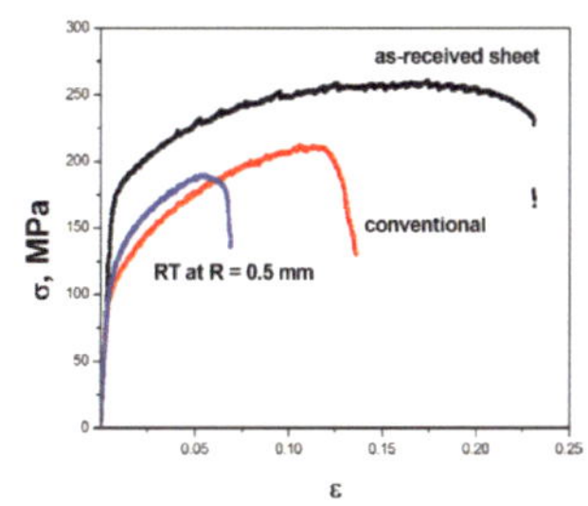

Figure 9. Tensile stress-strain curves for RT-FSW at R = 0 (conventional FSW), and 0.5 mm.

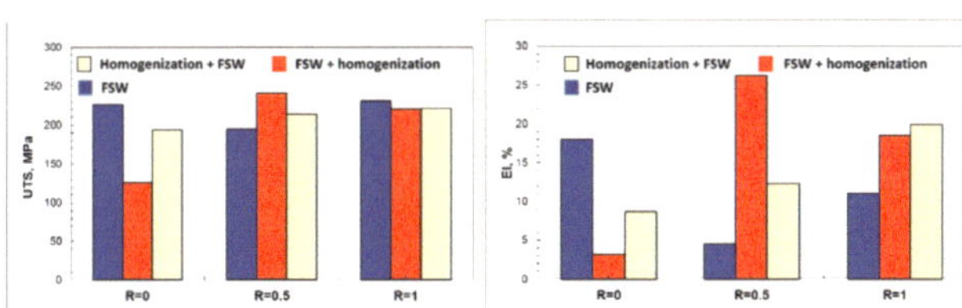

Figure 10. UTS and El of the AA5754 RT-FSW.

The tensile curves, irrespective of the specific FSW setup (T- and RT-type FSW), clearly showed the occurrence of serrated yielding, also termed the Portevin-Le Chatelier (PLC) effect, that is a common phenomenon in 5xxx aluminum alloys [36, 37]. The PLC phenomenon is driven by Mg solute atom cloud formation. This microstructure atomic-level evolution is actually responsible for the strain rate dependency of the observed serrated yielding phenomenon. The Mg solid solution, induced in the grains of the cold-rolled AA5754 sheets, effectively pins the dislocation sliding motion induced by the tensile test. This, in turns, generates the yielding phenomenon.

5.2.2. Microstructure of the joints at the different pin rotation radii

Figure 11 shows an overview of the FSW plate microstructure, in which the occurrence of a grain dynamic recrystallization process in the NZ is evident.

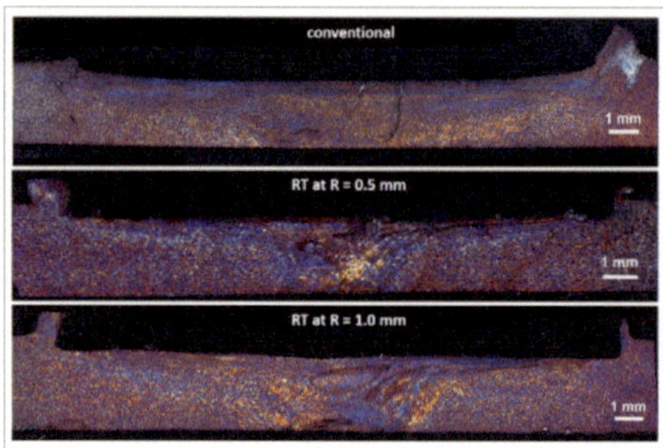

Figure 11. Montage of polarized optical micrographs (POM) RT-FSW at R = 0 (conventional FSW), 0.5, and 1 mm.

5.2.3. Microstructure modifications induced by pre- and post-welding annealing

The AA5754 was subjected to an annealing treatment at 415°C/3 h followed by furnace cooling, in one case prior FSW (AA5754-O state), and in another case, after FSW (post-weld annealing: PWA).

The microstructure of the annealed FSW AA5754-O sheets, obtained both under T-type and RT-type FSW configurations, is shown in **Figure 12**. As expected, the base material (BM) is fully recrystallized. It appeared that the equiaxed recrystallized mean grains did not change significantly in the BM, HAZ, and TMAZ, on either AS and RS of joint. This was found irrespective of the pin deviation extent. The NZ-grained structure, in the conventional (R = 0), and for 0.5 mm pin rotation deviation, appeared to be mixed, and characterized by the coexistence of fine equiaxed grains and stirred elongated grains (still remaining of the stirring effect induced by the FSW). It is actually a microstructure modification induced by concurring effect driven by the first recrystallization stage (due to the annealing treatment at 415°C/3 h), and by the following mechanical heat flow during the FSW. This latter is known to rise the temperature in the NZ aluminum alloys typically by 350–500°C [20, 27, 37]. The tool shoulder rotation, during welding, induces a large heat transfer and a high strain level at the top surface, which is considerably higher than that induced at the bottom surface. The bottom surface was in contact with a back-plate support, during the welding process. This indeed acted as a heat sink lowering the peak temperature and reducing the time to the peak temperature. Thus, in turns, grain growth in the bottom surface of the NZ was effectively slowed down during the

http://dx.doi.org/10.5772/64523

welding process. The NZ, obtained with $R = 1$ mm, still presented some oxide layers both in its center and near the TMAZ/NZ boundary.

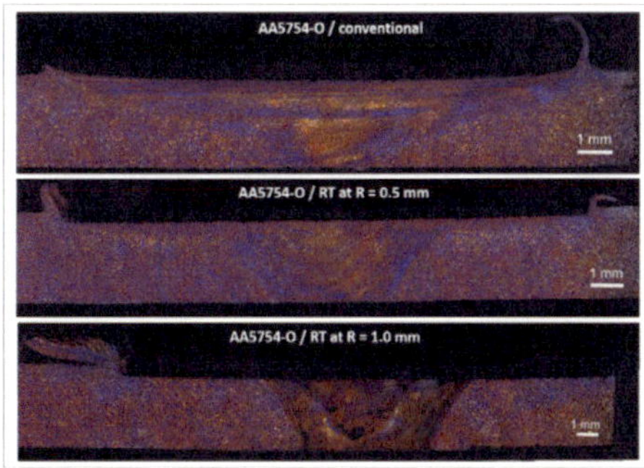

Figure 12. Montage of POM RT-FSW AA5754-O at $R = 0$ (conventional FSW), 0.5, and 1 mm.

The microstructure of the post-weld annealed (PWA) sheets, for the different rotational radii investigated, is shown in **Figure 13**. For $R = 0$ mm, fine equiaxed grains characterize the whole extension of the FSW sheet. These equiaxed grains had a mean size substantially same as the ones in the BM, and this was found in the HAZ and the TMAZ of the AS and RS. The only exception consisted in the grain size and morphology in upper welded zone, that is, the surface directly in contact with the shoulder, during the FSW. In this zone, very coarse irregular grains were induced to form by the stirring effect, and by the heat flow introduced in the aluminum plate by the tool shoulder. The depth extension of this coarse-grain region accounted for a minimum of one-third to a maximum of half of the whole sheet thickness. In particular, at $R = 0.5$ mm, fine recrystallized equiaxed grains characterized the whole extension of the BM, HAZ, and the TMAZ, in both the AS and the RS. More specifically, in this case, the NZ was characterized by the occurrence of very coarse irregular grains, mixed with fine recrystallized grains strips located throughout across the NZ, from the upper to the lower surface. The primary factors leading to the occurrence of abnormal grain growth process, during FSW, are associated with inhomogeneous gran deformation.

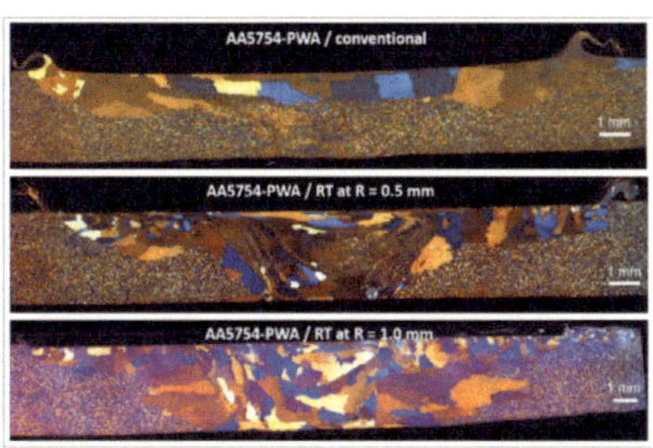

Figure 13. Montage of POM RT-FSW PWA at $R = 0$ (conventional FSW), 0.5, and 1 mm.

5.2.4. Defect and void formation during FSW

Traces of the presence of oxide layers (the lazy S-lines) are evident in the NZ microstructure (**Figures 12** and **13**). These, actually, follow the location of the fine grain strips. It thus appeared that the fine grains are formed where the oxide layers, the lazy S-line oxides, are present, and were formed at $R = 1$ mm of RT-type FSW. Thus, it appeared that the fine grain strips, at $R = 0.5$ mm, are being formed along already existing lazy S-line oxide, which formed during FSW.

5.2.5. Mechanical properties and hardness modifications induced by pre- and post-welding annealing

Typical stress-strain curves are shown in **Figure 14**. It appeared that the closest mechanical response to the unwelded annealed AA5754 sheet is obtained by welding with $R = 0.5$ mm in the PWA condition, where UTS differed only by 5%, and ductility differed by 30% with respect to the ductility of the unwelded annealed condition. In the other conditions, the UTS remained within a range of 14% of difference, with respect to the annealed sheet, with a ductility reduction ranging from 76 to 30%.

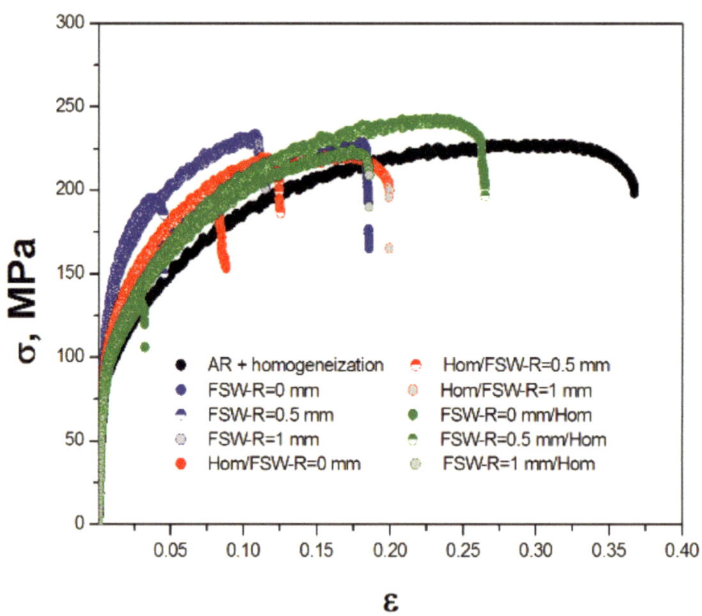

Figure 14. Tensile stress-strain curves for RT-FSW, in the AA5754-O stare and in the PWA condition, at $R = 0$ (conventional FSW), 0.5, and 1 mm.

Therefore, based on the microstructure evidence, and the obtained hardness and mechanical response, the use of a RT-type welding motion is justified when the plate is homogenized prior, or, even better, after FSW. Conversely, there is no need to deviate the pin, from its welding centerline, in the case of non-annealed AA5000 FSW.

http://dx.doi.org/10.5772/64523

6. Concluding remarks

In this contribution, two novel approaches and methodologies of friction stir welding on aluminum alloys were presented. The first approach consists of a double-side FSW (DS-FSW). The second approach is represented by a radial deviation of the rotating pin from its centerline, during FSW (RT-FSW). Both new methods were tested in a conventional pin and nonconventional pinless configuration. Several interesting achievements, from a technological point of view, were obtained and are here summarized.

DS-FSW:

DS-i: the elastic modulus and the hardness showed a larger uniformity across the sheet section, with respect to the FSW;

DS-ii: A better formability of the DS-FSW, compared to the conventional pin and pinless FSW, was obtained;

DS-iii: The DS-FSWed joints are characterized by LDH, and FLC values higher than those measured on conventional FSW.

RT-FSW:

RT-i: The RT setup, for a pin rotation radius of 0.5 mm, induced a low reduction of the mechanical response, compared to the conventional T setup FSW (i.e., with no pin deviation from the welding line). Accordingly, both the microstructure and the hardness profiles of all the characteristic welded zone were quite similar;

RT-ii: The post-weld annealing (PWA) showed the best mechanical response respect to the unwelded annealed AA5754 sheet. The best experimental setups were obtained setting a pin rotation radius R = 0.5 mm. In this configuration, UTS was ▪ 15% higher, and a ductility reduction of up to 30%, respect to the unwelded annealed sheet. In this condition, the microstructure of the NZ appeared to be characterized by very coarse grains. These coarse grains were generated by geometric dynamic recrystallization (GDR), which is induced by the combined effect of shoulder pressure (heat input), and post-welding annealing (PWA) thermal energy.

Author details

Marcello Cabibbo[1*], Archimede Forcellese[1] and Michela Simoncini[2]

*Address all correspondence to: m.cabibbo@univpm.it

1 DIISM – Department of Industrial Engineering and Mathematics, Marche Polytechnic University, Ancona, Italy

2 Università degli Studi e-Campus, Novedrate, Italy

References

[1] Davies RW, Grant GJ, Oliver HE, Khaleel MA, Smith MT. Forming-limit diagrams of aluminum tailor-welded blank weld material. Metall. Mater. Trans. A. 2001;32:275-283.

[2] Saunders FI, Wagoner RH. Forming of tailor-welded blanks. Metall. Mater. Trans. A. 1996;27:2605-2616.

[3] Merklein M, Johannes M, Lechner M, Kuppert A. A review on tailored blanks-production, applications and evaluation. J. Mater. Process. Tech. 2014;214:151-164.

[4] Sun Z, Ion JC. Laser welding of dissimilar metal combinations. J. Mater. Sci. 1995;30:4205-4214.

[5] Schubert E, Klassen M, Zerner I, Walz C, Sepold G. Light-weight structures produced by laser beam joining for future applications in automobile and aerospace industry. J. Mater. Process. Tech. 2001;115:2-8.

[6] Thomas W, Nicholas ED, Needham JC, Murch MG, Temple-Smith P, Dawes CJ. Friction stir butt welding: International Patent Application PCT/GB92/02203 and GB Patent Application 9125978.8, UK Patent Office, London, December 6, 1991.

[7] Mishra RS, Ma ZY. Friction stir welding and processing. Mater. Sci. Eng. R. 2005;50:1-78.

[8] AU: Data of Ref. [8] was duplicated in Ref. [15]. The latter was deleted, and references and their citations were renumbered accordingly.Su J-Q, Nelson TW, Mishra R, Mahoney M. Microstructural investigation of friction stir welded 7050-T651 aluminium. Acta Mater. 2003;51:713-729.

[9] Seidel TU, Reynolds AP. Visualization of the material flow in AA2195 friction-stir welds using a marker insert technique. Metall. Mater. Trans. A 2001;32:2879-2884.

[10] Nicholas ED, Thomas WM. A review of friction processes for aerospace applications. Int. J. Mater. Prod. Technol. 1998;13:45-54.

[11] Dawes CJ, Thomas WM. Friction stir process welds aluminum alloys. Welding J. 1996;75:41-45.

[12] Heinz B, Skrotzki B. Characterization of a friction-stir-welded aluminum alloy 6013. Metall. Mater. Trans. B 2002;33:489-498.

[13] Mahoney MW, Rhodes CG, Flintoff JG, Spurling RA, Bingel WH. The rate-controlling mechanism in superplasticity. Metall. Mater. Trans. A 1998;29:1955-1964.

[14] Jata KV, Sankaran KK, Ruschau JJ. Friction-stir welding effects on microstructure and fatigue of aluminum alloy 7050-T7451. Metall. Mater. Trans. A 2000;31:2181-2192.

[15] Khorrami MS, Kazeminezhad M, Kokabi AH. Microstructure evolutions after friction stir welding of severely deformed aluminum sheets. Mater. Des. 2012;40:364-372.

[16] Kumbhar NT, Sahoo SK, Samajdar I, Dey GK, Bhanumurthy K.Microstructure and microtextural studies of friction stir welded aluminium alloy 5052. Mater. Des. 2011;32(3):1657-1666.

[17] Oosterkamp A, Oosterkamp LD, Nordeide A. Kissing bond phenomena in solid-state welds of aluminum alloys. Weld. J. 2004;83(8):225s–231s.

[18] Forcellese A, Fratini L, Gabrielli F, et al. Formability of friction stir welded AZ31 magnesium alloy sheets. Mater. Sci. Forum 2010;638–642:1249–1254.

[19] AU: Please provide volume number for Ref. [19].M. Simoncini, M. Cabibbo, A. Forcellese. Development of double-side friction stir welding to improve post-welding formability of joints in AA6082 aluminum alloys. Proc. IMechE Part B: J. Eng. Man. 2015;229:1-11.

[20] M. Cabibbo, A. Forcellese, M. El Mehtedi, M. Simoncini. Double side friction stir welding of AA6082 sheets: microstructure and nanoindentation characterization. Mater. Sci. Eng. A 2014;590:209-217.

[21] M. Cabibbo, A. Forcellese, M. Simoncini, M. Pieralisi, D. Ciccarelli. Effect of welding motion and pre-/post-annealing of friction stir welded AA5754 joints. Mater. Des. 2016;93:146-159.

[22] Simoncini M, Forcellese A. Effect of the welding parameters and tool configuration on micro- and macro-mechanical properties of similar and dissimilar FSWed joints in AA5754 and AZ31 thin sheets. Mater. Des. 2012;41:50-60.

[23] Feistauer EE, Bergmann LA, Barreto LS, dos Santos JF. Mechanical behaviour of dissimilar friction stir welded tailor welded blanks in Al-Mg alloys for marine applications. Mater. Des. 2014;59:323-332.

[24] Movahedi M, Kokabi AH, Seyed Reihani SM, Najafi H, Farzadfar SA, Cheng WJ, Wang CJ. Growth kinetics of Al–Fe intermetallic compounds during annealing treatment of friction stir lap welds. Mater. Charact. 2014;90:121-126.

[25] Costa MI, Verdera D, Costa JD, Leitão C, Rodrigues DM. Influence of pin geometry and process parameters on friction stir lap welding of AA5754-H22 thin sheets. J. Mater. Proc. Technol. 2015;225:385-392.

[26] Kasman Ş, Yenier Z. Analyzing dissimilar friction stir welding of AA5754/AA7075. Int. J. Adv. Manuf. Technol. 2014;70:145-156.

[27] Costa MI, Verdera D, Leitão C, Rodrigues DM. Dissimilar friction stir lap welding of AA 5754-H22/AA 6082-T6 aluminium alloys: Influence of material properties and tool geometry on weld strength. Mater. Des. 2015;87:721-731.

[28] Polmear IJ. Light Alloys - Metallurgy of the Light Metals. 3rd ed. London: Arnold; 1995, 112-114.

[29] Canaday CT, Moore MA, Tang W, Reynolds AP. Through thickness property variations in a thick plate AA7050 friction stir welded joint. Mater. Sci. Eng. A 2013;559:678-682.

[30] Brown R, Tang W, Reynolds AP. Multi-pass friction stir welding in alloy 7050-T7451: Effects on weld response variables and on weld properties. Mater. Sci. Eng. A 2009;513–514:115-121.

[31] Cerri E, Leo P. Mechanical properties evolution during post-welding-heat treatments of double-lap friction stir welded joints. Mater. Des. 2011;32:3465-3475.

[32] Li JQ, Liu HJ. Characteristics of the reverse dual-rotation friction stir welding conducted on 2219-T6 aluminum alloy. Mater. Des. 2013;45:148-154.

[33] Forcellese A, Simoncini M. Plastic flow behaviour and formability of friction stir welded joints in AZ31 thin sheets obtained using the 'pinless' tool configuration. Mater. Des. 2012;36:123–129.

[34] Kim D, Lee W, Kim J, Kim C, Chung K. Formability evaluation of friction stir welded 6111-T4 sheet with respect to joining material direction. Int. J. Mech. Sci. 2010;52:612–625.

[35] Rhodes CG, Mahoney MW, Bingel WH, Spurling RA, Bampton CC. Effects of friction stir welding on microstructure of 7075 aluminum. Scripta Mater. 1997;36:69-75.

[36] Reynolds AP. Visualization of material flow in autogenously friction stir welds. Sci. Technol. Weld. Join. 2000;5:120-124.

[37] Imam M, Biswas K, Racherla V. On use of weld zone temperatures for online monitoring of weld quality in friction stir welding of naturally aged aluminium alloys. Mater. Des. 2013;52:730-739.

Chapter 3

A Mesh-Free Solid-Mechanics Approach for Simulating the Friction Stir-Welding Process

Kirk Fraser, Lyne St-Georges and Laszlo I. Kiss

Additional information is available at the end of the chapter

http://dx.doi.org/10.5772/64159

Abstract

In this chapter, we describe the development of a new approach to simulate the friction stir-welding (FSW) process using a solid-mechanics formulation of a mesh-free Lagrangian method called smoothed particle hydrodynamics (SPH). Although this type of a numerical model typically requires long calculation times, we have developed a very efficient parallelization strategy on the graphics processing unit (GPU). This simulation approach allows the determination of temperature evolution, elastic and plastic deformation, defect formation, residual stresses, and material flow all within the same model. More importantly, the large plastic deformation and material mixing common to FSW are well captured by the mesh-free method. The parallel strategy on the GPU provides a means to obtain meaningful simulation results within hours as opposed to many days or even weeks with conventional FSW simulation codes.

Keywords: Friction stir welding, Numerical simulation, Smoothed particle hydrodynamics, Coupled thermal-mechanical, GPU

1. Introduction

Friction stir welding (FSW) is a solid-state welding process that was patented in the UK by "The Welding Institute" (TWI) in 1991. In this process, illustrated in **Figure 1**, a non-consumable rotating tool is used and the workpieces are joined in a solid state, without fusing the materials. This tool is classically made up of a cylindrical shoulder and a cylindrical or conical pin. To perform a weld, the rotation of the tool is initiated, and then the tool is forced into the parts to be welded. When the shoulder reaches the surface of the material, an important amount of friction heat is generated along the contact surface. The increase in temperature softens the material and helps the workpieces to become highly plastic. Although significant heat is generated, the material nevertheless stays in the solid state, at about 0.8–0.9 times the melting point. The combined effect of the increased temperature and the pressure exerted by the tool allows the workpiece material to be mechanically mixed. The plates are then joined together in a solid state as the tool advances along the weld seam.

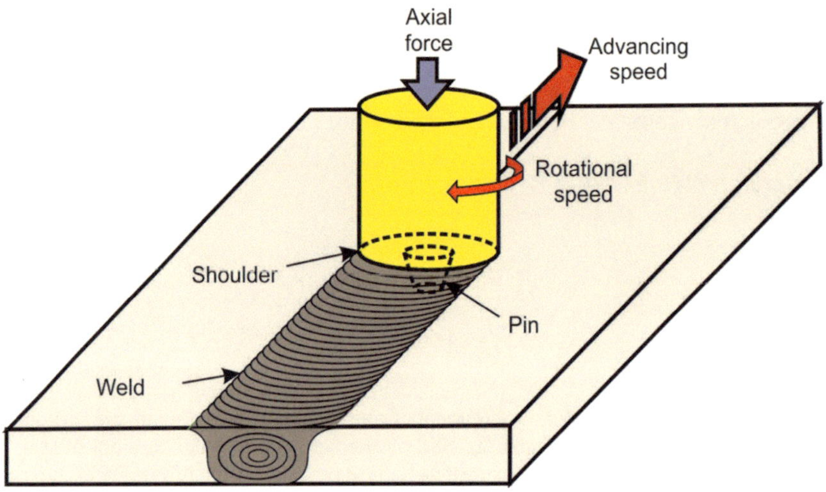

Figure 1. Friction stir-welding process.

FSW was initially developed and used to join aluminum alloys. However, since its invention, the application field of the process has been extended to weld various materials: copper, titanium, magnesium, steel, stainless steel, nickel, polymers, and lead.

To join two plates using the FSW process, a sequence of prescribed motions is performed. This sequence is normally divided into four different phases. Each phase plays a specific role in the welding process. These phases are illustrated in **Figure 2** and are identified as follows:

1) Plunge phase,

2) Dwell or stabilization phase,

3) Welding or advancing phase,

4) Tool removal or retraction phase.

During the plunge phase, the rotation of the welding tool is initiated and the tool plunges into the workpieces. During this phase, the material is relatively cold; only the pin is in contact with the workpiece. The axial force (also called forging force) and the torque applied to the tool are high, and in most cases, reach their highest values. At the end of the plunge phase, the pin has fully penetrated the workpiece and the shoulder is in contact with the surface. The rotation speed of the tool during the plunge and advance phase is frequently the same.

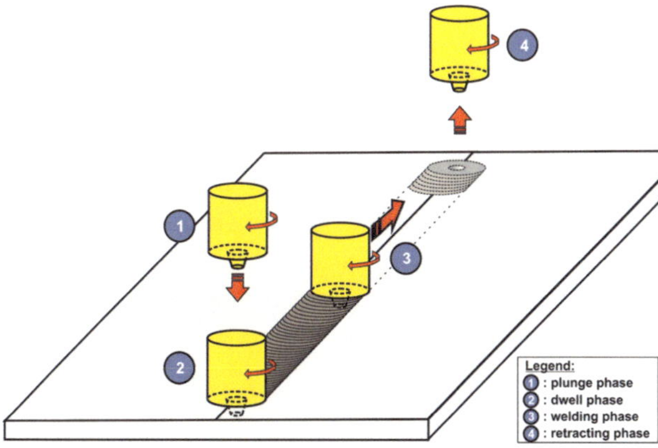

Figure 2. The four main phases of friction stir welding.

The dwell phase begins when the desired plunge depth has been achieved. The axial force **Figure 2** is maintained on the tool during this stabilization phase. The combined effect of the relative speed between the rotating tool and the material with the applied axial force generates heat due to friction at the tool-material interface. The tool is kept in place for a sufficiently long time to reach the temperature required for welding.

After the dwell phase, the tool starts to advance and accelerates to the prescribed translational velocity along the weld line. The acceleration may be fast if the dwell phase was sufficiently long and the temperature is high in the weld zone. However, too fast an acceleration can result in high mechanical stresses for both the tool and welding equipment, reducing their useful lifetime. Depending on the design of the tool and the specific process parameters, the FSW tool may be tilted slightly (a few degrees) to improve the quality of the weld.

In conventional arc-welding techniques, the material is physically melted to produce a weld. In FSW, numerous drawbacks associated with the presence of a liquid phase during welding are eliminated: solidification cracking is eradicated, and the distortions and the size of the heat-affected zone (HAZ) are reduced. Spatter, fume, and ultraviolet (UV) emissions are also eliminated. Compared to arc-welded parts, the FSW assemblies frequently exhibit higher mechanical properties in tension, compression, bending, and an increased life in fatigue. In addition, no flux, protective gas, or filler material is needed during welding. Finally, the thickness of FSW welds may go from few tenths of millimeters up to more than 70 mm in aluminum alloys.

However, the FSW process has certain limitations as well. In order to bring the material into the plastic state, the required torque and forces can be very high. The axial force applied on the tool can reach many kilonewtons (many tons of force). For this reason, the welding machine must be robust, typically leading to relatively expensive equipment. In order to have high-quality welds, it is also important to assure the appropriate clamping and support of the pieces to be welded. Further limitations of the FSW process are mostly related to geometrical factors. During welding, the tool shoulder must have constant and uniform pressure on the workpieces. Certain traditional types of welds such as the fillet weld cannot be accomplished without modification of the standard tool geometry.

There are two main classes of FSW tools: single and double shoulder. The tool shown in **Figure 2** belongs to the first category, while the double-shoulder tools have a pin located between two shoulders. These double-shoulder tools create high pressure in the weld zone by forcing the parts into a space slightly narrower than their thickness. This method eliminates the need for a solid backing plate that bears the axial force in case of single-shoulder technology. Furthermore, in the case of double-shoulder tools, the problem of insufficient penetration is eliminated and the temperature distribution is symmetrical about the center of the weld zone.

After its invention, FSW has been rapidly introduced in various fields: in marine and rail industries, automotive, aeronautic, aerospace, and fixed structures. Various types of materials are now welded, and composite welds (e.g., Al-Cu or Al-steel) are performed. There are also many variations on the standard FSW process. For example, using a procedure essentially similar to FSW, a method that is comparable to traditional resistance spot welding called Friction Stir Spot Welding (FSSW) has been developed. These two techniques can produce similar punctual welds, for various parts with similar geometry and thickness. To produce a weld, a rotating tool is plunged into the material. The axial motion stops when the shoulder touches the surface of the workpiece, the rotating tool stays there for a short period of dwell, and then it is extracted. FSSW has the benefit of being easy to mechanize with a robot, leading to excellent repeatability and improved weld quality compared to resistance spot welding. Another variation on the standard FSW process is the use of a tool with a retractable pin; this type of tool can be used to mitigate the presence of the hole left behind when the tool is retracted in phase 4. This process can be used to join parts where the presence of a hole at the end of the weld line is unacceptable.

The physical principle of FSW has also been used to improve the microstructure of the workpieces. In this technique, called friction stir processing (FSP), an FSW tool is used to modify the microstructure of the material. The principal improvements made by FSP are as follows:

- Creation of very fine microstructures to obtain super plasticity (nanograins can be produced);
- Homogenization of the microstructure to reduce segregation, eliminate porosity, and increase mechanical properties, ductility, and corrosion resistance;
- Introduction of particles to develop composite surface (metal matrix composite (MMC)) and modify the elasticity, wear resistance, thermal and electrical conductivity, or internal damping of the material.

The local modifications performed by FSP to the microstructure can be very beneficial in a zone of high stress, where a good ductility is needed, or where the fatigue life should be increased.

Numerical simulation of FSW is a popular field of research since the underlying physics is complex and requires the use of advanced multi-physics solvers. There are various numerical methods that can be used to simulate the friction stir-welding process. The finite difference method (FDM) and the finite element method (FEM) have certain applicability for studying the temperature distribution (heat transfer simulations). Lagrangian-based FEM typically will suffer excessive element distortion for processes that occur with large finite strains. The finite volume method (FVM) is also popular for studying the material flow and is strictly an Eulerian approach (cannot follow the evolution of each material point). Arbitrary Lagrangian Eulerian (ALE) is a meshed-based method that includes a material advection of the Lagrangian mesh within an Eulerian mesh. This allows for larger levels of plastic deformation to be studied. However, the method does have certain downfalls. Since the ALE scheme is highly dissipative, this makes simulating long processes (such as FSW) prone to precision error. The method also suffers from advection errors when the material movement is predominately out of the corner of an element (the classic ALE scheme advects material orthogonal to element faces). To date, mesh-free methods such as smoothed particle hydrodynamics (SPH) have shown the most potential to simulate the entire FSW process. Because SPH is meshfree, very large plastic deformation can be simulated without the problem of mesh distortion. Although the SPH method is computationally burdensome, the method can easily be adapted to run in parallel on the graphics processing unit (GPU) to significantly improve the calculation time.

Shi et al. [1] studied the effects of ultrasonic vibration to improve the weld quality using computational fluid dynamics (CFD). They validate their model by comparing predicted temperature and flow for experimental work. They note that the ultrasonic-assisted FSW process provides a larger flow region and allows for faster welding without the presence of defects. Since they use CFD, they are not able to follow the material history (Eulerian frame of reference). Furthermore, they cannot predict residual stresses or defects in the weld zone. Fraser et al. [2] have used FDM to predict the temperature distribution during the full FSW process. They use the results to find the optimal process parameters (based on an optimal temperature). Their method is efficient and was shown to correlate well with experimental work. The model is limited to temperature calculation and cannot be used to predict deformations, stresses, and defects.

Buffa et al. [3–5] used FEM to develop a hybrid model capable of determining the residual stresses in the resulting weld. They split the FSW process simulation into two phases. In the first phase, they model the plunge, dwell, and advance using a rigid viscoplastic model (fluid-based) that does not provide elastic stresses. Then, they switch to an elastic-plastic model to approximately calculate the resulting residual stresses during weld cooldown. They are able to obtain good correlation for the residual stresses. On the downside, their model does not allow for tracking defects since the welding phase is based on a fluid model.

Guerdoux and Fourmont [6] used the ALE method to study the different phases of the entire process. They used an elastoviscoplastic rate and temperature-dependant material model with the Hansel-Spittel rheological model. On the downside, the Hansel-Spittel model requires coefficient fitting from tensile tests and the coefficients are not commonly available. Grujicic et al. [6–8] as well as Chiumenti et al. [7] used ALE to simulate the FSW process and considered the effect of pin shape, contact friction, material and temperature flow. Their models are highly sophisticated, but are not able to predict residual stresses and defects. They noted that the calculation time is many weeks with their approach.

Bohjwani [8] used the SPH method to study the FSW process with the Johnson-Cook constitutive model in LS-DYNA. At the time, it was not possible to perform a coupled thermomechanical SPH simulation. As such, thermal softening is not taken into consideration. Timesli et al. [9] used the SPH method in two dimensions (2D) to simulate the FSW process. They have used the fluid formulation that directly calculated the deviatoric stress from the strain rate and a non-Newtonian viscosity (function of temperature). They showed that their model correlates well to an equivalent CFD model; however, they did not validate the model experimentally. Recently, Pan et al. [10] used the SPH method to solve the fully coupled thermomechanical problem for the FSW process in three dimensions (3D). Their approach gives detailed grain size, hardness, and microstructure evolution using the SPH method. However, they use a fluid-based formulation that does not allow the determination of elastic strains and stresses. Fraser et al. [11–13] have used the SPH method to simulate various FSW processes using a fully coupled thermos-mechanical SPH-FEM model. The tool is modeled with rigid FEMs and the workpieces with SPH. The model is able to predict temperatures, stresses, and defects all within a Lagrangian framework. This approach permits following the material point history throughout the entire welding process. Since the tool is modeled with FEMs, friction contact can be included.

In this chapter, we describe our approach toward simulating the entire FSW process using SPH on the GPU. In Section 2, we explain what SPH is and how the method can be used to solve large plastic deformation problems with an elastic-plastic formulation, including a description of our parallelization strategy on the GPU. Section 3 introduces the simulation model of a complex aluminum alloy joint. The simulation model will be used to show the power of the SPH method. A validation case is presented to show that the model is able to predict tool torque, force, and the temperature distribution, as well as the size and shape of the flash. Finally, Section 4 wraps up the chapter with concluding remarks and an outlook toward the future of FSW simulation.

2. Simulation theory

2.1. Smoothed particle hydrodynamics

Smoothed particle hydrodynamics is an advanced Lagrangian mesh-free simulation method. The numerical technique has applications in a wide variety of dynamic problems such as astrophysics, magnetohydrodynamics (MHD), computational fluid dynamics, and computational solid mechanics (CSM). The method was originally proposed by two independent research groups within the same year. Gingold and Monaghan [14] showed that the method could be used to simulate nonspherical stars, and Lucy [15] used the method to test the theory of fission for rotating protostars. One of the first groups to apply the SPH method to solid mechanics was Libersky and Petschek [16] in 1991.

What makes the SPH method meshfree is that the set of field equation (conservation equations for a solid body in this case) is solved by interpolation using a kernel, $W(r, h)$, from a set of j neighbor particles that are within the influence domain of a particle of interest, i. **Figure 3** gives a graphical representation of this concept. In this method, continuous field equations are "weakened" into a set of discrete ordinary differential equations. A continuous function is approximated by an interpolant through the use of a convolution integral:

$$f(\bar{x}) = \int f(\bar{x}')W(\bar{x} - \bar{x}', h)d\bar{x}' \tag{1}$$

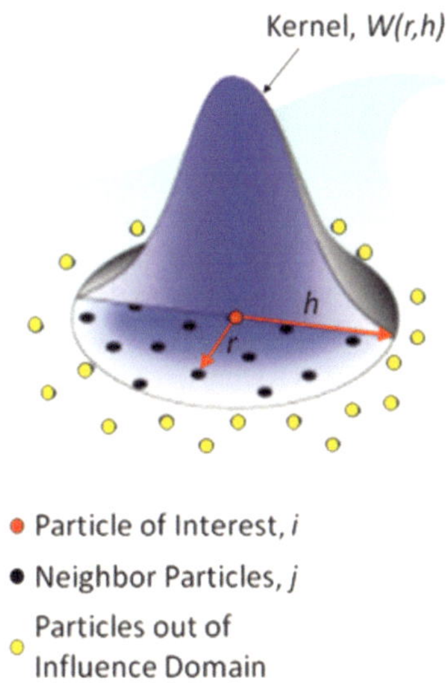

Figure 3. Smoothed particle interpolation.

is called the kernel, also commonly referred to as the smoothing function. It is a function of the spatial distance between the point at which the function is to be calculated (calculation point, $\bar{x}$), the interpolation location ($\bar{x}'$), and the smoothing length, h. The kernel is the key to the SPH method. The continuous SPH interpolation equation can then be written for a set of discrete material points:

$$f(x_i^\alpha) = \sum_{j=1}^{N_i} \frac{m_j}{\rho_j} f(x_j^\alpha) W(r,h) \tag{2}$$

x_i is the spatial location vector for particle i and x_j for the jth particle. m_j and ρ_j are the mass and density of a jth particle and $r = |x_i^\alpha - x_j^\alpha|$. The interpolation kernel, $W(r,h)$, will be written as W_{ij} throughout the rest of the paper. The sum is taken over the total number (N_i) of the j particles within the influence domain of i; these are termed the neighbors of the ith particle. As a general rule, we will use tensor notation to describe variables in continuum equations and indicial notation for the discrete SPH equations. Subscripts are reserved to indicate the ith or j^{th} particle, whereas superscripts follow the general rules of the Einstein notation. For example, the Cauchy stress tensor, $\bar{\bar{\sigma}}$, in this notation would be $\sigma_i^{\alpha\beta}$ for the ith particle.

Determining the neighbors list is a major part of the computational time in the SPH method. We have developed an efficient adaptive neighbor-searching algorithm (complete details in Fraser [17]). The adaptive search typically cuts the search time in half or better.

In the SPH method, the gradient of a vector function can be shown to be simply the function multiplied by the gradient of the smoothing function:

$$\nabla f\left(x_i^{\alpha}\right)=\sum_{j=1}^{N_i}\frac{m_j}{\rho_j}f\left(x_j^{\alpha}\right)\nabla_i W_{ij} \tag{3}$$

The evaluation of first derivatives is straightforward in the SPH method through the use of Eq. (3). The gradient of the smoothing function is given by

$$\nabla_i W_{ij}=\frac{\partial W_{ij}}{\partial \bar{x}_i}=\frac{dW\left(r,h\right)}{dR}\left(\frac{1}{h}\right)\left(\frac{1}{r}\right)\left(\bar{x}_i-\bar{x}_j\right) \tag{4}$$

The smoothing function is typically normalized using the ratio $R=\frac{r}{h}=\|\bar{x}_i-\bar{x}_j\|/h$. The available choices of smoothing functions are vast as this is an ongoing research topic. We have tested a number of different options such as the cubic spline by Monaghan [18], the quadratic function by Johnson and Beissel [19], the quintic function of Wendland [20], and the hyperbolic spline by Yang et al. [21], among others. Of those tested, we have found that the hyperbolic spline is well adapted for simulating friction stir welding with SPH. The function for simulations in three dimensions is defined as

$$W_{ij}=15/62\pi h3\begin{cases} R^3-6R+6, & 0\le R<1 \\ \left(2-R\right)^3, & 1\le R<2 \\ 0, & R\ge 2 \end{cases} \tag{5}$$

2.2. Coupled thermal mechanics SPH formulation for FSW

In this section, the solid-mechanics formulation of smoothed particle hydrodynamics that is used in this work is outlined. The formulation bears close resemblance to that of a fluid

http://dx.doi.org/10.5772/64159

approach; however, the main difference is the ability to account for elastic and plastic deformation. Liu and Liu [22] as well as Violeau [23] provided in-depth development of the SPH conservation equations.

2.2.1. Conservation equations

In order to simulate the FSW process, we must discretize conservation of mass, momentum, and energy using the SPH method previously outlined. We use the weakly compressible approach that is common for large deformation problems (e.g., see [24–27]). Fundamentally, for a system described by particles, mass is inherently conserved at the particle level. It follows then that mass would be conserved for a set of rigid particles (incompressible) that make up a system. On the other hand, for a system made up of non-rigid (compressible) particles, we must take into account the spatial and temporal change of mass, m, within an infinitesimal volume. A convenient measure of this change is the local density, $\rho = m/V$, of an element within the infinitesimal volume. The conservation of mass for a temporally changing compressible system is

$$\frac{d\rho}{dt} + \nabla \cdot \rho \bar{v} = 0 \tag{6}$$

where t is time and $\bar{v}$ is velocity. Using the definitions outlined in Eqs. (1)–(5), we can now write the discrete equation for Eq. (6) as

$$\frac{d\rho_i}{dt} = -\rho_i \sum_{j=1}^{N_i} \frac{m_j}{\rho_j} v_{ji}^{\beta} \frac{\partial W_{ij}}{\partial x_i^{\beta}} \tag{7}$$

where N_i is the number of neighbors of the ith particle and $v_{ji}^{\beta} = v_j^{\beta} - v_i^{\beta}$. There are other forms of conservation of mass in the SPH method; this form is found to be robust and has the added benefit that it provides improved results for a system with significant spatial variation of density such as in multi-phase problems. The continuum mechanics description of conservation of momentum for a solid body is

$$\frac{d\bar{v}}{dt} = \frac{1}{\rho} \nabla \cdot \bar{\bar{\sigma}} + \frac{1}{m} \bar{F}_{ext} + \bar{b} \tag{8}$$

Equation (8) describes the change in velocity (acceleration) of a material point in a solid body subject to internal forces due to stress, $\bar{\bar{\sigma}}$, external forces, $\bar{F}_{ext}$, (on the surface of the body), such as contact forces, and body forces, such as thermal expansion. Gravity is not considered in the formulation as its effects are not significant during the welding process. Now, we are ready to translate the momentum equation for a continuum to the discrete SPH form:

$$\frac{dv_i^\alpha}{dt} = \sum_{j=1}^{N_i} m_j \left(\frac{\sigma_i^{\alpha\beta}}{\rho_i^2} + \frac{\sigma_j^{\alpha\beta}}{\rho_j^2} \right) \frac{\partial W_{ij}}{\partial x_i^\beta} + \frac{1}{m_i} \left(F_{ext} \right)_i^\alpha + b_i^\alpha \tag{9}$$

This version of the momentum equation is commonly called the symmetric form since the pairwise particle interactions are balanced. Moreover, this form exactly conserves linear momentum.

In order to simulate the FSW process, we must take into account the change in energy in the system due to conversion of internal energy (plastic deformation) and frictional heating. The standard energy equation for a weakly compressible body takes on the same form as the heat diffusion equation:

$$\frac{\partial T}{\partial t} = \frac{1}{\rho C_p} \left(k \nabla^2 T + \dot{q} \right) \tag{10}$$

Equation (10) provides the temporal change in temperature, T, in a solid body due to the diffusion of thermal energy. C_p is the heat capacity and $\dot{q}$ takes into account heat generation and dissipation due to plastic deformation, frictional heating, convection, and radiation. The discrete SPH approximation of Eq. (10) is (see Cleary and Monaghan [28])

$$\frac{dT_i}{dt} = \frac{1}{\rho_i C_{pi}} \sum_{j=1}^{N_i} \frac{m_j}{\rho_j} \frac{\left(4k_i k_j\right)}{\left(k_i + k_j\right)} \frac{\left(T_i - T_j\right)}{\left|x_{ij}\right|^2} x_{ij} \frac{\partial W_{ij}}{\partial x_i^\beta} + \dot{q}_i \tag{11}$$

Although frictional heating, convection, and radiation are surface integrals, we have found that these terms can be approximated as volume integrals without any loss of precision for the FSW simulations. In this sense, the heat generation and dissipation take on the following form:

$$\dot{q}_i = \frac{1}{\rho_i C_{p_i}} \left(\sigma_i^{\alpha\beta} \dot{\bar{\bar{\varepsilon}}}_i^{\rho\,\alpha\beta} + \frac{\lambda_i}{V_i} \left(F_{T_i}^{\ \alpha} v_{T_{ij}}^{\ \alpha} \right) \right) + \frac{A_i}{m_i C_{p_i}} \left(h_{\text{conv}_i} \left(T_\infty - T_i \right) + \varepsilon_i \sigma_{\text{SB}} \left(T_{\text{surr}_i}^{\ 4} - T_i^4 \right) \right) \tag{12}$$

where $\dot{\bar{\bar{\varepsilon}}} p_i^{\alpha\beta}$ is the plastic strain rate tensor, $V_i = m_j / \rho_j$, $F_{T_i}^{\alpha}$ is the tangential force from sliding contact (we have used a constant coefficient of friction with the standard Coulomb friction law), $v_{T_{ij}}^{\ \alpha} = v_{T_i}^{\ \alpha} - v_{T_j}^{\ \alpha}$ is the relative tangential velocity at the contact surface, h_{convi} is the coefficient of convection, A_i is the equivalent surface area of a particle taken to be h^2, ε_i is the emissivity of the workpieces, σ_{SB} is the Stefan-Boltzmann constant, T_∞) is the temperature of the ambient air, and T_{surr} is the (average) temperature of the surroundings. The friction heat is distributed into the workpieces (i^{th} particle) and the tool (j^{th} finite element) by the λ_i parameter:

$$\lambda_i = \frac{\sqrt{k_i C_{pi} \rho_i}}{\sqrt{k_i C_{pi} \rho_i} + \sqrt{k_j C_{pj} \rho_j}} \tag{13}$$

Certainly, the heat loss and gain at the surface (convection, radiation, and friction heating) can be evaluated accordingly as surface integrals. However, we have found that the added complexity does not lead to improved precision for the FSW models that we have considered. In our experimental work, the surfaces of the workpieces are painted black to improve the quality of the image taken with an infrared camera. Note that for unpainted aluminum, the emissivity is very low (often less than 0.1); however, for a painted plate, the emissivity is ~0.95. Because of this, radiation effects are significant and should not be disregarded in the energy balance.

2.2.2. Stress and strain in SPH

The stress state can be updated in the material using a frame-indifferent objective stress rate equation. There are many different stress rate equations that can be used such as Truesdell, Green-Nahgdi, or the Jaumann rate equation (others exist). The Jaumann rate has a relatively simple formulation, thus making it unassuming to implement in a CSM code. The rate equation is

$$\dot{\bar{\bar{S}}} = 2G\left(\dot{\bar{\bar{\varepsilon}}} - \frac{1}{3} tr\left(\dot{\bar{\bar{\varepsilon}}}\right)\bar{\bar{\delta}}\right) + \dot{\bar{\bar{S}}}\,\bar{\bar{\Omega}}^T + \bar{\bar{\Omega}}\bar{\bar{S}} \tag{14}$$

$\dot{\bar{\bar{S}}}$ is the time rate of change of the deviatoric stress, G is the shear modulus of the material, $\dot{\bar{\bar{\varepsilon}}}$ and $\bar{\bar{\Omega}}$ are the strain rate and spin tensor, respectively, and $\bar{\delta}$ is the Kronecher delta. The rate equation can be transformed into the discrete SPH formulation by using the discrete form of the strain rate and spin tensors:

$$\dot{S}^{\alpha\beta} = 2G\left(\dot{\varepsilon}^{\alpha\beta} - \frac{1}{3}\delta^{\alpha\beta}\dot{\varepsilon}^{\gamma\gamma}\right) + S^{\alpha\gamma}\Omega^{\beta\gamma} + \Omega^{\alpha\gamma}S^{\gamma\beta} \tag{15}$$

The strain rate is found from

$$\dot{\varepsilon}_i^{\alpha\beta} = \frac{1}{2}\sum_{j=1}^{N_i}\left(\frac{m_j}{\rho_j} v_{ji}^{\alpha}\frac{\partial W_{ij}}{\partial x_i^{\beta}} + \frac{m_j}{\rho_j} v_{ji}^{\beta}\frac{\partial W_{ij}}{\partial x_i^{\alpha}}\right) - \beta_{expand\,i}\dot{T}_i\delta_{ij} \tag{16}$$

The $\beta_{\text{expand}_i}\dot{T}_i\delta_{ij}$ term takes into account the thermal strain rate and allows us to include thermal expansion. β_{expand} is the coefficient of volumetric expansion of the material. The SPH form of the spin rate is

$$\Omega_i^{\alpha\beta} = \frac{1}{2}\sum_{j=1}^{N_i}\left(\frac{m_j}{\rho_j}v_{ji}^{\alpha}\frac{\partial W_{ij}}{\partial x_i^{\beta}} - \frac{m_j}{\rho_j}v_{ji}^{\beta}\frac{\partial W_{ij}}{\partial x_i^{\alpha}}\right) \tag{17}$$

As the SPH method used is that of a weakly compressible approach, an equation of state is required to link the pressure, p, to the density, and speed of sound, c:

$$p_i = c^2\left(\rho_i - \rho_{i0}\right) \tag{18}$$

Plasticity is included in the simulation by using an elastic-perfectly-plastic–thermal-softening flow stress model of the form

$$\sigma_y\left(T\right) = \sigma_{y0}\left(1 - \left(\frac{T - T_R}{T_{melt} - T_R}\right)^m\right) \tag{19}$$

Here, σ_{y_0} is the room temperature yield strength, T_{R} and T_{m} are the room and melt temperature, respectively. m is the thermal-softening exponent. Plasticity is accounted for using the radial return algorithm (see [29–31] for further details).

2.3. Parallelization strategy on the GPU

Many types of engineering simulations require a large amount of computational time due to the complexity of the numerical model and/or the sheer size of the computational domain. In the case of friction stir welding, capturing all the aspect of the process requires a multi-physics approach that is very computationally burdensome. A typical FSW simulation can take many days or even weeks running on a single processing unit (sequential approach). For this reason, it is critical to be able to find an efficient means to run the simulation code in parallel. The idea is to split the domain into subregions and assigns them to individual processing units.

There are a number of different parallelization strategies that can be used. A popular method for small- to medium-sized models is to use a shared memory parallel (SMP) approach wherein each processor has its own set of tasks, but the processors share memory. In this sense, all the simulation data are stored in a common memory location. OpenMP is a very common directives-based programming language that can be used for SMP codes running on central processing units (CPUs).

http://dx.doi.org/10.5772/64159

For larger models, a different tactic is often employed with large number of CPUs, whereby the model and the data in memory are split up and assigned to individual compute "nodes." This approach is called distributed-memory parallel and requires the individual compute "nodes" to be linked by a network. A message-passing interface (MPI) can be used to provide the communication.

Another parallelization strategy that has become very popular is to use the graphics processing unit. Today's GPUs have hundreds and in most cases thousands of "cores". **Figure 4** shows a schematic of the architecture of a typical GPU. We can see that each multiprocessor is composed of a large number of "thread processors". The GPU has its own memory called global memory that is accessed by all the multiprocessors. For this reason, as much as possible of the code should be programmed on the GPU to limit the amount of data transfer between the CPU and GPU.

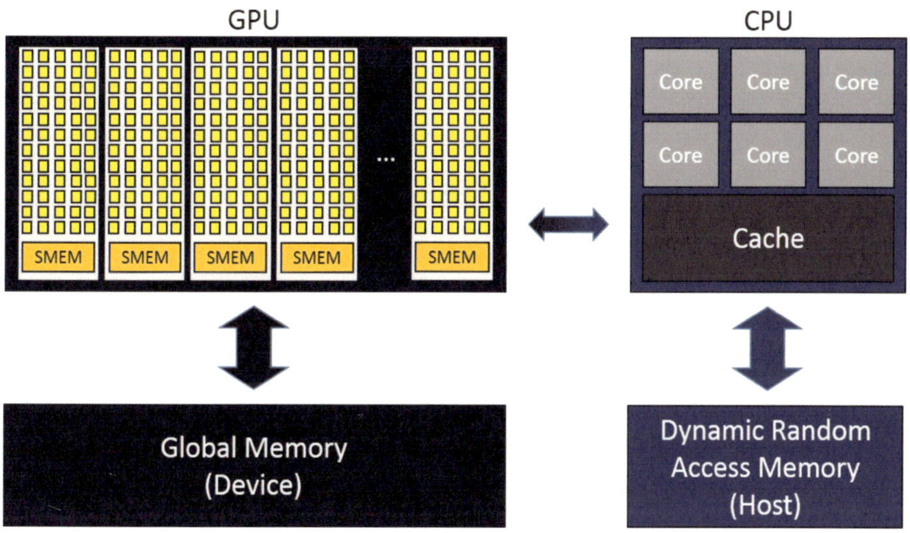

Figure 4. GPU architecture (adapted from NVIDIA [32] and Ruetsch and Fatica [33]).

In the case of simulating the FSW process with SPH, the GPU is ideally suited for parallelization. The large number of streaming multiprocessors on a GPU is perfect for the computationally heavy nature of SPH. SPH codes written to take advantage of the GPU can typically achieve speedup factors of 20–100× over an equivalent serial CPU (e.g., see Dalrymple et al. [34]). In some cases, speedup factors of over 150× are possible, although these are typically problems that are set up to fully exploit the architecture of a specific GPU.

Our parallelization strategy for the SPH code on the GPU is to assign each particle to a thread processor. In this sense, a thread will then carry out a set of calculations for a single particle. The number of threads that can run in parallel is hardware and code specific, but is typically in the multi-thousand range. Certainly, there are different parallelization strategies for SPH on the GPU; however, we have found this approach to be straightforward and efficient.

3. Simulation of a complex FSW joint

To date, most of the work on simulating the FSW process has been focused on a simple butt-joint geometry model. Such a model is sufficient for academic research. However, for real engineering applications, the numerical model should be robust enough to be able to simulate complex geometries within a reasonable timeframe. In this section, we describe the FSW simulation model and results for a complex geometry. The case considered is of an aluminum alloy bridge deck that is fabricated by extrusion in multiple sections and joined using FSW. The joint geometry can be seen in **Figure 5**. One of the drawbacks of using extruded sections is that the parts tend to fit together with some undesirable qualities for FSW. In this case, the two workpieces join together with a ~0.5-mm step at the top surface of the joint (as shown in **Figure 5**). The left-most workpiece is slightly thicker than the other, and, as such, poses a challenge for FSW. The tool will have to push down an extra 0.5 mm in order to come into contact with the lower of the two surfaces. This in turn causes the formation of a significant flash on the thicker workpiece. The overall height of the joint is 100 mm, the three vertical members are 3 mm thick, the thicker plate (left side of step in image) is 3.7 mm thick, and the thinner plate is 3.2 mm thick.

Figure 5. Complex joint.

3.1. Model description

The complex joint geometry is modeled by a combination of SPH for the workpieces and rigid finite elements for the tool. Since the tool is made of hardened steel, it can safely be approximated as a rigid body. The simulation model is shown in **Figure 6**; here, we can see the rigid tool and the two workpieces including the step at the top surface. The mesh size for the finite elements is 0.6 mm in the pin and shoulder region. Large elements are used outside of this region since contact with the workpieces is only during flash formation.

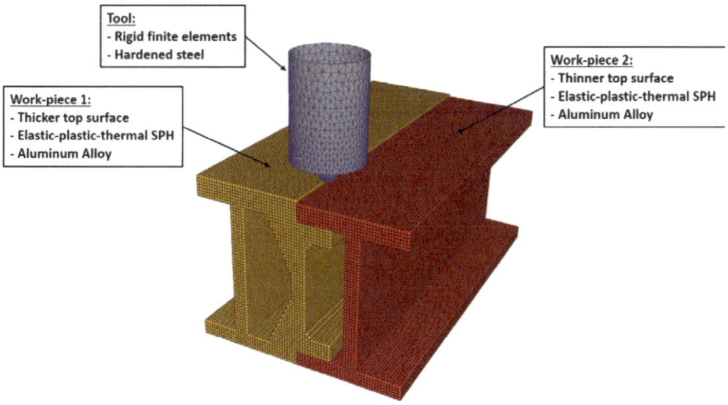

Figure 6. FSW joint simulation model.

The entire joint geometry is modeled with elastic-plastic-thermal SPH elements to allow for an improved prediction of the thermal expansion and the stresses in the joint during the welding process. The vertical member below the weld seam carries 90% of the forge force during the welding process. With our modeling approach, the stresses and the possibility that the member could collapse can be evaluated. The tool interacts with the workpieces through a penalty-based contact algorithm that we have developed for FSW (full details in Fraser et al. [35]). The tool has a shoulder diameter of 15 mm, an average pin diameter of 6 mm, and a pin depth of 3.8 mm. The simulation model is composed of only a small region of interest of the actual bridge deck. Convection (10 W/m^{2}K) is included in the model as well as radiation (the surface of the workpieces was painted black, an emissivity of 0.95) using a novel adaptive thermal boundary condition algorithm (see Fraser et al. [36]). The material parameters of the aluminum alloy used in the simulation are shown in **Table 1**.

Mechanical			Workpieces Thermal		
Parameter	Value	Units	**Parameter**	Value	Units
Density, ρ	2700.0	Kg/m^3	**Conductivity, k**	175.0	W/mK
Initial yield, σ_{y_0}	240.0	MPa	**Heat capacity, C_p**	895.0	J/kgK
Shear modulus, G	26.3	GPa	**Tool thermal**		
Room temperature, T_R	20.0	°C	**Conductivity, k**	55.0	W/mK
Melt temperature, T_{melt}	605.0	°C	**Heat capacity, C_p**	485.0	J/kgK
Softening exponent, m	1.34	-	**Density, ρ**	7850.0	Kg/m^3
Speed of sound, c	4722	m/s			

Table 1. Thermal-physical properties of the aluminum alloy.

We have used a uniform grid particle distribution of 0.6 mm to discretize the workpieces. This spacing allows for a sufficient number of particles through the thickness without incurring excessive calculation penalty. The time step size is selected based on the Courant-Friedrichs-Lewy (CFL) criteria, $dt_{min} = \text{CFL}[h/(v_{max}+c)]$. For this FSW model, we found that CFL = 0.7 was acceptable, leading to $dt_{min} = 9.8 \times 10^{-8}$). The small time step size is one of the major drawbacks of using a solid-mechanics approach. Nevertheless, the time step size is required in order to capture the propagation of elastic stresses within the aluminum.

The model is run as two distinct phases: plunge and advance. The dwell phase was not part of the process as a ramp-up procedure to full advance speed was used in the experiment. A well-defined ramp-up is good practice to limit the forces and torque on the tool and can replace the dwell phase. The plunge speed is 25 mm/min and the full advance speed is 1250 mm/min with 2100 rpm. The ramp-up is performed linearly for an initial tool displacement of 40 mm; after this point, the tool speed is constant at 1250 mm/min.

Because of the 0.5-mm step, excessive amounts of flash are produced as the tool advances. The flash has to be removed following the welding phase and requires a significant amount of work for the welding technician. In order to attempt to reduce the quantity of flash produced, we investigate three cases as follows:

Case 1- As performed in experiment—Full depth plunge (4.3 mm) until the tool shoulder contacts the lower workpiece surface with a counterclockwise tool rotation. This simulation case uses the same process parameters as the production run. The model serves as the validation case using temperature, force, torque, and flash height.

Case 2- Variation 1—Partial depth plunge (4.2 mm) with a counterclockwise tool rotation.

Case 3- Variation 2—Full depth plunge (4.3 mm) until the tool shoulder contacts the lower workpiece surface with a clockwise tool rotation.

Case 1 represents the actual process parameters used in the experiment. This case is used to validate the tool force and torque, as well as the temperature distribution and history. Cases 2 and 3 are variations on case 1. In case 2, we attempt to reduce the quantity of flash by plunging less (4.2 as opposed to 4.3 mm). This will have the effect of limiting the volume of material that is sheared off the top surface of the thicker plate. In case 3, the flash formation will be reduced by operating the FSW tool with a clockwise rotation. This results in the advancing side being on the surface of the thicker plate. This will increase the weld temperature and help to move more material to the lower side of the step, ultimately creating a superior weld compared to cases 1 and 2.

3.2. Simulation results

The three cases were run in SPHriction-3D; in this section, we present the results from the three different cases. The production process parameters correspond to case 1 and are used to validate the model. A video of the results for the three cases is available here: https://www.youtube.com/watch?v=eLOQILkUx-A.

http://dx.doi.org/10.5772/64159

The temperature distribution results for the three cases are shown in **Figure 7** at different times during the simulation. We can see that the maximum temperature for case 2 is lower than for the other two cases. This is because the tool plunges 0.1 mm less, in turn decreasing the forge force and the heat generated due to Eq. (12). The ultimate result is that the quality of the weld in case 2 is significantly lower than in the other cases. Of the three cases, the best weld quality is obtained from case 3. Since the tool rotates clockwise, the advancing side is on the surface of the thicker workpiece. This helps to move the hot material to the thinner workpiece at the front of the tool. This is a favorable situation compared to having the hot material move around the back of the tool (as in cases 1 and 2). We have every advantage to have the advancing side on the thicker workpiece since the pressure is higher there. This causes the workpieces to heat up more uniformly than is possible in either case 1 or 2.

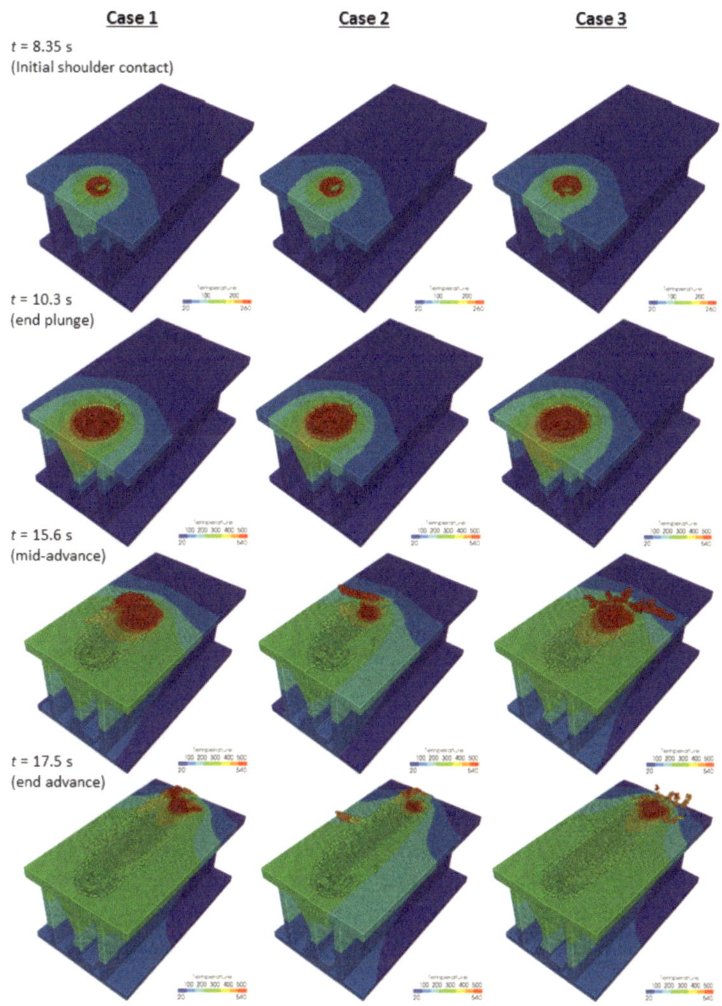

Figure 7. Temperature and deformation results for the three cases.

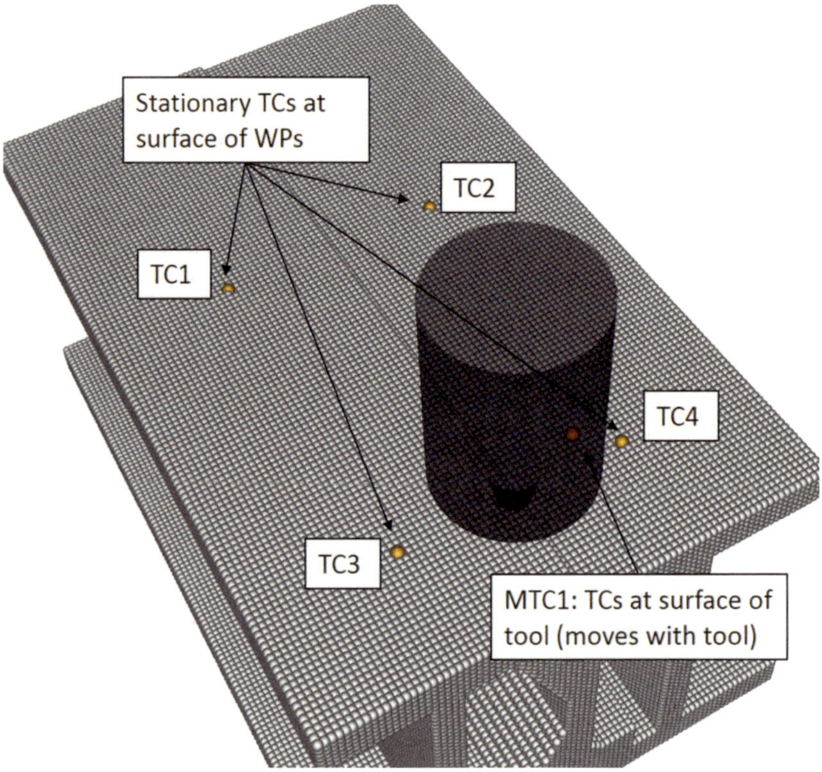

Figure 8. Temperature measurement points in the simulation model.

We have used four measurement points (TCs) for the temperature histories as shown in **Figure 8**. TC1 and TC2 are placed at the middle of the workpiece (along the weld direction). TC3 and TC4 are placed in line with the tool axis during the plunge phase. The four TCs are at the surface of the workpieces and located 11.5 mm from the interface of the two workpieces. MTC1 is a moving temperature measurement point that is located on the underside of the tool and follows the tool as it rotates and advances. MTC1 is located 6 mm from the tool axis on the underside of the tool shoulder.

The temperature was measured experimentally at two points on the surface of the thicker workpiece (at locations TC1 and TC3) using data obtained from an infrared camera (IRcam). Due to the filming angle available with the IRcam (restricted access to work area), temperatures on the thinner workpiece could not be evaluated. **Figure 9** shows that there is a good agreement between the experimental and simulation results. The simulation model has a tendency to slightly overpredict the temperature. Since we have used the perfectly-plastic-thermal-softening model presented in Eq. (19), there is an overprediction of the plastic deformation and in turn an increase in the heat generated as shown in Eq. (12). Furthermore, the heat capacity and thermal conductivity of the aluminum alloy at high temperature are not known. These parameters play an important role in the coupled thermal-mechanical model.

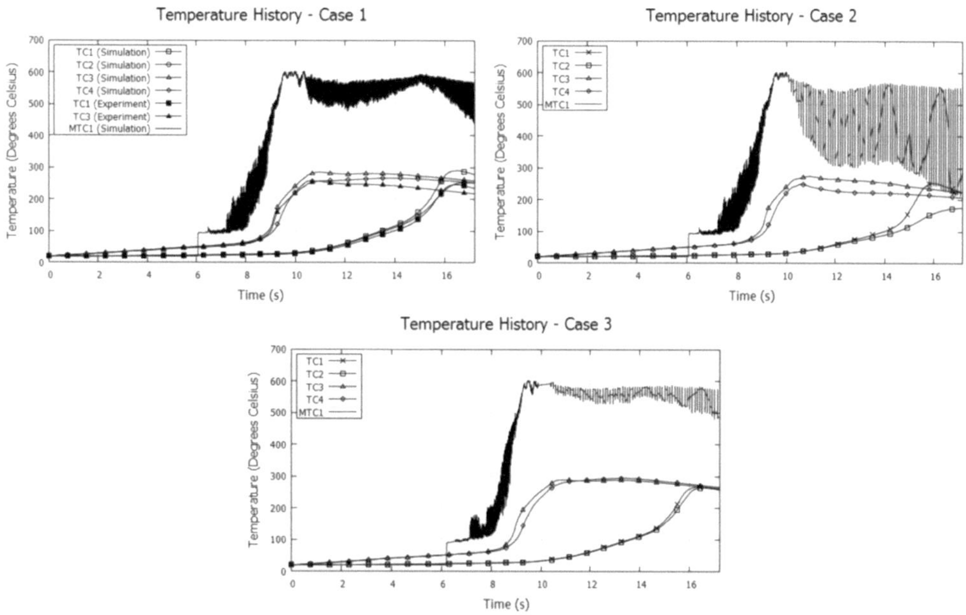

Figure 9. Temperature history results for the three cases.

The relative difference between the TCs on the thicker and thinner plates gives a good means of diagnosing the quality of the weld. If there is a large difference in the temperature reading, we can conclude that the pressure is higher on one side of the weld than the other. This leads to an unfavorable temperature distribution and the weld quality suffers. Case 2 is an excellent example of such a situation. Notice the large difference in temperature in TC3 and TC4. Since the plunge depth was insufficient, there is not enough pressure on the thinner plate, leading to a decrease in temperature.

The temperature results at MTC1 are also an excellent indication of the weld quality. Since MTC1 follows the tool as it rotates and advances, large temperature fluctuations are suggestive of inadequate process parameters. We can see that the variation in temperature at MTC1 for case 3 is significantly less than in other two cases. The experimental setup that we used did not allow us to embed thermocouples in the workpiece or in the tool. Using an IRcam is beneficial in cases such as this since holes do not need to be drilled in the aluminum or in the tool. The surfaces to be filled should be painted a light coat of flat black paint that can easily be removed with light buffing following welding. Temperature measurements with an IRcam provide a very powerful diagnosis tool in the laboratory or in the hands of an FSW technician at a commercial company. The images obtained can help the technician or an engineer to understand whether their chosen process parameters are adequate and if not give good hints as to why. For example, if the IRcam shows a significantly higher surface temperature on the advancing side than the retreating side, the tool is likely advancing too fast for the chosen rpm. During the plunge phase, the IRcam can again be used to determine whether the plunge speed is too high (surface temperature too low) or low (surface temperature too high) (**Figure 10**).

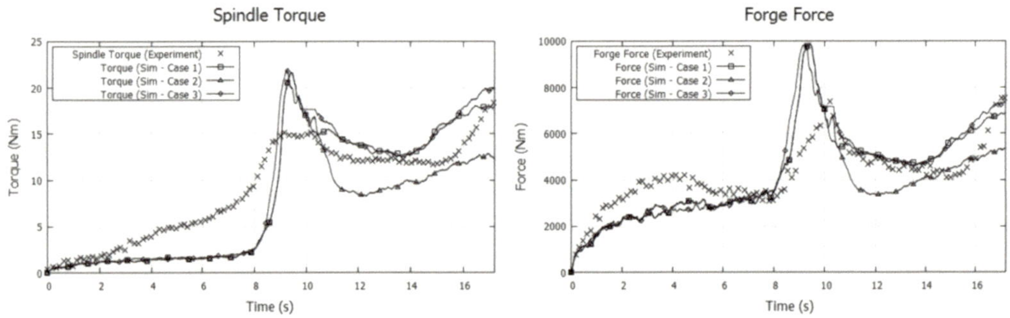

Figure 10. Spindle torque and forge-force comparison.

We can conclude that the weld zone has a more uniform temperature distribution. This leads to favorable welding conditions and results in improved weld quality. Of particular interest is the strong oscillation at MTC1 for case 2. Near the end of the simulation, there is a peak-to-peak temperature change of over 300°C. The temperature on the thinner plate is too low to allow the aluminum material to flow and the weld is essentially incomplete. This can be verified by investigating the plastic strain contours in the weld zone as shown in **Figure 12**. We can see that case 3 is the only one of the three in which the mechanically effected zone spans the entire diameter of the tool. In case 1, the welded zone gets narrower as the tool advances. For case 2, the welded zone spans no more than half the tool diameter from the edge of the tool pin on the thinner plate into the thicker plate.

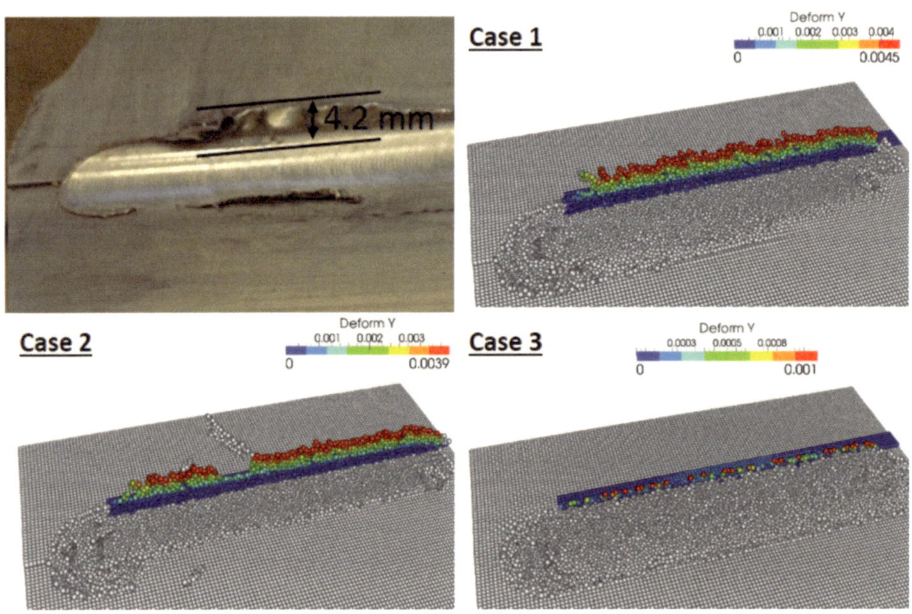

Figure 11. Flash height comparison at the end of advancing phase.

http://dx.doi.org/10.5772/64159

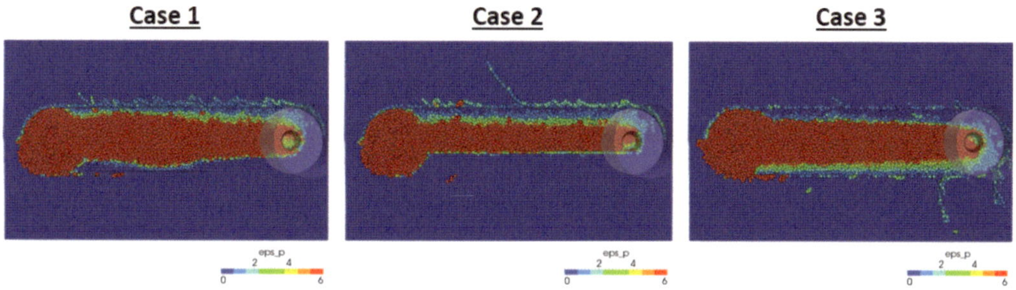

Figure 12. Plastic strain at the end of advancing phase showing the effective weld zone.

A comparison of the spindle torque and the forge force is shown in **Figure 10**. The inertia of the spindle plays a strong role in the experimentally measured torque. Because the plates being welded are very thin, the maximum process torque does not exceed 25 Nm and the average torque during the advancing phase is ~20 Nm. However, the no-load torque measured was ~10 Nm, accounting for almost half of the typical process torque. In the simulation models, the inertial effects of the spindle are not taken into consideration. The simulation torque is calculated by taking the cross-product of the contact forces and the distance vector between the tool axis and an SPH element. For this reason, the torque trends line up well with the experimental data, though the magnitude is diffident (**Figure 11**).

A good correlation between the forge force from experiment and simulation was obtained. The inertial effects do not play an important role here, leading to a better prediction than was obtained with the torque. There are other factors that lead to a reduction in the precision of the predicted torque and forge force such as the thermophysical properties of the material, the chosen friction law, differences in how the FSW machine and simulation model control the position, and rpm of the tool, as well as discrepancies between the actual geometry of the workpieces and the tool compared to their idealization in the simulation model.

Nevertheless, the simulation model provides an excellent understanding of how a change in process parameters effects the torque and forces. We can see that the tool torque and forge force for case 2 are lower than those for cases 1 and 3. This is an intuitive result as the plunge depth is shallower, leading to less contact pressure (**Figure 12**).

A flash height of 4.2 mm was measured experimentally; case 1 predicts a flash height of 4.5 mm, 3.9 mm for case 2, and less than 1 mm for case 3. The flash heights are shown in **Figure 11**; notice that the wavy pattern of the flash is well represented in the simulation model for case 1. The images have been cleaned up to show only the continuous flash line by omitting sporadic "flash flakes" that typically do not require much effort to remove. Clearly, the flash produced in case 3 is significantly less than that in other two cases. The reason is entirely due to the change in tool rotation. Flash lines will most commonly be laid down on the retreating side of the weld. By ensuring that the advancing side is on the thicker side, the material is "ripped" from the thicker side and transported to the retreating side. Because of the height change, the flash is not able to attach to the thinner side and creates intermittent "flakes" that can be removed in less time than is possible in the case of a continuous flash line on the thicker side (as in cases 1 and 2).

4. Conclusions

In this work, we have presented our approach toward simulating the entire FSW process using a solid-mechanics approach. By using a mesh-free numerical method such as SPH, the large plastic deformation encountered during FSW can be easily calculated. Mesh-based methods struggle to capture all the physics of the process due to discretization errors as the mesh distorts. The fully coupled elastic-plastic-thermal code is able to predict temperature, stress, and deformation histories. Because of the mesh-free Lagrangian nature of SPH, the model is able to predict defects (free surface changes) in a way that other numerical methods cannot. The prediction of defects is an invaluable feature for an engineer working on the design of the joint geometry to be welded. Optimal process parameters can then be chosen that lead to no-weld defects. In this manner, the design engineer can find the fastest rate of advance that can be used to increase the overall profit margin during a high-volume production run.

One of the major advantages of using a solid-mechanics approach compared to a fluid approach is that the simulation models are able to capture the elastic stresses and strains. **Figure 13** shows the effective stress in the joint at the end of the plunge phase. This is the point when the forge force reaches its maximum value. This is of great interest to a joint designer who is interested to know if the joint will withstand the forge force during the welding process. If the vertical members under the weld seam are too thin, they will likely undergo significant plastification and could collapse. This certainly would be disastrous for the finished product. Other benefits of including the elastic stresses and strain are the ability to more precisely predict defect size and shape, as well as residual stresses and deformation following a cooldown phase.

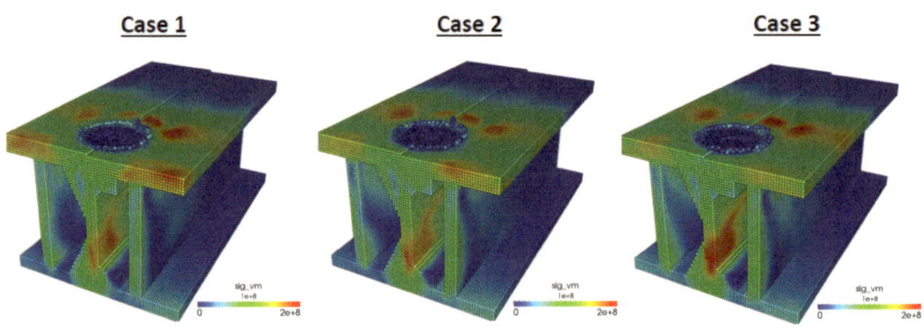

Figure 13. Stress state at the end of plunge phase.

Looking toward the future of numerical simulation of FSW, we can see that as the performance of GPUs continues to improve, larger and more complex simulation models will be possible. We are currently working on a multi-GPU parallelization strategy that will allow tens and even hundreds of millions of SPH elements to be simulated. This approach requires the use of a highly optimized communication strategy between the GPUs (e.g., using MPI). We are currently working on various developments in the code, such as follows:

- Improved contact models with different friction treatments (such as including the shear limit and/or a stick-slip behavior);
- Wear prediction at the surface of the tool using Archard's model;
- Improved thermo-physical material representations that more accurately model the behavior of the aluminum alloy during the FSW process;
- An implicit mesh-free collocation approach that will permit efficient simulation of long duration phases such as cooling.

Since the simulation code is developed using a highly optimized parallel-processing strategy, complex 3D-joint geometries can be simulated within a reasonable period. In this work, the three simulation models were run simultaneously on a personal workstation with three individual GPUs. The cost of such a computer is less than five thousand dollars in today's market. Because of the parallel strategy, a cluster with many GPUs can be used with 100% efficiency (as long as an individual GPU has enough memory for each simulation model). In the sense of optimization, a company with access to a GPU compute cluster (say eight or more GPUs) could run parametric models (e.g., varying the rpm and advance speed) simultaneously. The obtained data sets would provide the required information to construct a response surface and find the optimal advancing speed and rpm.

Acknowledgements

The authors would like to thank NVIDIA for donating the GeForce GTX Titan Black GPU that was used to perform the simulations. We would also like to thank the Portland Group (PGI) for having generously provided a license for PGI Visual Fortran (PVF) with CUDA Fortran. The project is supported in part by funding from FRQNT, CQRDA, GRIPS, REGAL, RTA, and CURAL. We declare no conflicts of interest associated with this work.

Author details

Kirk Fraser[1,2*], Lyne St-Georges[1] and Laszlo I. Kiss[1]

*Address all correspondence to: Kirk.Fraser@PredictiveEngineering.com

1 Université du Québec à Chicoutimi (UQAC), Saguenay, QC, Canada

2 Predictive Engineering, Saguenay, QC, Canada

References

[1] Shi L, Wu CS, Liu XC. Modeling the effects of ultrasonic vibration on friction stir welding. Journal of Materials Processing Technology. 2015;222:91–102.

[2] Fraser K, St-Georges L, Kiss L. Optimization of friction stir welding tool advance speed via monte-Carlo simulation of the friction stir welding process. Materials. 2014;7:3435–52.

[3] Buffa G, Ducato A, Fratini L. Numerical procedure for residual stresses prediction in friction stir welding. Finite Elements in Analysis and Design. 2011;47:470–6.

[4] Buffa G, Fratini L, Shivpuri R. Finite element studies on friction stir welding processes of tailored blanks. Computers & Structures. 2008;86:181–9.

[5] Buffa G, Hua J, Shivpuri R, Fratini L. A continuum based fem model for friction stir welding—model development. Materials Science and Engineering: A. 2006;419:389–396.

[6] Guerdoux S, Fourment L. A 3D numerical simulation of different phases of friction stir welding. Modelling and Simulation in Materials Science and Engineering. 2009;17.

[7] Chiumenti M, Cervera M, Agelet de Saracibar C, Dialami N. Numerical modeling of friction stir welding processes. Computer Methods in Applied Mechanics and Engineering. 2013;254:353–69.

[8] Bohjwani S. Smoothed particle hydrodynamics modeling of the friction stir welding-process: University of Texas at El Paso; 2007.

[9] Timesli A, Zahrouni H, Braikat B, Moufki A, Lahman H. Numerical model based on SPH to simulate friction stir welding. Revue de Mechanique Appliquee et Theorique. 2011;2:537–46.

[10] Pan W, Li D, Tartakovsky AM, Ahzi S, Khraisheh M, Khaleel M. A new smoothed particle hydrodynamics non-Newtonian model for friction stir welding: Process modeling and simulation of microstructure evolution in a magnesium alloy. International Journal of Plasticity. 2013;48:189–204.

[11] Vigh LG, St-Georges L, Kiss LI, Fraser K. FSW hegesztett aluminium palyalemez—Technologia, analizises meretezes. Femszerkezetek. 2014;III:31–35.

[12] Fraser K, St-Georges L, Kiss LI. Prediction of defects in a friction stir welded joint using the smoothed particle hydrodynamics method. Proceedings of the 7th Asia Pacific IIW International Congress on Recent Development in Welding and Joining Methods. Singapore; 2013.

[13] Fraser K, St-Georges L, Kiss LI. Smoothed particle hydrodynamics numerical simulation of bobbin tool friction stir welding. 10th International Friction Stir Welding Symposium. Beijing, China: TWI; 2014.

[14] Gingold RA, Monaghan JJ. Smoothed particle hydrodynamics: theory and application to non-spherical stars. Montly Notices of the Royal Astronomical Society. 1977;181:375–89.

[15] Lucy LB. A numerical approach to the testing of the fission hypothesis. Astronomy J. 1977;82:1013–24.

[16] Libersky L, Petschek AG. Smoothed particle hydrodynamics with strength of materials. The Next Free Lagrange Conference. New York, NY: Springer-Verlag; 1991.

[17] Fraser K. Adaptive smoothed particle hydrodynamics neighbor search algorithm for large plastic deformation computational solid mechanics. 13th International LS-DYNA Users Conference. Dearborn, MI: LSTC; 2014.

[18] Monaghan JJ. An introduction to SPH. Computer Physics Communications. 1988;48:89–96.

[19] Johnson GR, Beissel SR. Normalized smoothing functions for sph impact computations. International Journal for Numerical Methods in Engineering. 1996;39:2725–41.

[20] Wendland H. Piecewise polynomial, positive definite and compactly supported radial functions of minimal degree. Advances in Computational Mechanics. 1995;4(1):389–96.

[21] Yang X, Liu M, Peng S. Smoothed particle hydrodynamics modeling of viscous liquid drop without tensile instability. Computers and Fluids. 2014;92:199–208.

[22] Liu GR, Liu MB. Smoothed Particle Hydrodynamics : A Meshfree Particle Method. Hackensack, NJ: World Scientific; 2003.

[23] Violeau D. Fluid Mechanics and the SPH Method: Theory and Applications. Oxford, UK: Oxford University Press; 2012.

[24] Cleary PW. Extension of SPH to predict feeding, freezing and defect creation in low pressure die casting. Applied Mathematical Modelling. 2010;34:3189–201.

[25] Cleary PW, Prakash M, Das R, Ha J. Modelling of metal forging using SPH. Applied Mathematical Modelling. 2012;36:3836–55.

[26] Das R, Cleary, P.W. Modelling plastic deformation and thermal response in welding using smoothed particle hydrodynamics. 16th Australasian Fluid Mechanics Conference. Australia 2007.

[27] Das R, Cleary PW. Effect of rock shapes on brittle fracture using smoothed particle hydrodynamics. Theoretical and Applied Fracture Mechanics. 2010;53:47–60.

[28] Cleary PW, Monaghan JJ. Conduction modelling using smoothed particle hydrodynamics. Journal of Computational Physics. 1998;148:227–264.

[29] Banerjee B. An evaluation of plastic flow stress models for the simulation of high temperature and high strain-rate deformation of metals. ActaMaterialia. 2006;58:6810–6827.

[30] Brannon RM. Geometric Insight into Return Mapping Plasticity Algorithms. New Mexico: University of New Mexico; 2002.

[31] Mitsoulis E. Flows of viscoplastic materials: models and computations. In: Binding DM, Hudson NE, Keunings R, editors. Reology Reviews. London, UK: Brit. Soc. Rheol.; 2007. p. 135–78.

[32] Awile O, Büyükkeçeci F, Reboux S, Sbalzarini IF. Fast neighbor lists for adaptive-resolution particle simulations. Computer Physics Communications. 2012;183:1073–81.

[33] Ruetsch G, Fatica M. CUDA Fortran for Scientists and Engineers. Waltham, MA, USA: Elsevier Inc.; 2014.

[34] Valdez-Balderas D, Domínguez JM, Rogers BD, Crespo AJC. Towards accelerating smoothed particle hydrodynamics simulations for free-surface flows on multi-GPU clusters. Journal of Parallel and Distributed Computing. 2013;73:1483–93.

[35] Fraser K, Kiss LI, St-Georges L. Hybrid thermo-mechanical contact algorithm for 3D SPH-FEM multi-physics simulations. 4th International Conference on Particle-Based Method. Barcelona, Spain; 2015.

[36] Fraser K, St-Georges L, Kiss LI. Adaptive thermal boundary conditions for smoothed particle hydrodynamics. 14th Internation LS-DYNA Conference. Detroit, MI, USA; 2016.

Chapter 4

Gas Tungsten Arc Welding with Synchronized Magnetic Oscillation

Thiago Resende Larquer and Ruham Pablo Reis

Additional information is available at the end of the chapter

http://dx.doi.org/10.5772/64158

Abstract

The search for improvements in mechanized/automated welding techniques has been intense due to skilled labour shortage. In this line, the combination of operational modes (polarity and/or metal transfer mode) within a process has gained attention, since it expands the adjustability of the arc energy. By combining this feature with arc motion, the arc energy delivered to the workpiece can be optimally distributed. Therefore, this work exploits the synchronization between arc magnetic oscillation and gas tungsten arc welding (GTAW) process to control weld bead formation. A system was devised to control the magnetic oscillation and a welding power source synchronously. Characterization of the synchronized magnetic oscillation was carried out based on high-speed filming and electrical data. The welding process was then synchronized with the magnetic oscillation varying the level of welding current according to the arc time-position, being the effect on weld bead width considered for analysis. Welding without oscillation and with unsynchronized magnetic oscillation were taken as references. The synchronized magnetic oscillation made possible to achieve larger weld bead width on the side with higher current level and longer lateral stop time and vice versa. This technique might be beneficial to applications where extreme weld bead control is required.

Keywords: magnetic oscillation, arc welding, GTAW

1. Introduction

The search for improvements in mechanized/automated welding processes has been in evidence for quite a long time, with remarkable recurrence and intensity nowadays due to shortage of qualified workforce. One way to exploit the mechanical/automated welding processes with more efficiency (productivity) has been through the combination of operating modes (polarity and/or metal transfer mode) within a single process, in addition to the possibility of combining welding current levels. With the combination of operating modes (possible with modern power sources) and current levels in the same welding operation, it is possible to vary the energy of the process, both thermal (heat delivered to the base metal) and mechanical (arc pressure and impact of droplets on the base metal, the latter in the case of consumable electrodes). In this line, an interesting approach, not much exploited yet, but promising, is to use this feature to distribute the energy of welding optimally into the workpiece to control the weld bead formation (molten material from both electrode and workpiece). This could be done by synchronizing the welding operating modes and/or current levels with the position of the arc/torch. The arc position can be changed mechanically (by moving the torch) or magnetically. The magnetic deflection of arcs (deviation of arc coupling with the workpiece by external magnetic fields) is a relatively versatile and inexpensive technique. The arc magnetic oscillation is composed of a series of magnetic deflections (pendulum-like movement of the arc when subjected to a variable and/or alternating magnetic field). Once the electromagnet is positioned/mounted relative to the arc/torch and thereby the direction of the magnetic flux lines is defined (longitudinal for lateral/transversal oscillation and transversal for longitudinal oscillation), the extension of the arc movement in each position depends on the magnetic field level applied and the time spent in each position depends on the application time of the magnetic field. As shown in **Figure 1**, the direction of deflection (left and right or forward and backward in relation to the welding travel speed direction) depends on the direction of magnetic flux lines produced by the electromagnet; the inversion of arc positions/direction of deflection is given by the inversion of the electromagnet control signal (voltage/current).

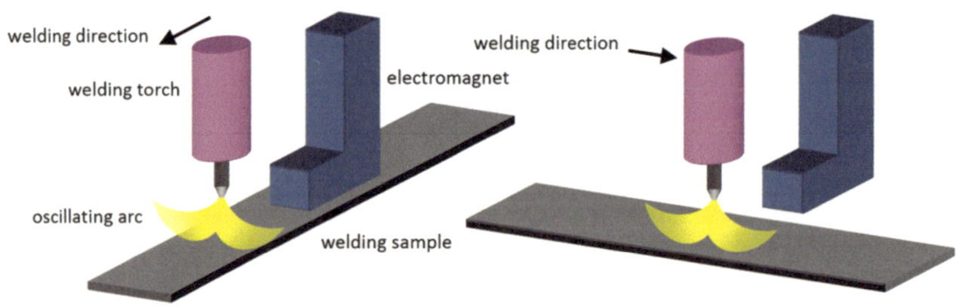

Figure 1. Direction of magnetic deflection: on the left hand side—magnetic field parallel/aligned with the welding direction generates transversal/lateral arc deflection; on the right hand side—magnetic field transversal to the welding direction generates longitudinal arc deflection.

This work aims to better exploit the potential of magnetic oscillation. The overall goal is to synchronize the magnetic oscillation of the arc with the welding process (current levels), gas tungsten arc welding (GTAW) in this case, and evaluate the potential of this technique to control/modify the welding results, specifically in terms of the weld beads' external geometry. With the synchronization proposed, it would be possible to choose the thermal and mechanical energy of the arc (current levels) for each of its positions. The setting of arc position and time at each position with prechosen energies (welding current levels) is controlled by the wave-

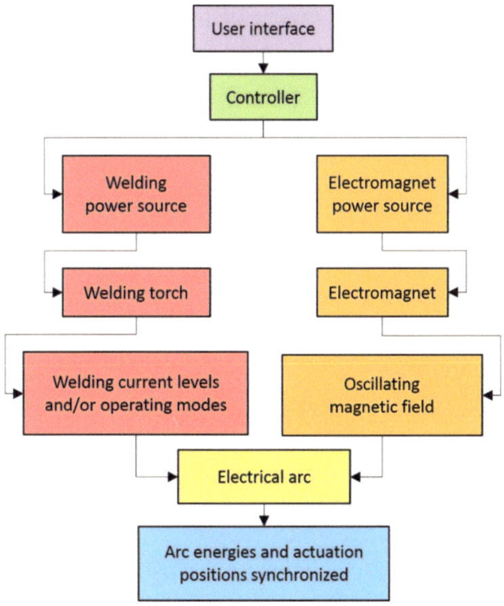

Figure 2. Schematic description of the idea of synchronizing the arc positions with its energy levels using magnetic oscillation.

form (amplitude and time) of the voltage/current signal applied to the electromagnet. A short schematic description of the idea of synchronizing the arc positions with its energy levels by the use of magnetic oscillation is shown in **Figures 2** and **3**. Examples of motivating applications for the development of magnetic oscillation synchronized with welding processes are

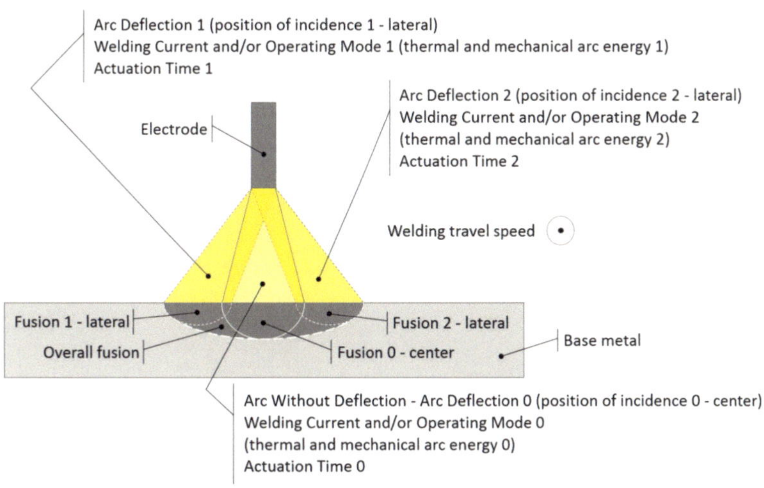

Figure 3. Schematic welding transverse cross section with the arc at the centre, left, and right positions with different levels of thermal and mechanical energy in each position as result of synchronizing the arc positions with its energy levels using magnetic oscillation in the case of transverse arc deflection.

the ability to act differently on the geometry of the weld beads (molten and heat-affected zones), affect grain size for improvement of weld properties, allow weld pool control for out-of-position welding operations, and facilitate narrow gap welding, root passes, among others.

2. Bibliographic review

Magnetic fields are intrinsic to the welding arc. As it is widely known, self-induced magnetic field is the basis for the formation of plasma jet, which has beneficial effects on the penetration of the weld bead, for example [1, 2]. On the other hand, external magnetic fields can be used to oscillate arcs, replacing mechanical devices for coating operations, for example. The idea of using magnetic fields to oscillate welding arcs is not new. It was designed and patented in 1960 by Greene [3]. Currently there are commercial systems to magnetically oscillate welding arcs, and alternating current sources are used to control the oscillation. General recommendations for construction of electromagnets for welding arc deflection are found in the literature [4].

The magnetic deflection of welding arcs can occur in various ways. The most commonly known is certainly the magnetic arc blow [1, 5, 6]. Another is the deflection of arcs in double wire gas metal arc welding (GMAW) [7]. Additionally, there is the case of deflection caused by external magnetic fields, such as those used in equipment to deflect welding arcs [8].

It is important to review the basic electromagnetic effect that governs the magnetic deflection phenomenon. If an electric charge travels within a magnetic field, it is subjected to a magnetic force of magnitude proportional to its velocity and magnetic field strength (**Figure 4**). The direction and orientation of the force are determined by the left-hand rule; place the index finger of the left hand in the direction of the magnetic field lines and the middle finger in the direction of conventional electrical current. In this case, the thumb when oriented perpendicular to the index finger points in the direction of the force to which the electric charge is subjected. A charged particle, stationary or moving parallel to the magnetic field lines, will not suffer any magnetic-induced force due to this field. However, a charged particle travelling, not parallel, through a magnetic field will have its direction of movement changed, that is, its trajectory will undergo magnetic deflection.

With the illustration of **Figure 4** in mind, **Figure 5** sequentially shows an arc deflection can be obtained by applying an external magnetic field on it. In accordance with the principles of electromagnetism, a linear conductor in the presence of a magnetic field is subjected to a force

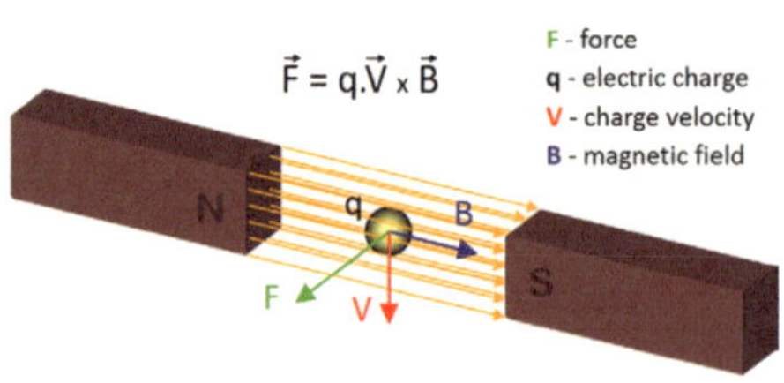

Figure 4. Force produced on a positive electrical charge moving through a magnetic field (the force direction points the opposite way if the electrical charge is negative).

proportional to the conductor length within the magnetic field, the electrical current flowing through this conductor and the magnetic flux density. Therefore, in welding, in a simplified manner, if a current *I* is flowing from the electrode to the workpiece through an arc of length *La* and this arc is in the presence of a magnetic field *Be* (externally produced by an electromagnet, for example), a force *F* acting in the arc (perpendicular to the magnetic field and current flow) is generated.

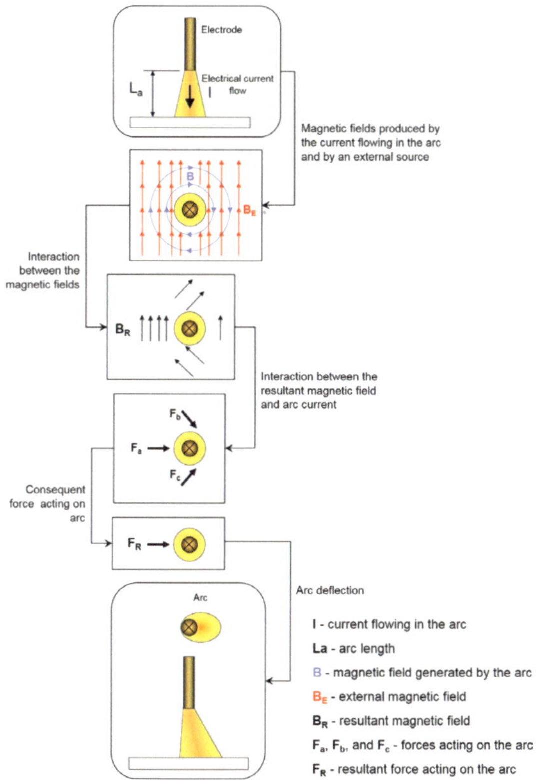

Figure 5. Diagrammatic explanation on how the deflection of an arc in the presence of an external magnetic field takes place.

2.1. Advantages and limitations of arc magnetic oscillation

Perhaps the main advantage of using magnetic oscillation is the virtually unlimited capability to create arc deflection patterns, either sideways or forward and backward relative to the direction of welding. Manufacturers of magnetic oscillation systems commonly point arc stabilization, arc positioning, heat distribution control, undercut minimization, porosity reduction, improved penetration, and uniform side melting in joints as advantages of this technique. In practical terms, the magnetic deflection is more adequate for high-frequency movements and with greater precision (no mechanism inertia, etc., typical of mechanical devices). Despite the fact that magnetic oscillation can be used in favour of welding, some issues may arise related to the use of magnetic fields along electric arcs. Perhaps the destabilization of the arc in the presence of strong magnetic fields is the main disadvantage of using magnetic fields to oscillate welding arcs. These instabilities in the arc may even lead to its extinction, even temporally. Problems with arc instability and interruption in double wire GMAW are mentioned in the literature [7, 9–11]. The main reason for this phenomenon is linked to magnetic fields generated by the arcs operating adjacent to each other and the "stiffness" presented by these arcs. Magnetic fields up to 50 Gauss have been used to oscillate welding arcs without problems [8], although manufacturers build systems to operate up to 600 Gauss. Of course, what really matters is the value of the magnetic field acting effectively on the arc. In practical terms of magnetic oscillation, there may be limitations on the range (extension) of the arc deflection, since the arc is attached on one end (electrode) and moves on the other (workpiece) such as a pendulum.

2.2. Applications of magnetic oscillation

Several studies have been conducted to explore the application of magnetic oscillation to control the weld bead geometry and hence mitigate defects, as well as to improve mechanical properties of the weld as a result of grain refinement, for instance. A study analyzed the effect of frequency and amplitude of GTAW arc oscillation on the mechanical properties of the welded material [12]. The results demonstrated a grain refinement as compared with welds realized with constant and pulsed currents, both without arc oscillation. The obtained hardness was higher due to the grain refinement and low segregation of phases. Another study investigated the grain refinement in aluminium alloys [13]. The results concluded that by magnetically oscillating the arc it is possible to disturb the profile of solidification of the weld pool, causing the grain refining of the molten zone. Magnetic oscillation has been successfully used in GTAW for grain refinement of titanium alloys [14]. Another work used the transverse magnetic oscillation in GTAW with filler metal, and by extending the amplitude of the magnetic field, the authors obtained an increase of the weld bead width and were able to reduce penetration [15]. Another study used transverse magnetic oscillation in GMAW for narrow gaps and the authors obtained good penetration and melting uniformity on both sides of the groove [16]. A more recent work used a system to synchronize the electrode polarity with the torch position in GMAW for hardfacing, that is, the synchronized oscillation (weaving) was with a mechanical device [17]. In this case, negative polarity was used in the centre of the weld bead (high melting rate and welding speed, low dilution and penetration) and positive polarity was used on the sides of the weld bead to facilitate overlapping of the next bead, avoiding lack of fusion defects. According to the authors, the process was satisfactory for carrying out hardfacing with little penetration, surface smoothness, and good aspect ratio (width/height). In addition, the weld beads showed no discontinuities and had an excellent visual appearance with few spatters. Therefore, the synchronization between magnetic oscillation (arc position) and the welding process (level of current and/or operating mode) may have potential in similar situations.

3. Methodology and results

The synchronization between the magnetic oscillation of the arc and the welding process, GTAW in this case, was assessed in two parts; characterization of arc deflection and magnetic oscillation synchronized with GTAW. Concerning the arc deflection characterization, high-speed filming and electrical signal data were used to evaluate the GTAW arc behaviour during magnetic oscillation and to verify if the synchronization system was working properly. In addition, some general consideration on the effect of the synchronization on the weld bead formation was carried out. In the assessment of the magnetic oscillation synchronized with GTAW, transverse/lateral oscillation to the direction of welding was employed with three arc stop positions synchronized with three welding current levels and three actuation times (one for each position), as illustrated in **Figure 3**. To support the analysis of this combination, electrical signals from the electromagnet and the welding process, including electrical transients, were assessed along with weld surface appearance as well as measurements related to the width of the resulting weld beads. All welds were produced as bead-on-plate tests in 250 X 60 X 3 mm mild carbon steel and argon was used at 14 l/min as shielding gas. The arc length (electrode to workpiece distance) was always kept at 6.5 mm (this setting is a little above the value conventionally used for welding, but was adopted to increase the arc deflection and therefore boost any related effect). A Th2 tungsten electrode with 4 mm diameter and 60 degrees sharp was used. The welding travel speed used was always 200 mm/min, unless stated differently. The magnetic flux density acting on the arc was estimated with a Gaussmeter by conducting measurements for different electromagnet voltages for an electromagnet-to-GTAW-electrode (arc centre) distance of 15 mm and with the electromagnet placed 3 mm above the test sample (as the actual welding tests) (the measurements are shown in **Figure 6**), but with no arc (no welding). **Figure 7** illustrates the general equipment used during the tests. As shown, the test samples were replaced by a stationary water-cooled copper block to facilitate high-speed filming (no weld pool formation). It is worth saying that the welding power source employed (IMC DIGIPlus A7) allows to switch between up to six pre-set welding programs (welding modes and/or current levels) by an external control input, the same used to control the electromagnet and then synchronize the magnetic oscillation with the welding process.

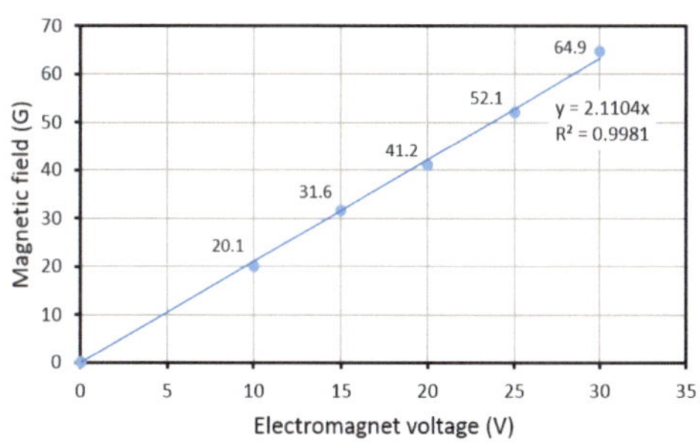

Figure 6. Magnetic field "acting on the arc" versus electromagnet voltage for an electromagnet-to-GTAW-electrode (arc centre) distance of 15 mm and with the electromagnet placed 3 mm above the test sample.

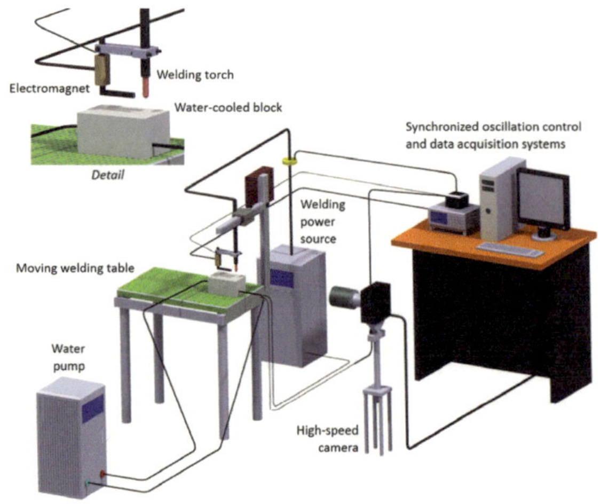

Figure 7. Illustration of general equipment used for testing GTAW with synchronized magnetic oscillation.

3.1. Characterization of arc deflection

3.1.1. Magnetic deflection response time

In order to verify the responsiveness of the synchronized magnetic oscillation system, three tests were carried out as shown in **Table 1**, and high-speed images of the GTAW arc analyzed. By the images of the deflected arc (**Figure 8**), it is possible to see that with the lateral (left and right) stop times set in 50 ms the arc reached virtually the same deflection levels obtained with 200 ms of lateral (left and right) stop times. Thus, 50 ms was considered sufficient for the electromagnet coil current (controlled by the electromagnet voltage) to reach the level required to take the arc to the expected deflection range (around 12 mm). On the other hand, the reduction of the lateral (left and right) stop times to only 5 ms made the arc reach levels significantly reduced (to about 8 mm), indicating that in this case the electromagnet coil current had not reached the level required to lead the arc to the expected deflection range. As the manufacturer of the welding power source recommends a dwell time in each welding mode or current level of at least 100 ms and considering that 50 ms allowed the expected level of arc deflection, 100 ms will be the minimum allowed for the actuation time of the combinations of arc position and welding current level.

Test	Welding current (A)	Left stop time (ms)	Right stop time (ms)
1	200	50	50
2	200	5	5
3	200	200	200

Table 1. Tests to assess the responsiveness time of the synchronized magnetic oscillation system (electromagnet voltage = ±20 V—tests without central stop time).

http://dx.doi.org/10.5772/64158

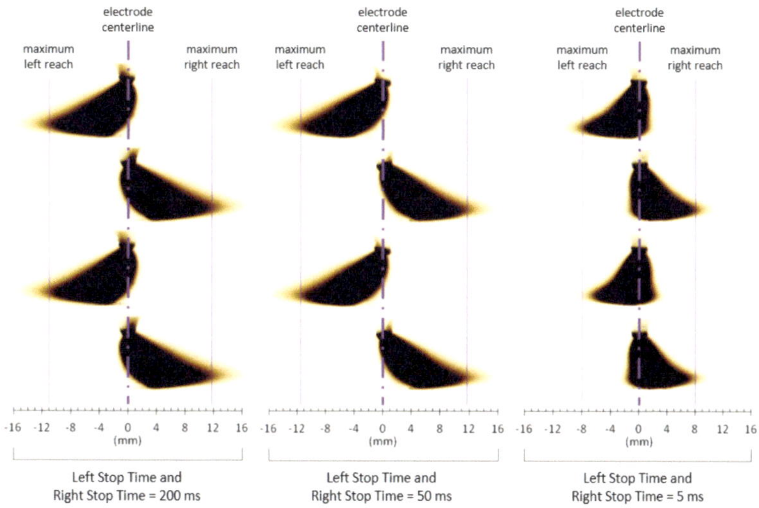

Figure 8. Maximum arc deflection reach for different lateral stop times (the arc image colours are inverted for better visualization).

3.1.2. Synchronism between magnetic oscillation of the arc and welding current level

A test was carried out to demonstrate visually the synchronism between the arc melting capacity (represented by the welding current level) and the position-time of application of this melting capacity (determined by the magnetic oscillation). **Table 2** lists the parameters used in this test.

Test	Left stop time (ms)	Central stop time (ms)	Right stop time (ms)	Left welding current (A)	Central welding current (A)	Right welding current (A)
4	250	500	750	100	200	300

Table 2. Test to demonstrate the synchronism between magnetic oscillation (arc time-position) and welding current level (electromagnet voltage = ±20 V).

Figure 9 demonstrates how the GTAW arc resultant from test 4 changes its position due to the magnetic field controlled by the electromagnet voltage as well as how it changes its "volume" due to the welding current applied in each position. The arc starts deflected to the left (left stop position) showing small "volume" level due to the small welding current used (low ionization degree). Past the short application time of low current, when the electromagnet voltage goes to zero, the welding current changes to an intermediate level, which is evident by the increase in arc "volume". The arc then quickly reaches the position without deflection (central stop position) with the welding current kept at the intermediate level. Next, after being centralized for relatively long time, the electromagnet voltage goes to the same level previously used in the left stop position deflection, but this time with reverse sign (negative). At this moment, the welding current rises to a high level (the arc clearly further increases in "volume"). The arc is then quickly deflected to the right side (right stop position) maintaining this high welding current level. Next, after the arc spends an even longer time at this high current level and in this position, the electromagnet voltage is set back to zero and the welding current switches back to the intermediate level (the arc decreases in "volume"). The arc then quickly returns to the state of no deflection (central stop position) keeping the level of intermediate welding current. Once again, elapsing the time without arc deflection and at the intermediate welding current, the electromagnet voltage returns to the level programmed with a positive sign, which makes the welding current return to the low level. The arc is then quickly deflected to the left again (left stop position), beginning a new cycle of magnetic oscillation synchronized with the welding current. The short time required for the arc to stabilize at each oscillation position

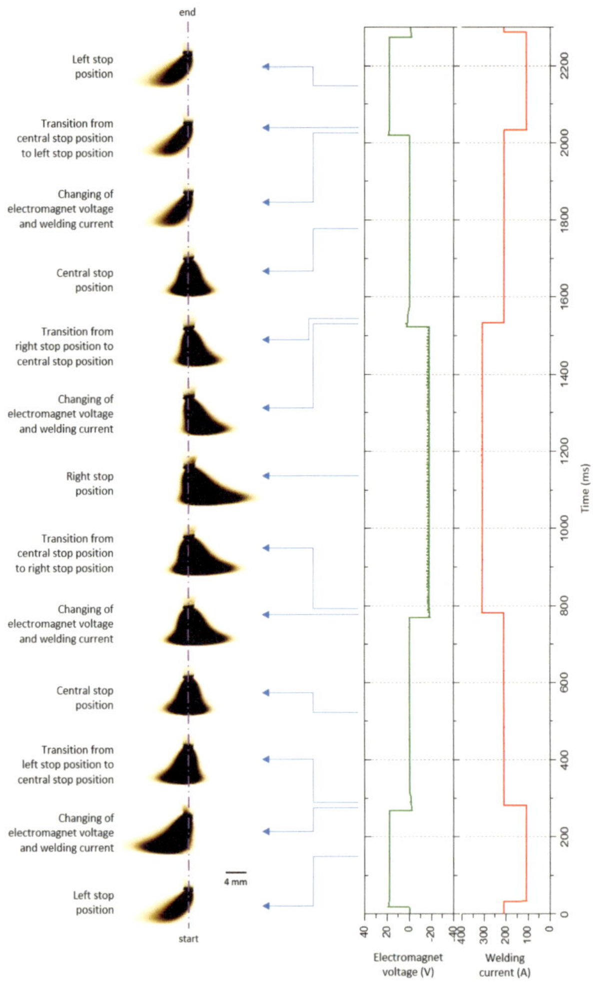

Figure 9. Sequence of high-speed images of a GTAW arc with synchronized magnetic oscillation (the arc image colours are inverted for better visualization).

http://dx.doi.org/10.5772/64158

(transition between deflections) reflects the behaviour of the electromagnet coil current, which induces the magnetic field for deflection changes and slightly lags the electromagnet voltage, here used as control signal. It is believed that this arc stabilization time could be reduced by using a current source for the electromagnet control. Finally, it is also observed in **Figure 9** that the electromagnet voltage and welding current levels as well as the dwell times at these levels were according to plan (**Table 2**).

3.1.3. General effect of the synchronization on weld bead formation

In order to demonstrate in a simple way the effect of the synchronization between magnetic oscillation of the arc and welding current on weld bead formation, three tests were performed (**Table 3**), with the object of evaluation being the surface appearance of the resulting weld beads. The arc stop times and current levels were set in such a way their product resulted always in 37.5 A.s in each arc position.

Test	Welding speed (mm/min)	Left stop time (ms)	Central stop time (ms)	Right stop time (ms)	Left welding current (A)	Central welding current (A)	Right welding current (A)	Average welding current (A)
5*	180	250	250	250	150	150	150	154.4
6	180	150	350	150	250	107	250	154.1
7	200	150	350	150	250	107	250	152.3

* Test with constant current and magnetic oscillation—without synchronization.

Table 3. Tests to assess the general effect of the synchronism between magnetic oscillation of the arc (arc time-position) and welding current level on weld bead formation (electromagnet voltage = ±30 V; oscillation frequency = 1 Hz).

By comparing the weld beads resulted from tests 5 (constant current magnetic oscillation) and 6 (synchronized magnetic oscillation), shown in **Figure 10**, it can be noted that, for the same oscillation frequency and amplitude (voltage applied to the electromagnet) and the same average welding current and welding speed, the condition with constant current magnetic oscillation (conventional oscillation— test 5) did not result in a weld bead with continuous lateral melting. On the other hand, by using the synchronized magnetic oscillation (test 6), there was lateral melting continuity as the molten marks of the arc merged on both sides of the weld bead. Thus, it is demonstrated, by the visual appearance of the weld surface, that the magnetic oscillation enhances the ability to adjust the shape of the weld bead. In conventional arc oscillation, more lateral melting could be attained by increasing the arc stop time on each side and reducing it at the centre, but this would certainly result in increased "melting waves", leaving the weld beads "zigzag" shaped as the arc would stay too long on one side before returning to the centre and then to the other side. The formation of lateral "melting waves" could be overcome by decreasing the welding travel speed, but with sacrifice in productivity. With the synchronized oscillation approach, the arc stop times and the welding current levels can be combined to keep the overall melting capacity of the arc (average welding current).

Thus, it is possible to avoid the formation of lateral "melting waves" on the welds without reducing the welding speed (productivity loss).

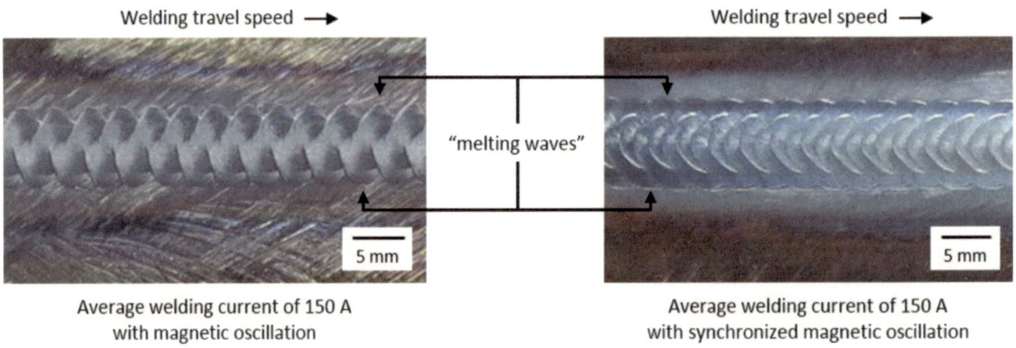

Figure 10. Superficial appearance of weld beads resulted from tests 5 (left hand side) and 6 (right hand side).

Test 7 (**Table 3**) was carried out to show more clearly the arc action at each position using the synchronized magnetic oscillation. The current levels and times at each arc stop position were the same as in test 6, but there was a small increase in welding speed to cause a larger spacing between the arc action marks. As shown in **Figure 11**, the so-called arc action marks are represented by melting edges left by the arc at each stop position. The marks denote something like typical pulsing current marks in Pulsed GTAW, but displaced both longitudinally (as would be in Pulsed GTAW) and transversely/laterally to the weld bead axis. It is possible to note the formation of large marks on the sides (high current) and small marks in the centre (low current). These arc action marks tend to become more evident (spaced) for low frequencies and high amplitudes of arc oscillation and for high welding speeds.

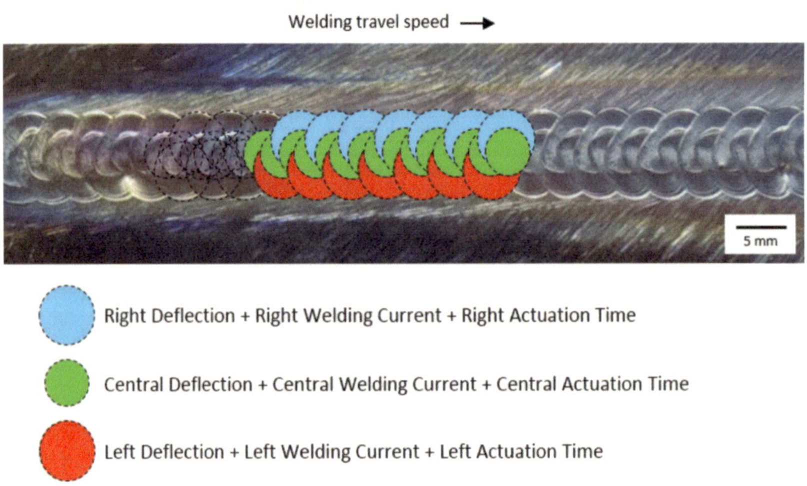

Figure 11. Demonstration of arc action marks in synchronized magnetic oscillation.

3.2. Magnetic oscillation of the arc synchronized with GTAW

The effect of the synchronization between the magnetic oscillation of the arc and the GTAW process was assessed based on the combination of tests showed in **Figure 12**. Two average welding current levels, two oscillation frequencies, and two oscillation amplitudes (electromagnet voltages) were tested for the synchronized approach and compared to similar situations without synchronization and even without arc oscillation. In the synchronized case, lateral/transversal arc oscillation was employed as illustrated in **Figure 3**. For the tests with magnetic oscillation (with and without synchronization), the arc stop times (left, central and right) were as shown in **Table 4**. In order to assess the effect of welding current change in each arc stop position, different current levels were used for each position according to **Table 5**, but always keeping the average welding currents to 150 and 200 A as shown in **Figure 12**. For the tests in pulsed mode, a different approach was used compared to conventional pulsed GTAW. The same three different current levels were applied in sequence and with the same actuation times as in the cases with synchronized oscillation for comparison. All results were assessed in terms of weld bead formation, specifically the effect on the weld width.

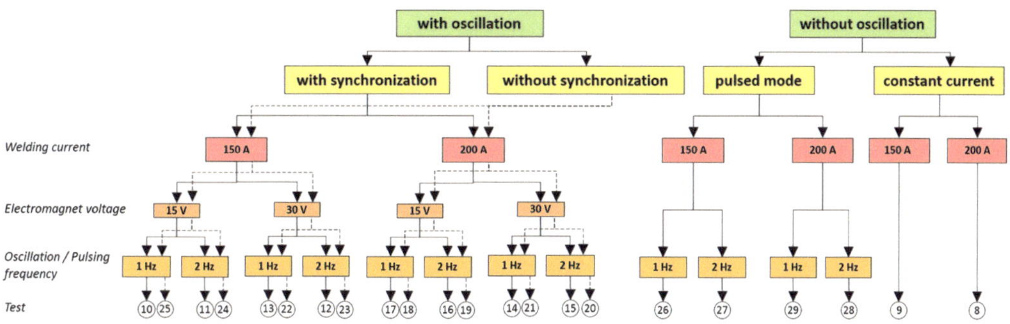

Figure 12. Flowchart of tests with magnetic oscillation of the arc synchronized with GTAW.

Left stop time (ms)	Central stop time (ms)	Right stop time (ms)	Oscillation frequency (Hz)
300	250	200	1
150	125	100	2

Table 4. Arc stop times for the tests with magnetic oscillation and oscillation frequencies.

Left welding current (A)	Central welding current (A)	Right welding current (A)	Average welding current (A)
210	111	158	150
280	148	210	200

Table 5. Welding current levels for each position of the arc and average currents.

Test	Electromagnet voltage				Welding current and Arc Voltage		
	Left Voltage (V)	Central Voltage (V)	Right Voltage (V)	Frequency [Hz]	Current [A]	Voltage [V]	Frequency [V]
			Without oscillation and with constant current				
8	–	0.07	–	–	205.1	14.53	–
9	–	0.07	–	–	154.5	12.86	–
			With oscillation and without synchronization (with constant current)				
10	−12.98	0.30	13.16	1.01	154.4	13.14	–
11	−13.03	0.53	13.23	1.97	154.4	13.03	–
12	−27.47	0.78	27.72	1.98	154.4	12.89	–
13	−27.51	−0.24	27.72	1.00	154.4	12.86	–
14	−27.49	−0.24	27.70	1.00	204.9	13.87	–
15	−27.54	−0.61	27.78	2.00	204.9	13.76	–
16	−13.03	0.51	13.23	1.98	204.8	14.22	–
17	−13.01	0.15	13.19	1.01	204.8	14.20	–
			With oscillation and with synchronization				
18	−13.02	-0.18	13.21	0.99	203.7	14.62	1.00
19	−13.04	0.41	13.22	1.99	206.4	14.15	2.04
20	−27.52	−0.60	27.74	2.01	202.9	14.18	2.03
21	−27.48	−0.28	27.68	1.00	204.7	14.35	1.00
22	−27.50	0.25	27.69	0.99	153.9	13.33	1.01
23	−27.57	−0.56	27.81	1.99	154.8	13.55	1.98
24	−13.06	−0.38	13.24	2.00	154.2	13.56	2.00
25	−13.02	−0.17	13.20	1.00	155.6	13.39	1.00
			Without oscillation and with pulsed mode				
26	–	0.07	–	–	153.9	13.43	1.01
27	–	0.07	–	–	156.3	13.27	1.98
28	–	0.07	–	–	204.5	14.05	1.98
29	–	0.07	–	–	205.1	14.19	1.01

Table 6. Resultant average electrical parameters from the GTAW tests.

3.2.1. Resultant electrical parameters and oscillograms

Table 6 shows the electrical parameters resulted from the GTAW tests and collected by the data acquisition system. All the electrical parameters, including those resulting from the synchronized magnetic oscillation tests were reasonable according to plan, showing the proper operation of the electromagnet control system and the welding power source. **Figures 13** and **14** present examples of electrical oscillograms resulted from tests with synchronization between magnetic oscillation and welding current level. It is possible to note that the welding current and the arc voltage followed the electromagnet voltage changes, evidencing the synchronism.

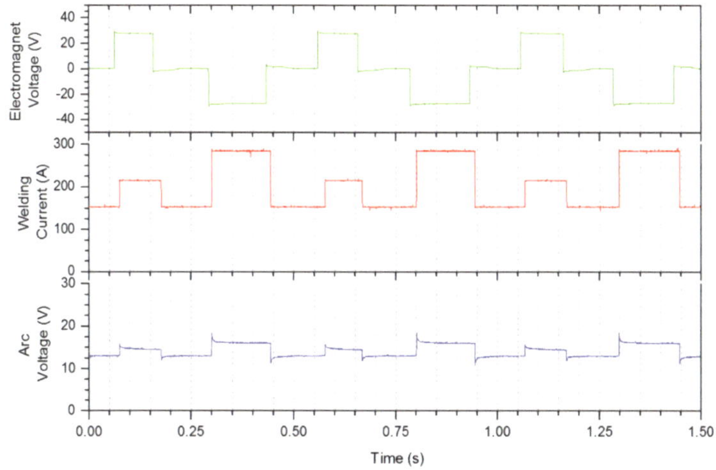

Figure 13. Electrical oscillogram from test 20—with oscillation and with synchronization.

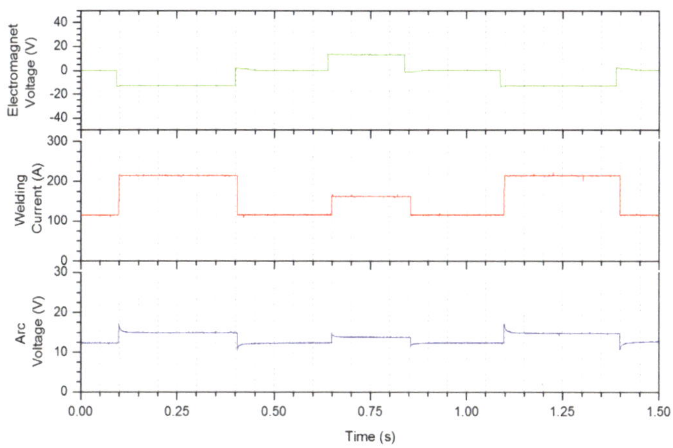

Figure 14. Electrical oscillogram from test 25—with oscillation and with synchronization.

3.2.2. *Effect on weld bead width*

The effect of the synchronization between magnetic oscillation of the arc and the welding current level on the change in parameters of weld bead width was evaluated. **Figure 15** shows how the width parameters of the resulting GTAW weld beads were measured. The centre of each welding movement (GTAW electrode route—baseline A) was obtained by markings previously made on the specimens. Three width measurements in different regions (beginning, middle, and end) for each specimen were carried out. The average total, right, and left widths of the weld beads then were taken for analysis. The standard deviation, not shown in the following graphs for visualization issues, was generally very low, which corroborates to the synchronized magnetic oscillation system robustness.

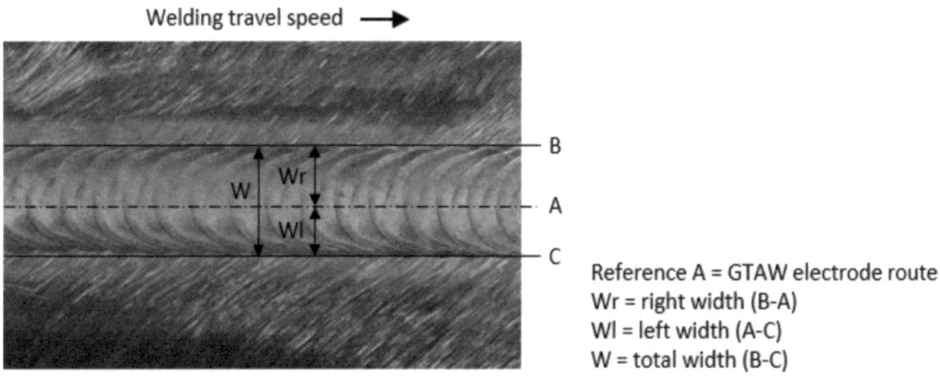

Figure 15. Weld bead width parameters analyzed.

Figure 16 shows the total width of the weld beads versus the electromagnet voltage (arc deflection) for the different GTAW configurations tested with an average welding current of 150 A. The lowest total width was reached for the case with constant current without oscillation. In the pulsed mode cases both frequencies result in weld beads of total width slightly larger and quite similar, since the arc reached higher current levels, increasing the size of the weld pool, at least in the bead surface. The two pulsing frequencies did not result in different total widths probably because the molten puddles (arc action marks) of the low levels of current were superimposed by the molten puddles (arc action marks) of the high levels of current, with the highest levels defining the total width of the bead, at least for the welding speed used. The pulsed mode resulted in intermediate levels of total width. As expected, the larger the magnetic deflection (electromagnet voltage) used, the greater the total width of the weld beads. The synchronized configuration resulted in greater total width compared to the constant current configuration with oscillation. This result can be explained as for the same average current, the lateral welding currents for the synchronized oscillation cases, especially on the left side, were superior to the central current and thus the likely tendency was the spreading of the molten metal surface (weld bead). The oscillation frequencies of 1 Hz provided larger total widths than those of 2 Hz. With 1 Hz the arc stays longer in each stop position in each deflection, giving more time for the current action in each deflection. The largest total width was achieved in the case of the synchronized oscillation at 1 Hz and with large arc defection (electromagnet voltage of 30 V).

The following graphs show the partial widths (left and right) of the weld beads to better analyze the effect of synchronizing the current level with the arc position for an average current of 150 A. The partial width values for constant current without oscillation and for the pulsed current cases are not shown here, as these conditions are transversely symmetrical to the welding direction and there is no significant difference between left and right widths.

Figure 17 shows the left width of the weld beads versus the electromagnet voltage (arc deflection) for the different GTAW configurations with an average welding current of 150 A. Generally, the left width tends to increase with the increase of arc deflection, particularly for the oscillation frequency of 1 Hz. It is clear that the synchronized configurations significantly

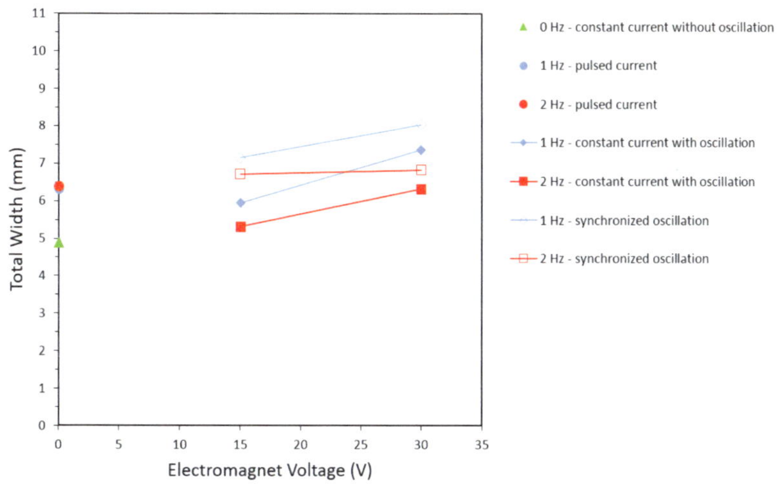

Figure 16. Total width of weld beads versus electromagnet voltage (arc deflection) for different configurations of GTAW with an average welding current of 150 A.

increased the left width, with greater effect for the frequency of 1 Hz. For the case of 2 Hz, with shorter times of current action in each of the arc positions, the resulting left width remained largely unchanged, and even tended to decrease in the synchronized case, with increase in electromagnet voltage. It may be that with 2 Hz the current action time to promote melting has been so short that the effect of deflection increase (expected increase in width) was attenuated. As in the total width analyses, the largest left width took place in the case of synchronized oscillation at 1 Hz and with large arc defection (electromagnet voltage of 30 V).

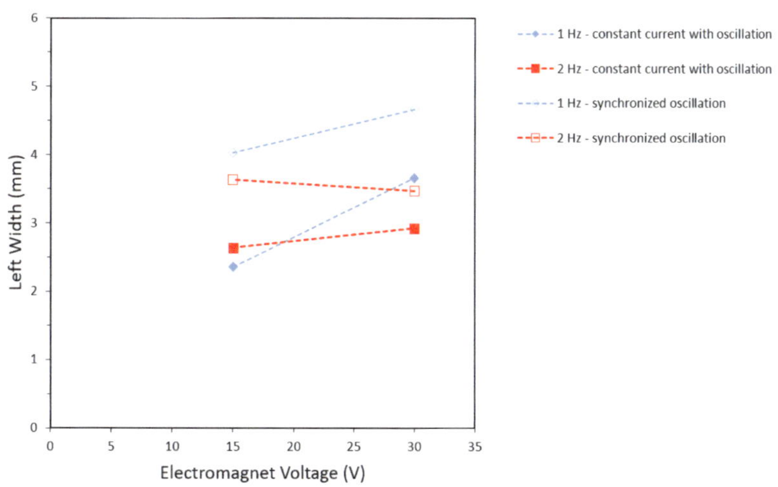

Figure 17. Left width of weld beads versus electromagnet voltage (arc deflection) for different configurations of GTAW with an average welding current of 150 A.

Figure 18 shows the right width of the weld beads versus the electromagnet voltage (arc deflection) for the different GTAW configurations with an average welding current of 150 A. Generally, the right width tends to increase slightly with the voltage electromagnet increase. Since the welding currents on the right side were practically the same, the right width values almost did not change comparing the cases with synchronized oscillation with those with constant current and oscillation.

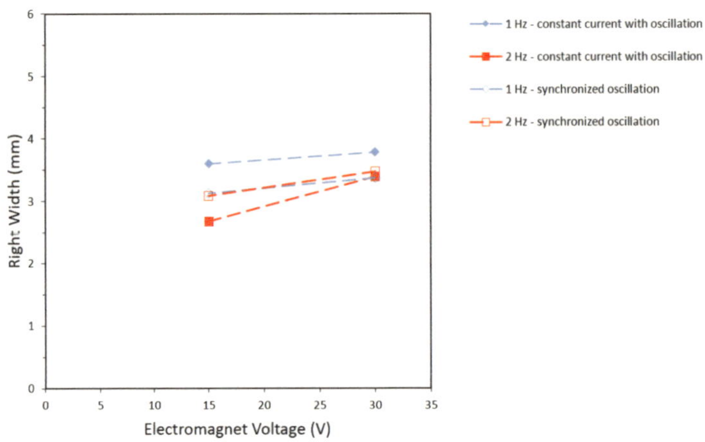

Figure 18. Right width of weld beads versus electromagnet voltage (arc deflection) for different configurations of GTAW with an average welding current of 150 A.

By comparing the effect on the total, left and right widths with the average welding current of 150 A, it is evident, especially for the oscillation frequency of 1 Hz, that the total width of the weld beads was mainly defined by left width. This indicates that the synchronized magnetic oscillation system was able to control the formation of the weld bead (at least in terms of surface width) as desired. That is, the highest welding current level and longest arc stop time on the left side of the oscillation led to increase of the left width, which, in turn, led to increase of the total width of the weld bead. Regarding the effect of the central welding current in terms of width, its main function is to "join" the two lateral arc deflections and melting capacities. This good control of the weld puddle could be exploited, for example, in the welding of dissimilar materials, in joining materials of different thicknesses, for root passes, narrow gaps, etc., always trying to direct more or less heat/melting capacity according to the arc position and need.

The following graphs are related to the GTAW configurations with an average welding current of 200 A. **Figure 19** shows the total width of the weld beads versus the electromagnet voltage (arc deflection) of all GTAW configurations tested with this average current level. As has occurred with the average current of 150 A, the smallest total width for 200 A took place with constant current without oscillation. In the pulsed current cases, the total width exhibited higher levels, the effect being slightly more pronounced with the frequency of 1 Hz, probably because the time the arc stays in the high current level is longer for this frequency. In the cases of synchronized oscillation and in those with constant current with oscillation, for the frequency of 1 Hz, the greater the magnetic deflection (electromagnet voltage) used, the greater the total width. The opposite took place for the frequency of 2 Hz (more pronounced for the constant current oscillation case), i.e., the overall width decreased with the electromagnet voltage increase. This unexpected result might have occurred due to the high current levels used (for the 200 A average welding current) as they make it more difficult for the arc to deflect —the higher the current flowing through the arc, the smaller its magnetic deflection [8]. In this case, to surpass this effect, even higher electromagnet voltages would be necessary, which were not attempted due to limitations in the electromagnet voltage/coil current allowed by the synchronized oscillation system. The synchronized oscillation configuration resulted in larger total widths compared with the constant current with oscillation configuration, as in the case of the average current of 150 A. This result can be explained since, for the same average welding current (in this case 200 A), the currents in the synchronized oscillation, especially on the left side, were significantly superior to the central current, which led to spreading of the weld puddle. Therefore, in general, the oscillation frequency of 1 Hz provided larger widths than those with 2 Hz, this fact being more pronounced for the high electromagnet voltage level (30 V). With the low frequency (1 Hz), the arc stays longer in each stop position for each deflection, giving more time for the current action in each deflection. Similar to the occurrence for the 150 A average current, the largest total width for the 200 A average current was achieved in the case of the synchronized oscillation at 1 Hz and with large arc defection (electromagnet voltage of 30 V).

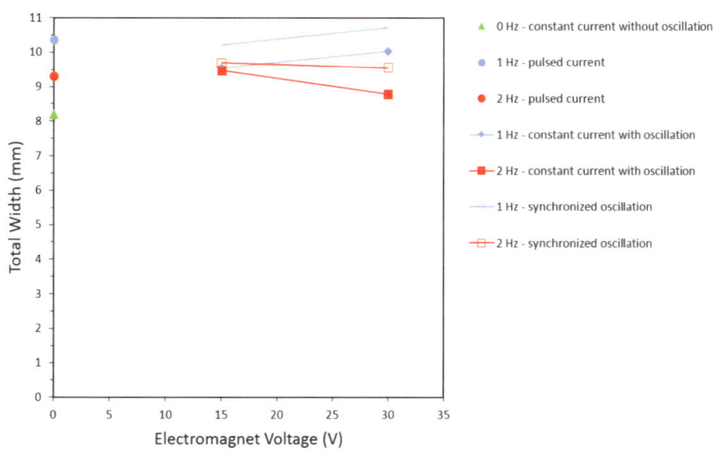

Figure 19. Total width of weld beads versus electromagnet voltage (arc deflection) for different configurations of GTAW with an average welding current of 200 A.

The following graphs show partial widths (left and right) of the weld beads for an average current of 200 A, also to better analyze the effect of synchronizing the current level with the arc position. The partial width values for constant current without oscillation and for the pulsed current cases are not shown here as they were shown earlier, as these conditions are transversely symmetrical to the welding direction and there is no significant difference between left and right widths.

Figure 20 shows the left width of the weld beads versus the electromagnet voltage (arc deflection) for the different GTAW configurations tested with an average welding current of 200 A. Here the left width also tended to increase with the arc deflection (electromagnet voltage) increase, particularly for the oscillation frequency of 1 Hz. However, for 2 Hz the left width practically remained unchanged with the increase in the electromagnet voltage, tending particularly in the synchronized oscillation case to a small decrease, probably because at 2 Hz the current action times in each of the arc stop positions were shorter. In this case also, the largest left width was obtained for the case of synchronized oscillation at 1 Hz and with large arc deflection (electromagnet voltage of 30 V).

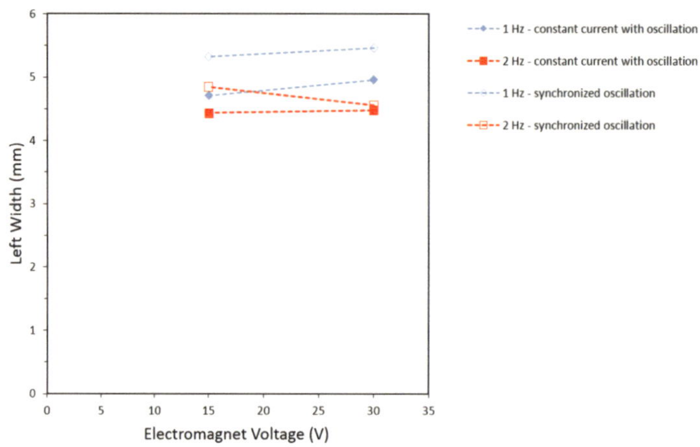

Figure 20. Left width of weld beads versus electromagnet voltage (arc deflection) for different configurations of GTAW with an average welding current of 200 A.

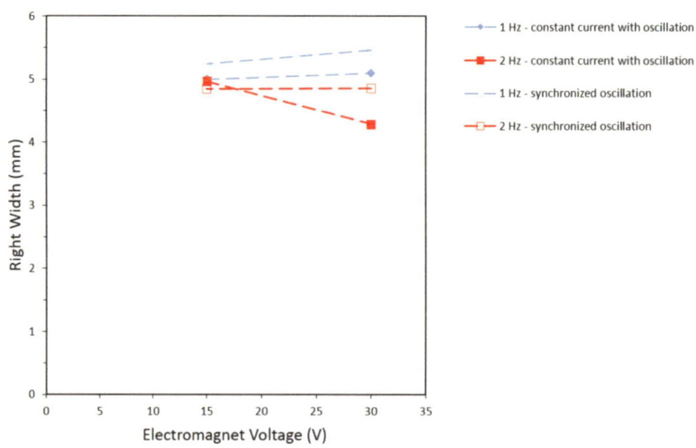

Figure 21. Right width of weld beads versus electromagnet voltage (arc deflection) for different configurations of GTAW with an average welding current of 200 A.

http://dx.doi.org/10.5772/64158

Figure 21 shows the right width of the weld beads versus the electromagnet voltage (arc deflection) for the different GTAW configurations tested with an average welding current of 200 A. Since the welding currents on the right side were practically the same, the right width values almost did not change comparing the synchronized oscillation to the constant current with oscillation cases. The right width resulting from the constant current with oscillation configuration at 2 Hz was the only one that showed unexpected result—decreased with the electromagnet voltage increase, collaborating to reduce the total width—and further investigation will be needed to clarify this fact.

The analysis of the width parameters for the average welding current of 200 A indicates that the most satisfactory results (greatest control of the weld puddle and of the formation of the weld bead) were obtained with synchronized magnetic oscillation at a frequency of 1 Hz. It is worth recalling that this good control could be exploited, for example, in the welding of dissimilar materials, joining of materials of different thicknesses, root pass, in narrow gaps, etc., always seeking to drive more or less heat/melting capacity according to the arc position and need.

By comparing the results from the average welding current levels used (150 and 200 A), the total, left and right width, values were larger with 200 A as expected, since increases in the current give the arc more melting capacity. However, the increase in width values with the electromagnet voltage increase was more pronounced for 150 A—arcs with low current are easier to deflect [8]. In general, the synchronized oscillation configurations resulted in the largest widths for both average welding currents used, with 1 Hz oscillations favouring larger values compared to 2 Hz.

4. Conclusions

According to the conditions used and tests performed, the main findings were:

a. Regarding the characterization of arc deflection

- From the evaluation of the synchronism between magnetic oscillation and welding current level and by analyzing the synchronizing device response time, it was observed the efficiency of the synchronization system developed;
- From the examination of the general effect of the synchronization on weld bead formation, more flexibility to optimize the arc melting capacity was verified in each arc position during oscillation.

b. Regarding the magnetic oscillation of the arc synchronized with the GTAW process

- In general the electrical parameters, including those resulting from the synchronized magnetic oscillation tests, were according to plan, showing the proper/synchronous functioning of the electromagnet control system and of the welding power source;

- The synchronized oscillation configuration generally resulted in the largest values of weld bead width. Furthermore, the larger the magnetic deflection used, the greater the total width resulted;
- Analysis of width parameters for both average welding currents used indicated that the best results (control over the weld pool and over the weld bead formation) were obtained with the synchronized magnetic oscillation of the arc at a frequency of 1 Hz;
- For both average current levels tested, the left side width, where higher levels of current and longer times of arc action were employed, influenced more on the increase of the total width, demonstrating that the synchronized magnetic oscillation technique was able to control the formation of the weld beads. That is, the higher current and actuation time on the left side of the weld pool led to left width increase, resulting in increase of the total weld bead width.

5. Future developments

Aiming to further develop and evaluate the synchronized magnetic oscillation technique, the following ideas are proposed:

- Improve the synchronized oscillation system to have the capacity to deflect the arc with different and higher intensities at each stop position;
- Evaluate the synchronized magnetic oscillation within a broad frequency range to better exploit its capacity to control the weld bead formation;
- Evaluate the synchronized magnetic oscillation with GTAW and GMAW in applications such as the welding of dissimilar materials, joining of materials of different thicknesses, in root pass, in narrow gap, in hardfacing, for grain refinement, for out-of-position welding, etc., always seeking to drive more or less heat/melting capacity according to the position of the arc and need;
- Perform longitudinal magnetic oscillation synchronized with GTAW and GMAW to verify the effects on factors such as weld bead geometry and maximum welding speed allowed;
- Record the magnetic oscillation synchronized with GMAW using high-speed filming to verify possible effects on the transfer of metal droplets from the electrode to the workpiece.

Acknowledgements

The authors thankfully acknowledge the financial support received from CNPq (project 458428/2014-0), FAPEMIG (project APQ-00858-14), and CAPES and the infrastructure used in the Centre for Research and Development of Welding Processes at Federal University of Uberlândia, Brazil.

Author details

Thiago Resende Larquer[1] and Ruham Pablo Reis[2*]

*Address all correspondence to: ruhamreis@mecanica.ufu.br

1 Federal Institute of Education, Science and Technology of Triângulo Mineiro, Patos de Minas, Minas Gerais, Brazil

2 Federal University of Uberlândia, Uberlândia, Minas Gerais, Brazil

References

[1] Lancaster, J. F. The Physics of Welding. 2nd ed. Oxford: Pergamon Press; 1986. 340 p.

[2] SCOTTI, A.; PONOMAREV, V. GMA Welding – better understanding, better performance (in Portuguese). 1st ed. São Paulo: Artliber; 2008. 284 p.

[3] GREENE, W. J. Magnetic oscilation of wedling arc, US Patent 2,920,183. New York: United States Patent Office; 1960.

[4] BLUNT, F. J.; RIBTON, C. N Design of electromagnets for arc oscillation in surfacing applications, for MIG and TIG processes. TWI Industrial Member Report 541/1996. Cambridge: TWI; 2000.

[5] Houldcroft, P.; John, R. Welding and Cutting. 1st ed. Cambridge: Woodhead-Faulkner Limited; 1988. 232 p.

[6] Reis, R. P.; Souza, D.; Scotti, A. Models to Describe Plasma Jet, Arc Trajectory and Arc Blow Formation in Arc Welding. Welding in the World. 2011;55(3):24–32. DOI: 10.1007/BF03321283

[7] Ueyama, T.; Ohnawa, T.; Tanaka, M.; Nakata, K. Occurrence of Arc Interference and Interruption in Tandem Pulsed GMA Welding-Study of Arc Stability in Tandem Pulsed GMA Welding (Report 1). Quarterly Journal of the Japan Welding Society. 2005;23(4): 515–525. DOI: 10.2207/qjjws.23.515

[8] Kang, Y. H.; Na S. J. A Study on the Modeling of Magnetic Arc Deflection and Dynamic Analysis. Welding Journal. 2002;81(1):8s–13s.

[9] Reis, R. P.; Scotti, A.; Norrish, J.; Cuiuri, D. Investigation on Welding Arc Interruptions in the Presence of Magnetic Fields: Welding Current Influence. IEEE Transactions on Plasma Science. 2012;40(3):870–876. DOI: 10.1109/TPS.2012.2182781

[10] Reis, R. P.; Scotti, A.; Norrish, J.; Cuiuri, D. Investigation on Welding Arc Interruptions in the Presence of Magnetic Fields: Arc Length, Torch Angle and Current Pulsing

Frequency Influence. IEEE Transactions on Plasma Science. 2013;41(1):133–139. DOI: 10.1109/TPS.2012.2230650

[11] Reis, R. P.; Souza, D.; Ferreira Filho, D. Arc Interruptions in Tandem Pulsed Gas Metal Arc Welding. Journal of Manufacturing Science and Engineering. 2015;137(1): 011004-011004-9. DOI: 10.1115/1.4028681

[12] Kumar, A.; Shailesh, P.; Sundarrajan, S. Optimization of Magnetic Arc Oscillation Process Parameters on Mechanical Properties of AA 5456 Aluminum Alloy Weldments. Materials and Design. 2008;29(10):1094–1913. DOI: 10.1016/j.matdes.2008.04.044

[13] Fasching, A. A.; Edwards, G. R.; David, S. A. Grain Refinement and Hydrogen Embrittlement in Iron Aluminide Alloy FA129. Science and Technology of Welding and Joining. 1997;2(4):167–173. DOI: 10.1179/stw.1997.2.4.167

[14] Sundaresan, S.; Ram, G. D. J. Use of Magnetic Arc Oscillation for Grain Refinement of Gas Tungsten Arc Welds in a–ß Titanium Alloys. Science and Technology of Welding and Joining. 1999;4(3):151–160. DOI: 10.1179/136217199101537699

[15] Chen, X. Q.; Smith, J. S.; Lucas, J. Microcomputer Controlled Arc Oscillator for Automated TIG Welding. Journal of Microcomputer Applications. 1990;13(4):347–360. DOI: 10.1016/0745-7138(90)90034-5

[16] Kang, Y. H.; Na S. J. Characteristics of Welding and Arc Signal in Narrow Groove Gas Metal Arc Welding Using Electromagnetic Arc Oscillation: Experiments Produce Optimum Parameters for Obtaining Uniform and Sufficient Groove Face Penetration. Welding Journal. 2003;82(5):93s–99s.

[17] Dutra, J. C.; Bidese, E.; Bonacorso, N. G.; Gonçalves E.; Silva, R. H. Improving Surfacing Performance with GMAW: A Method of Synchronizing Polarity is Used for Applications that Require Minimal Dilution. Welding Journal. 2013;92(5):42–47.

Chapter 5

A Comprehensive Model of the Transport Phenomena in Gas Metal Arc Welding

Junling Hu, Zhenghua Rao and Hai-Lung Tsai

Additional information is available at the end of the chapter

http://dx.doi.org/10.5772/64160

Abstract

A comprehensive two-dimensional gas metal arc welding (GMAW) model was developed to take into account all the interactive events in the gas metal arc welding process, including the arc plasma, melting of the electrode, droplet formation, detachment, transfer, and impingement onto the workpiece, and the weld-pool dynamics and weld formation. The comprehensive GMAW model tracks the free surface using the volume of fluid method and directly modeled the coupling effects between the arc domain and the metal domain, thus eliminating the need to assign boundary conditions at the interface. A thorough investigation of the plasma arc characteristics was conducted to study its effects on the dynamic process of droplet formation, detachment, impingement, and weld-pool formation. It was found that the droplet transfer and the deformed electrode and weld-pool surfaces significantly influence the transient distributions of current density, arc temperature, and arc pressure, which in turn affect the droplet formation, droplet transfer, and weld-pool dynamics.

Keywords: GMAW, arc plasma, weld-pool dynamics, metal transfer, droplet formation

1. Introduction

Gas metal arc welding (GMAW) is the most widely used joining process due to its ability to provide high-quality welds for a wide range of ferrous and non-ferrous alloys at low cost and high speed. As shown in **Figure 1**, GMAW is an arc-welding process that uses arc plasma between a continuously fed filler metal electrode and the workpiece to melt the electrode and the workpiece. The melted filler metal forms droplets and deposits on the partially melted workpiece to form a weld pool. The weld pool solidifies to bond the workpieces after the arc moves away. A shielding gas is fed through the gas nozzle to protect the molten metal from nitrogen and oxygen in the air. GMAW is also commonly known as metal inert gas (MIG) since inert gasses argon and helium are often used as a shielding gas. An active shielding gas containing oxygen and carbon dioxide is also used and thus the GMAW process is also called metal active gas (MAG). Direct current is usually used with the filler wire as the anode electrode to increase wire melting rate. GMAW can be easily adapted for high-speed robotic, hard automation, and semiautomatic welding applications.

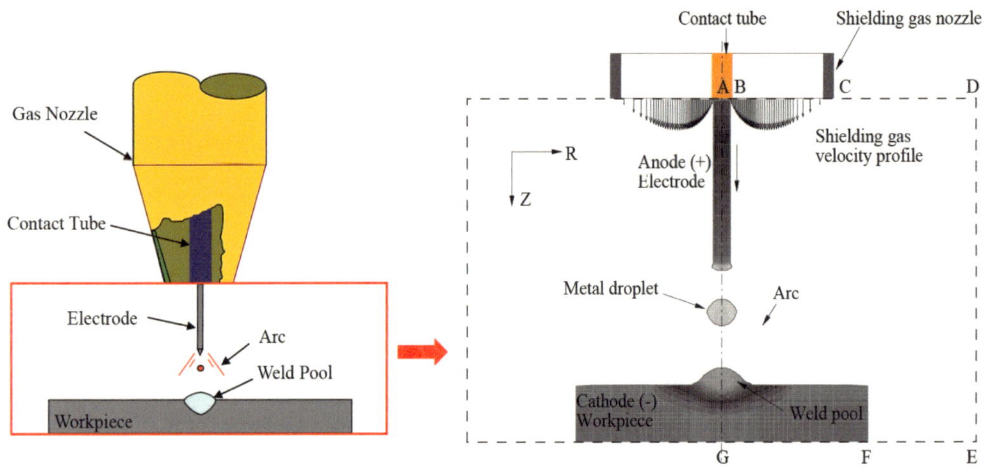

Figure 1. Schematic representation of a GMAW system with the computational domain shown inside the frame.

GMAW is a complex process with three major coupling events: (1) the evolution of arc plasma, (2) the dynamic process of droplet formation, detachment, and impingement onto the weld pool, and (3) the dynamics of the welding pool under the influences of the arc plasma and the periodical impingement of droplets. The stability of the GMAW process and the weld quality depend on many process parameters, such as welding current, welding voltage, wire feed speed, wire material and wire size, arc length, contact tube to workpiece distance, workpiece material and thickness, shielding gas properties, shielding gas flow rate, welding speed, etc. Selection of these welding-processing parameters relies on extensive experimentation and is an expensive trial-and-error process. Therefore, tremendous research efforts have been devoted to developing mathematical models of the GMAW process in order to reveal the underlying welding physics and provide key insights of process parameters for process optimization and defect prevention. Due to the complexity of the welding process and the associated numerical difficulty, many numerical models in the literature have simplified the GMAW process and only focused on one or two events. Many works on droplet formation [1–8] and weld-pool dynamics [9–29] have not included the arc plasma. More works now have been devoted to study the arc plasma and its influence on the metal transfer [30–41] and weld-pool dynamics [41–49].

http://dx.doi.org/10.5772/64160

In these simplified models, the droplet formation is considered as an isolated process in the electrode. The influence of the arc plasma is considered as boundary conditions with assumed distributions, such as linear current density distributions [1–3] or Gaussian distributions for the current density and heat flux [6–8].

The effects of droplet impingement on the weld pool have been significantly simplified as boundary conditions in the modeling of the weld-pool dynamics by many researchers [9–21]. The weld-pool surface was assumed to be flat [9–14] or modeled with boundary-fitted coordinates [13–15]. The dynamic impingement of a droplet onto the weld pool has been omitted [13], treated as a liquid column [14] or cylindrical volumetric heat source [15–19] acting on the weld pool in many weld-pool models. Only recent models [20–29] have simulated the dynamic interaction of droplets impinging onto the weld pool including both heat transfer and fluid flow effects and tracked the deformed weld pool free surface. However, they all applied assumed current, heat flux, and arc pressure boundary conditions at the weld-pool surface and also approximated the droplet impingement with the assumed droplet shape, volume and temperature, and impinging frequency and velocity.

In almost all aforementioned studies, the interaction of arc plasma with electrode melting, droplet generation and transfer, and weld-pool dynamics was not considered. Linear or Gaussian current density and heat flux were assumed as boundary conditions at the electrode surface [1–6] and weld-pool surface [15–29]. However, the surface of the workpiece is highly deformable, and the profile of the electrode changes rapidly, which greatly influence the arc plasma flow and thus change the current, heat flux, and momentum distribution at the surfaces of electrode and workpiece. Furthermore, the arc plasma can be dramatically distorted when there are free droplets between the electrode tip and the surface of the weld pool as observed in experimental studies [30–32]. Several models [33–39] have been developed to study the dynamic interaction of the arc plasma with the droplet formation. However, the droplet was eliminated when it was detached from the electrode tip or when it reaches the workpiece. The weld-pool dynamics was also omitted and the workpiece was treated as a flat plate. Some recent models [40–43] included the arc plasma, the filler wire, and the workpiece to study the direct interactions of the three domains. However, they are not completely coupled models since the droplet transfer in the arc still relies on an empirical formulation to calculate the plasma drag force in [43] or the droplet impingement is not simulated in [40–42].

The authors developed a fully coupled comprehensive GMAW model [44–52] to include the entire welding process—the arc plasma evolution, the electrode melting, the droplet formation and detachment, the droplet transfer in the arc, the droplet impingement onto the weld pool, and the weld-pool dynamics and solidification. The volume of fluid (VOF) technique was used to track the interface of the arc plasma and the metal. The temperature, pressure, velocity, electric, and magnetic fields are calculated in the entire computational domain, including the arc, filler wire, and the workpiece without using assumed heat, current, and pressure distributions at the interfaces. In the following sections, the comprehensive mathematical model is first presented to model the GMAW physics, and then the computational results are presented to show the evolution of the arc plasma and its dynamic interaction with the droplet formation, detachment, transfer, and impingement, and the weld-pool dynamics.

2. Mathematical model

2.1. Governing equations

The computational domain is shown in **Figure 1**, which has an anode region, an arc region, and a cathode region. The governing equations for the arc, the electrode, and the workpiece can be written in a single set based on the continuum formulation given by Diao and Tsai [53]:

Mass continuity

$$\frac{\partial}{\partial t}(\rho)+\nabla\cdot(\rho \boldsymbol{V})=0 \tag{1}$$

Momentum

$$\begin{aligned}&\frac{\partial}{\partial t}(\rho u)+\nabla\cdot(\rho \boldsymbol{V} u)=\nabla\cdot(\mu_l\frac{\rho}{\rho_l}\nabla u)-\frac{\partial p}{\partial r}-\frac{\mu_l}{K}\frac{\rho}{\rho_l}(u-u_s)-\frac{C\rho^2}{K^{1/2}\rho_l}|u-u_s|(u-u_s)\\&-\nabla\cdot(\rho f_s f_l \boldsymbol{V}_r u_r)-J_z\times B_\theta\end{aligned} \tag{2}$$

$$\begin{aligned}&\frac{\partial}{\partial t}(\rho v)+\nabla\cdot(\rho \boldsymbol{V} v)=\nabla\cdot(\mu_l\frac{\rho}{\rho_l}\nabla v)-\frac{\partial p}{\partial z}-\frac{\mu_l}{K}\frac{\rho}{\rho_l}(v-v_s)-\frac{C\rho^2}{K^{1/2}\rho_l}|v-v_s|(v-v_s)\\&-\nabla\cdot(\rho f_s f_l \boldsymbol{V}_r v_r)+\rho g\beta_T(T-T_0)+J_r\times B_\theta\end{aligned} \tag{3}$$

where $\boldsymbol{V}$ is the velocity vector, and u and v are the velocities in the r and z directions, respectively; $\boldsymbol{V}_r$ is the relative velocity vector between the liquid phase and the solid phase. The subscripts s and l refer to the solid and liquid phases, respectively, and the subscript 0 represents the initial condition. g is the gravitational acceleration, p is the pressure, ρ is the density, μ is the viscosity, β_T is the thermal expansion coefficient, T is the temperature, J_r and J_z are current densities in the respective r and z directions and B_θ is the self-induced electromagnetic field. K is the permeability function, C is the inertial coefficient, and f is the mass fraction.

Energy

$$\begin{aligned}&\frac{\partial}{\partial t}(\rho h)+\nabla\cdot(\rho \boldsymbol{V} h)=\nabla\cdot(\frac{k}{c_s}\nabla h)+\nabla\cdot(\frac{k}{c_s}\nabla(h_s-h))-\nabla\cdot(\rho(\boldsymbol{V}-\boldsymbol{V}_s)(h_l-h))\\&-\Delta H\frac{\partial f_l}{\partial t}+\frac{J_r^2+J_z^2}{\sigma_e}-S_R+\frac{5k_b}{e}(\frac{J_r}{c_s}\frac{\partial h}{\partial r}+\frac{J_z}{c_s}\frac{\partial h}{\partial z})\end{aligned} \tag{4}$$

where h is the enthalpy, k is the thermal conductivity, c is the specific heat, S_R is the radiation heat loss, k_b is the Stefan-Boltzmann constant, σ_e is the electrical conductivity, and e is the electronic charge.

Current continuity

$$\nabla^2\phi = \frac{1}{r}\frac{\partial}{\partial r}(r\frac{\partial \phi}{\partial r}) + \frac{\partial^2 \phi}{\partial z^2} = 0 \tag{5}$$

Ohm's law

$$J_r = -\sigma_e \frac{\partial \phi}{\partial r}, \; J_z = -\sigma_e \frac{\partial \phi}{\partial z} \tag{6}$$

Maxwell's equation

$$B_\theta = \frac{\mu_0}{r}\int_0^r J_z r dr \tag{7}$$

where ϕ is the electrical potential and μ_0 is the magnetic permeability.

The continuum model [53] included the first- and second-order drag forces and the interaction between the solid and liquid phases due to the relative velocity in the mushy zone ($0<f_l<1$ and $0<f_s<1$), which are represented by the corresponding third to fifth terms in the right-hand side of Eqs. (2) and (3). The energy flux due to the relative phase motion in the mushy zone is represented as the second term in the right-hand side of Eq. (4). The enthalpy method is used for phase change during the fusion and solidification processes. The enthalpy for the solid and liquid phases can be expressed as

$$h_s = c_s T, \; h_l = c_l T + (c_s - c_l)T_s + H \tag{8}$$

where H is the latent heat of fusion.

Continuum density (ρ), specific heat (c), thermal conductivity (k), velocity ($\boldsymbol{V}$), and enthalpy (h) are defined as follows:

$$\rho = \rho_s g_s + \rho_l g_l, \; c = f_s c_s + f_l c_l, \; k = k_s g_s + k_l g_l$$

$$f_l = \frac{\rho_s g_s}{\rho}, \quad f_s = \frac{\rho_l g_l}{\rho} \tag{9}$$

$$V = f_s V_s + f_l V_l, \quad h = f_s h_s + f_l h_l$$

where g is the volume fraction of the solid or liquid phase.

The permeability function is assumed to be analogous to fluid flow in porous media employing the Carman-Kozeny equation [54, 55]

$$K = \frac{g_l^3}{c_1 (1 - g_l)^2}, \quad c_1 = \frac{180}{d^2} \tag{10}$$

where d is proportional to the dendrite dimension. In this study, it is assumed to be a constant and is on the order of 10^{-2} cm.

The inertial coefficient, C, is calculated from [56]

$$C = 0.13 g_l^{-3/2} \tag{11}$$

2.2. Arc region

The arc region includes the arc plasma column and the surrounding shielding gas. The arc plasma is assumed to be in local thermodynamic equilibrium (LTE) [57]. The plasma properties, including enthalpy, density, viscosity, specific heat, thermal conductivity, and electrical conductivity, are calculated from an equilibrium composition [57, 58]. The influence of metal vapor on plasma material properties [37–42] is not considered in the present study. The plasma is also assumed to be optically thin, thus the radiation may be modeled as a radiation heat loss per unit volume represented by S_R in Eq. (4) [57, 58].

2.3. Metal region and tracking of free surfaces

The metal region includes the electrode, droplet in the arc, and the workpiece. The dynamic evolution of the droplet formation of the electrode tip, the droplet transfer in the arc, and the weld-pool dynamics require precise tracking of the free surface of the metal region. The volume of fluid method is used to track the moving free surface [59]. A volume of fluid function, $F(r,z,t)$, is used to track the location of the free surface. This function represents the volume of fluid per unit volume and satisfies the following equation:

$$\frac{dF}{dt} = \frac{\partial F}{\partial t} + (V \cdot \nabla) F = 0 \tag{12}$$

The average value of F in a cell is equal to the volume fraction of the cell occupied by the metal. A zero value of F indicates that a cell contains no metal, whereas a unit value indicates that the cell is full of metal. Cells with F values between zero and one are partially filled with metal.

2.4. Forces at the interface of the arc plasma and metal regions

The molten metal is subject to body forces and surfaces forces at the interface of the arc plasma and metal regions. The body forces include gravity, buoyancy force, and electromagnetic force. The surface forces include arc plasma shear stress, arc pressure, surface tension due to surface curvature, and Marangoni shear stress due to temperature difference. The surface forces are included as source terms to the momentum equations according to the CSF (continuum surface force) model [59–61]. Using F of the VOF function as the characteristic function, the surface forces are transformed to the localized body forces and added in the free surface cells.

The arc plasma shear stress is calculated from the velocities of the arc plasma cells at the free surface

$$\vec{\tau}_{ps} = \mu \frac{\partial \mathbf{V}}{\partial \vec{s}} \tag{13}$$

where μ is the viscosity of the arc plasma. The arc pressure at the metal surface is also obtained from the results in the arc region.

Surface tension pressure is normal to the free surface and can be expressed as [60]

$$p_s = \gamma \kappa \tag{14}$$

where γ is the surface tension coefficient. The free surface curvature κ is given by

$$\kappa = -\left[\nabla \cdot \left(\frac{\vec{n}}{|\vec{n}|} \right) \right] = \frac{1}{|\vec{n}|} \left[\left(\frac{\vec{n}}{|\vec{n}|} \cdot \nabla \right) |\vec{n}| - (\nabla \cdot \vec{n}) \right] \tag{15}$$

where $\vec{n}$ is the surface normal, calculated as the gradient of the VOF function

$$\vec{n} = \nabla F \tag{16}$$

The temperature-dependent Marangoni shear stress is in a direction tangential to the local free surface and is given by [7]

$$\vec{\tau}_{Ms} = \frac{\partial \gamma}{\partial T} \frac{\partial T}{\partial \vec{s}} \tag{17}$$

where $\vec{s}$ is a tangent vector of the local free surface.

2.5. Energy terms at the interface of the arc plasma and metal regions

2.5.1. Plasma-anode interface

The anode sheath region at the plasma-electrode interface is a very thin region, about 0.02-mm thick [57], and is at nonlocal thermal equilibrium. The very thin region is treated as a special interface by adding energy source terms, S_a in the metal region and S_{ap} in the arc region:

$$S_a = \frac{k_{eff}(T_{arc} - T_a)}{\delta} + J_a\phi_w - \varepsilon k_b T_a^4 - q_{ev}H_{ev} \tag{18}$$

$$S_{ap} = -\frac{k_{eff}(T_{arc} - T_a)}{\delta} \tag{19}$$

where T_{arc} and T_a are the respective arc plasma and metal temperature at the plasma-anode interface, k_{eff} is the harmonic mean of the thermal conductivities of the arc plasma and the anode materials, δ is the thickness of the anode sheath region and is taken as 0.1 mm according to the maximum thickness observed by experiments [62], ϕ_w is the work function of the anode material, J_a is the anode current calculated as the square root of J_r^2 and J_z^2, ε is the metal surface emissivity and k_b is the Stefan-Boltzmann constant, H_{ev} is the latent heat of vaporization of metal vapor, and q_{ev} is the mass rate of evaporation at the metal surface. The mass rate of evaporation of metal, q_{ev} for steel can be expressed as [63]

$$\log(q_{ev}) = A_v + \log P_{atm} - 0.5\log T \tag{20}$$

$$\log P_{atm} = 6.121 - \frac{18836}{T} \tag{21}$$

The four terms in Eq. (18) take into account thermal conduction, electron heating associated with the work function of the anode material, black-body radiation heat loss, and evaporation heat loss, respectively, at the metal surface. The energy equation for the plasma region only considers the cooling effects through conduction.

2.5.2. *Plasma-cathode interface*

Similarly, energy source terms S_c and S_{cp} are added to the corresponding metal and arc regions at the plasma-cathode interface, taken into account the conduction, radiation, and evaporation terms:

$$S_c = \frac{k_{eff}(T_{arc} - T_c)}{\delta} - q_{ev}H_{ev} - \varepsilon k_b T_c^4 \tag{22}$$

$$S_{cp} = -\frac{k_{eff}(T_{arc} - T_c)}{\delta} \tag{23}$$

where T_c is the metal surface temperature at the cathode surface, k_{eff} is the harmonic mean of the thermal conductivities of the arc plasma and the cathode materials, and δ is the thickness of the cathode and is taken as 0.1 mm.

2.6. External boundary conditions

The computational domain for a two-dimensional (2D) axisymmetric GMAW system is shown as ABCDEFGA in **Figure 1**. The external boundary conditions are listed in **Table 1**. Symmetrical boundary condition is assigned along the centerline AG.

	AB	BC	CD	DE	EF	FG	GA
u	0	0	0	$\frac{\partial(\rho u)}{\partial r}=0$	0	0	0
v	v_w	Eq. (24)	$\frac{\partial(\rho v)}{\partial z}=0$	0	0	0	$\frac{\partial v}{\partial r}=0$
h	$T = 300$ K	$T = 300$ K	$T = 300$ K	$T = 300$ K	$T = 300$ K	$T = 300$ K	$\frac{\partial T}{\partial r}=0$
ϕ	$-\sigma\frac{\partial \phi}{\partial z} = \frac{I}{\pi R_c^2}$	$\frac{\partial \phi}{\partial z}=0$	$\frac{\partial \phi}{\partial z}=0$	$\frac{\partial \phi}{\partial r}=0$	$\phi=0$	$\phi=0$	$\frac{\partial \phi}{\partial r}=0$

Table 1. Boundary conditions on the outer boundaries.

The velocity boundary takes into account the wire feed rate at AB, shielding gas inlet at BC, open boundaries at CD and DE, and non-slip wall condition at EF. The inflow of shielding gas from the nozzle at BC is represented by a fully developed axial velocity profile for laminar flow in a concentric annulus [64]:

$$v(r) = -\frac{2Q}{\pi}\frac{R_n^2 - r^2 + (R_n^2 - R_w^2)\dfrac{\ln(r/R_n)}{\ln(R_n/R_w)}}{R_n^4 - R_w^4 + \dfrac{(R_n^2 - R_w^2)^2}{\ln(R_n/R_w)}} + V_w\frac{\ln(R_n/r)}{\ln(R_n/R_w)} \tag{24}$$

where Q is the shielding gas flow rate, V_w is the wire feed rate, R_w and R_n are the radius of the electrode and the internal radius of the shielding gas nozzle, respectively.

The temperature boundaries along AD, DE, and EG are set as the room temperature. The boundary conditions for current flow include a zero voltage at the bottom of the workpiece FG, uniform current density along AB specified as $J_z = -\sigma_e \frac{\partial \phi}{\partial z} = \frac{I}{\pi R_w^2}$, and zero current flow along the other surfaces.

3. Numerical methods

At each time step, the calculation involves separate calculations in the arc region and the metal region, the coupling of the two regions through the interface boundary conditions described in Sections 2.4 and 2.5, and updating the arc and metal regions after obtaining the new free surface using the VOF method, Eq. (12), in the metal region.

The arc plasma region uses a fully implicit formulation and an upwind scheme for the combined convection/diffusion coefficients, and the SIMPLE algorithm [65] for the velocity and temperature fields. The metal region uses the method developed by Torrey et al. [59] to calculate the velocity and temperature fields.

The computational domain is 5 cm in radius and 3.04 cm in length. A nonuniform grid system is used with finer meshes near the electrode tip, in the arc column and the weld pool, where a fine mesh of 0.01 cm is used. Time step size is set as 5×10^{-6} s for a stable numerical solution.

4. Results and discussion

In this chapter, the comprehensive model [44, 45] is used to simulate a spot GMAW welding of a mild steel workpiece with a mild steel electrode under a constant current of 220 A shielded by argon. The electrode has a diameter of 0.16 cm and the workpiece is a mild steel disk with a 3-cm diameter and a 0.5-cm thickness. The contact tube is set flush with the bottom of the gas nozzle and has a contact tube to workpiece distance of 2.54 cm. The wire feed rate is 4.5 cm/s and the initial arc length is 0.8 cm. The shielding gas flow rate is 24 l/min and the inner diameter of the nozzle is 1.91 cm.

The temperature-dependent material properties of argon and the radiation loss term (SR) in Eq. (4) are taken from [58] and are plotted in **Figure 2**. **Table 2** lists the properties of the solid and liquid mild steel taken from [7] and other parameters used in the computation.

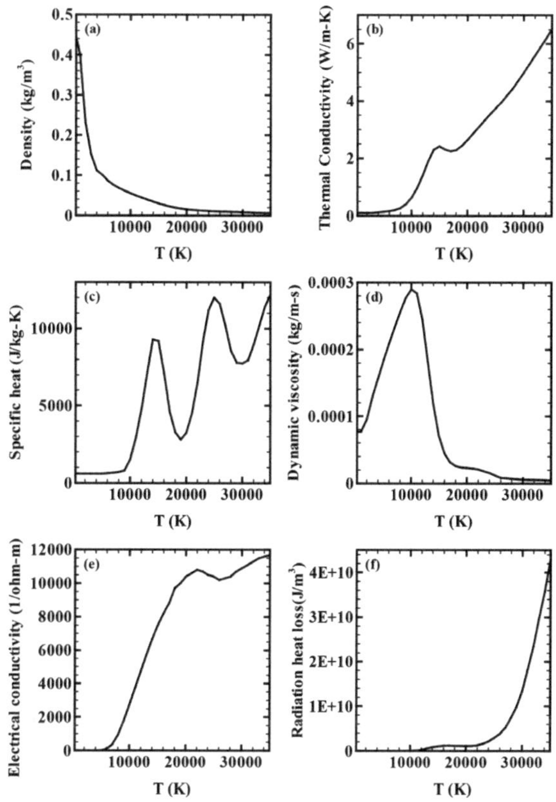

Figure 2. Temperature-dependent material properties of argon and the volume radiation heat loss taken from [58].

Nomenclature	Symbol	Value (unit)
Constant in Eq. (20)	A_v	2.52
Specific heat of solid phase	c_s	700 (J kg^{-1} K^{-1})
Specific heat of liquid phase	c_l	780 (J kg^{-1} K^{-1})
Thermal conductivity of solid phase	k_s	22 (W m^{-1} K^{-1})
Thermal conductivity of liquid phase	k_l	22 (W m^{-1} K^{-1})
Density of solid phase	ρ_s	7200 (kg m^{-3})
Density of liquid phase	ρ_l	7200 (kg m^{-3})
Thermal expansion coefficient	β_T	4.95×10^{-5} (K^{-1})
Radiation emissivity	ε	0.4
Dynamic viscosity	μ_l	0.006 (kg m^{-1} s^{-1})
Latent heat of fusion	H	2.47×10^5 (J kg^{-1})

Nomenclature	Symbol	Value (unit)
Latent heat of vaporization	H_{ev}	7.34×10^6 (J kg^{-1})
Solidus temperature	T_s	1750 (K)
Liquidus temperature	T_l	1800 (K)
Ambient temperature	T_∞	300 (K)
Vaporization temperature	T_{ev}	3080 (K)
Surface tension coefficient	γ	1.2 (N m^{-1})
Surface tension temperature gradient	$\partial\gamma/\partial T$	10^{-4} (N m^{-1} K^{-1})
Work function	ϕ_w	4.3 V
Electrical conductivity	σ_e	7.7×10^5 (Ω^{-1}m^{-1})

Table 2. Thermophysical properties of mild steel and other parameters.

4.1. Arc plasma evolution

Figure 3 shows the distributions of arc plasma temperature and pressure before and after the first droplet is detached and transferred to the workpiece. The shape of the electrode and workpiece are marked with thick lines. At $t = 100$ ms, the first droplet is formed at the electrode tip and the workpiece is still flat before a weld pool is formed. The arc shown in **Figure 3(a)** has a bell-shaped envelope with a maximum temperature of 19,300 K underneath the droplet. The high-temperature arc covers the droplet and expands as it moves toward the workpiece. The arc pressure contours at $t = 100$ ms in **Figure 3(b)** has two high-pressure regions. One is underneath the droplet caused by the pinch effect of the electromagnetic force, and the other is near the workpiece due to the stagnation of the plasma flow impinging onto the workpiece. The velocity field and streamlines in **Figure 4** show that shielding gas flows downward from the gas nozzle along the electrode surface and then is drawn to the electrode around the electrode tip. The ionized shielding gas around the electrode tip is pinched by the radially

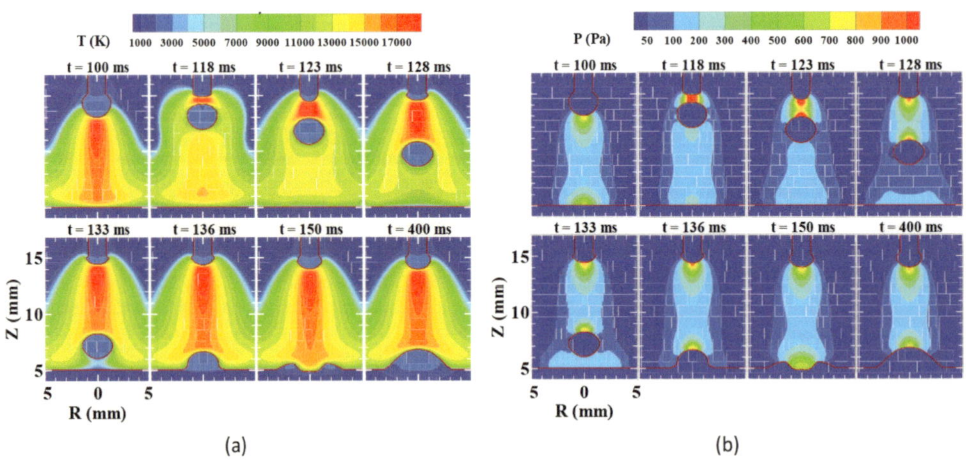

Figure 3. Arc plasma evolution during the first droplet formation, detachment, transfer, and impingement onto the workpiece: (a) temperature distributions in the arc plasma and (b) pressure distributions in the arc plasma.

inward and axially downward electromagnetic force toward the workpiece. The maximum axial velocity in the arc column is found to be 230 m/s on the axis. The corresponding current density distribution in **Figure 5** clearly shows that current diverges from the electrode tip and converges at the cathode in the workpiece, which results in inward and downward electromagnetic forces around the droplet and the inward and upward electromagnetic forces near the workpiece.

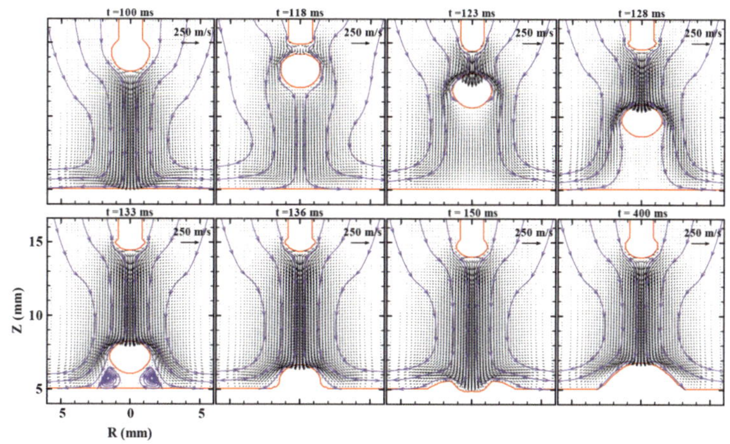

Figure 4. The corresponding velocity distributions in the arc plasma for the cases shown in **Figure 3**.

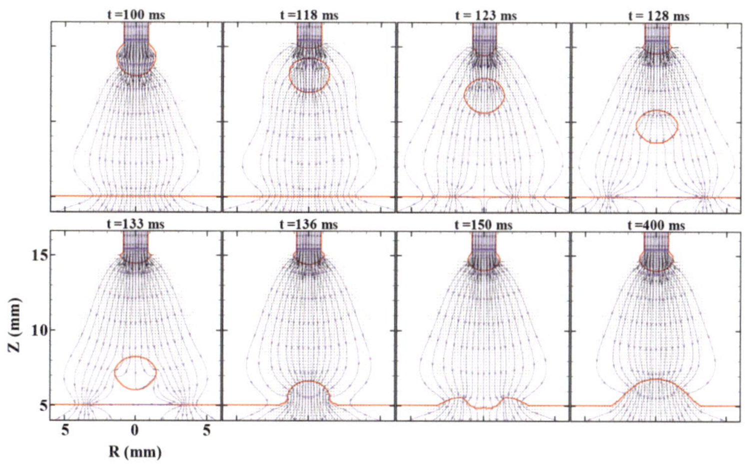

Figure 5. The corresponding current distributions in the arc plasma for the cases shown in **Figure 3**.

After the droplet is detached from the electrode at t = 118 ms, a new arc plasma is struck between the electrode tip and the top surface of the detached droplet. During the transfer of the detached droplet to the workpiece, the existence of the moving droplet greatly distorts the arc shape and flow pattern. Between t = 118 and 133 ms, current flow through the moving droplet decreases and the temperature of the arc plasma underneath the droplet also decreases. The arc plasma above the droplet is driven by the electromagnetic force and accelerates above the electrode. The high-velocity arc plasma flow impinges onto the top surface of the relatively slow-moving droplet and then flows around it. The flow pattern of the arc plasma around the droplet is similar to a flow around a sphere, including a high-pressure region formed at the droplet top surface due to the impingement and a low-pressure wake region below the droplet. The pressure drag is the main driving force for the droplet acceleration in the arc plasma. These arc plasma transport phenomena are confirmed by the experimental results of [30–32], but are significantly different from the numerical results in [43]. The unified GMAW model in [43] predicted current tended to flow through the detached droplet and a strong arc plasma flow formed beneath the droplet. However, the flow pattern in [43] failed to push the detached droplet downward in the arc plasma and thus an empirical equation was used to calculate the arc plasma drag force.

The first droplet reaches the workpiece around $t = 136$ ms, and a weld pool with an oscillating surface forms at the workpiece. The current distribution at the workpiece is greatly influenced by the weld-pool surface shape. The current tends to converge on the projected area at the workpiece, which may be at the workpiece center as in the cases of both $t = 136$ and 400 ms or not at the center as that of $t = 150$ ms.

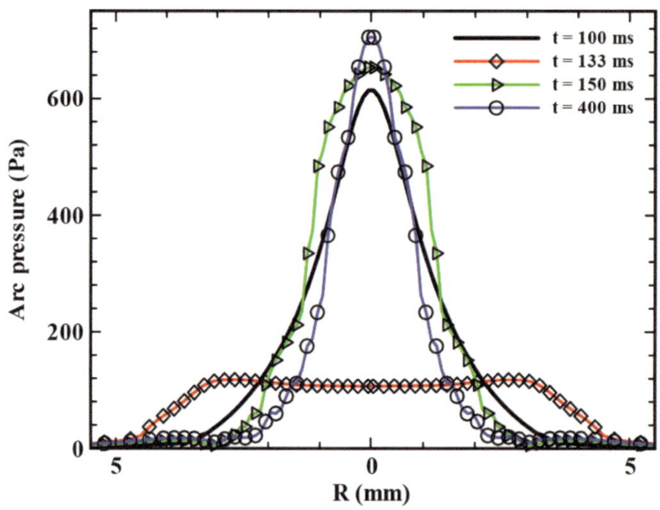

Figure 6. Arc pressure distributions along the radial direction at the workpiece surface.

In many of the weld-pool models [7–29], the arc pressure distribution at the center of the workpiece surface was assumed to be a Gaussian distribution with a fixed amplitude and distribution radius. However, the arc pressure distribution at the workpiece surface changes dramatically during the welding process as shown in **Figure 6**. Both the magnitude and distribution region varies with the evolution of the electrode and weld-pool surfaces and the presence of the detached droplet. Low arc pressure with a flat-top distribution is found at the

http://dx.doi.org/10.5772/64160

weld-pool surface at t = 133 ms when there is a detached droplet in the arc column, whereas high and concentrated arc pressure distribution is observed at t = 400 s when the weld-pool surface is projected and the arc column has no detached droplet. These distributions significantly deviate from the assumed Gaussian distribution in many weld-pool models [7–29]. Similarly, their assumed Gaussian distributions of current and heat flux cannot represent the dynamic boundary conditions at the weld-pool surface.

4.2. Droplet formation and transfer

Droplet formation is determined by the concentrated heating due to the recombining electrons at the electrode surface and the flow pattern within the droplet caused by a balance of forces acting on the droplet, which includes electromagnetic force, surface tension force, gravity, arc pressure, and plasma shear stress. To clearly illustrate the heat transfer and fluid flow within the droplet at the electrode tip, the distributions of temperature, velocity, electrical potential, current, and electromagnetic force within the droplet at t = 100 ms are drawn in **Figure 7**. A vortex flow forms in the droplet with a downward flow along the centerline and an upward flow at the surface. The fluid circulation enhances the mixing of cold fluid at the center with the hot surface fluid. The downward flow is caused by the inward and downward electromagnetic force at the upper part of the droplet near the melt line marked as a dashed line. The electromagnetic force and current density are determined by the electrical potential distribution. Current slightly diverges in the upper part of the droplet and converges in the lower part and then flows out of the droplet surface from the lower part of the droplet. The current flow pattern results in an electromagnetic force that is radially inward and axially downward at the upper part and upward at the bottom part. The upward electromagnetic force, surface tension, and arc pressure at the droplet bottom change the fluid to flow upward along the surface. At the balance of electrons heating, arc plasma heating, evaporation and radiation cooling, and convection cooling, the maximum temperature at the droplet surface is found to be 2936 K, which is close to the experimental result of [58].

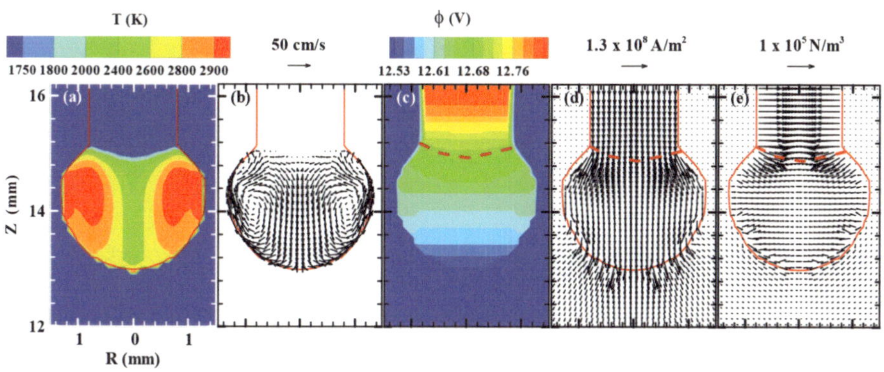

Figure 7. Distribution of physical variables within the droplet at t = 100 ms. (a) Temperature, (b) velocity, (c) electrical potential, (d) current density, and (e) electromagnetic force.

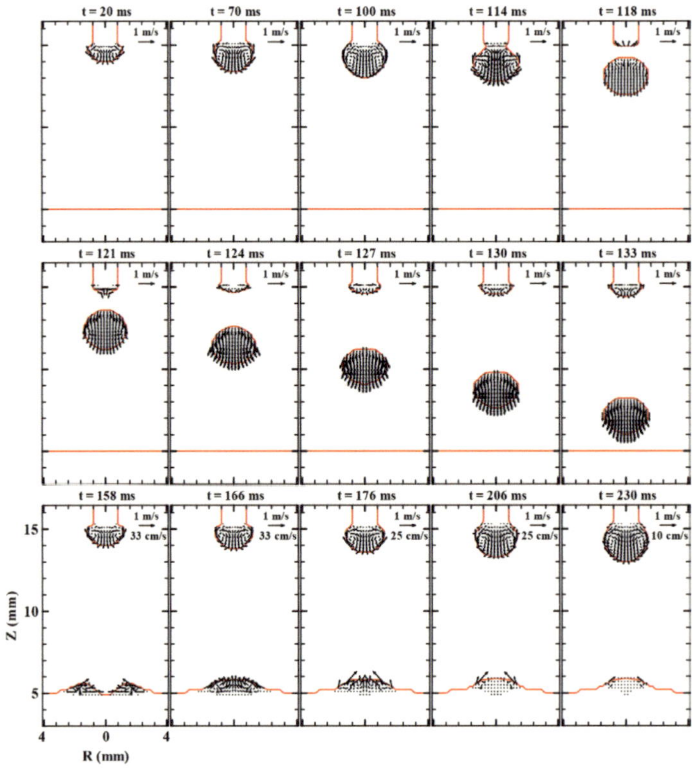

Figure 8. Velocity distributions in the metal showing droplet generation, detachment, transfer in the arc, and impingement onto the weld pool.

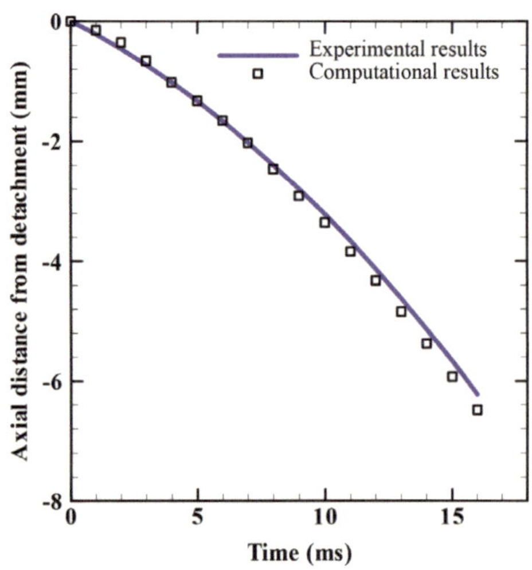

Figure 9. Comparison of droplet flight trajectory with experiment results [32].

The first droplet formation is shown in **Figures 3–5** and **8** from t = 20 to 116 ms. A round droplet forms at the electrode tip and grows larger. After a neck is formed at t = 116 ms, the velocity within the droplet increases due to the increased electromagnetic pinch force at the neck. After the first droplet is detached at t = 118 ms, the second droplet begins to form at the electrode tip. The droplet is detached and transferred to the workpiece from t = 118 to 133 ms. The detached droplet is accelerated by the plasma arc and gravity and reaches the workpiece with an axial velocity of about 60 cm/s. The center positions of the first detached droplet shown in **Figure 8** are plotted as a function of time and compared with the flight trajectory taken by Jones et al. [32] in **Figure 9**. The droplet trajectory from the computation matches the experimental results. The droplet acceleration obtained by taking the second derivative of the curve is found to be 24 m/s^2, which is comparable to 21 m/s^2 in [32].

4.3. Weld-pool dynamics and solidification

Figures 3 and **8** show the first droplet impingement onto the workpiece and form a weld pool. The weld pool grows wider and deeper with more droplets deposited into it. The final weld-

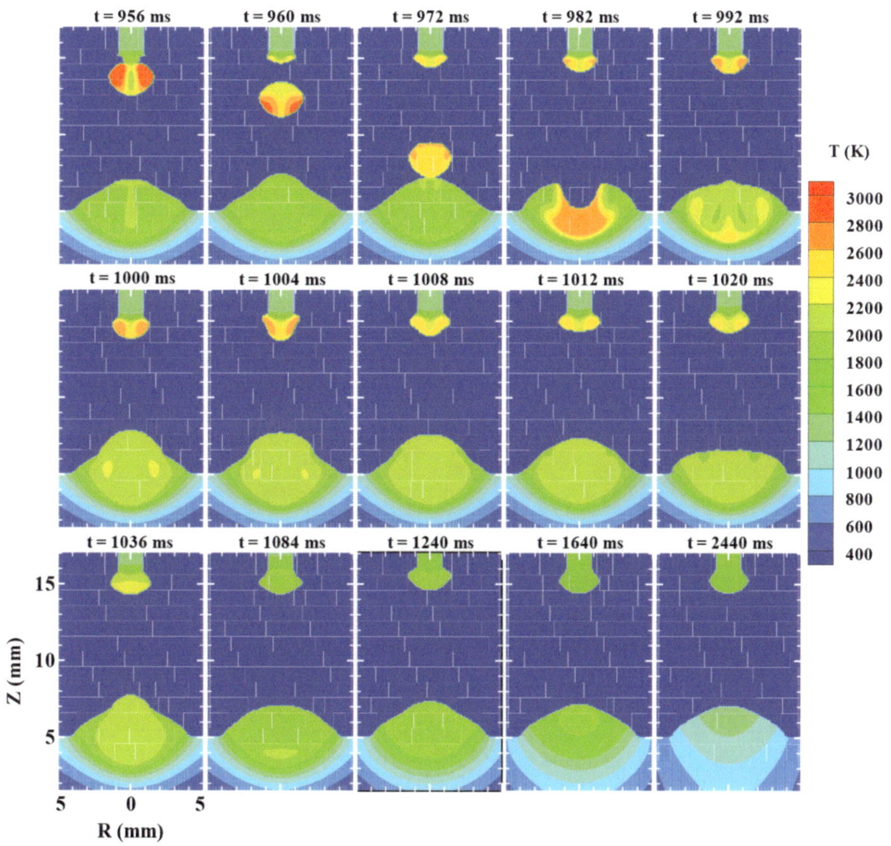

Figure 10. A sequence of temperature distributions in the metal showing droplet generation, detachment, transfer in the arc, impingement onto the weld pool, and weld-pool dynamics.

pool shape and the resulting final weld shape are determined by weld-pool dynamics subject to periodic droplet impingement and several important forces, including electromagnetic force, arc pressure, plasma shear stress, surface tension, and gravity force. As shown in **Figures 10** and **11**, a droplet is ready to be detached from the electrode tip at $t = 956$ ms. Two vortices formed in the weld pool with an inward flow at the weld-pool surface and a downward flow at the center. The inward flow at the weld-pool surface is driven by the surface tension and the downward flow is mainly by the arc pressure force. When the arc pressure at the weld-pool surface decreases due to a droplet stuck in the arc column, as shown in **Figures 12** and **13**, the fluid at the weld-pool center rises up at $t = 960$ ms. A crater is formed after a droplet impinges onto the weld pool at $t = 982$ ms. The high-temperature filler metal carried by the droplet reaches the bottom of the weld pool. The crater is then filled up by the surrounding fluid, and the weld pool first oscillates at high amplitude, then the oscillation gradually subsides. A sequence of experimental images is given in Ref. [43] showing the weld-pool oscillation after a droplet impinges onto the weld pool, which can be seen by the up-and-down movement of the weld-pool surface.

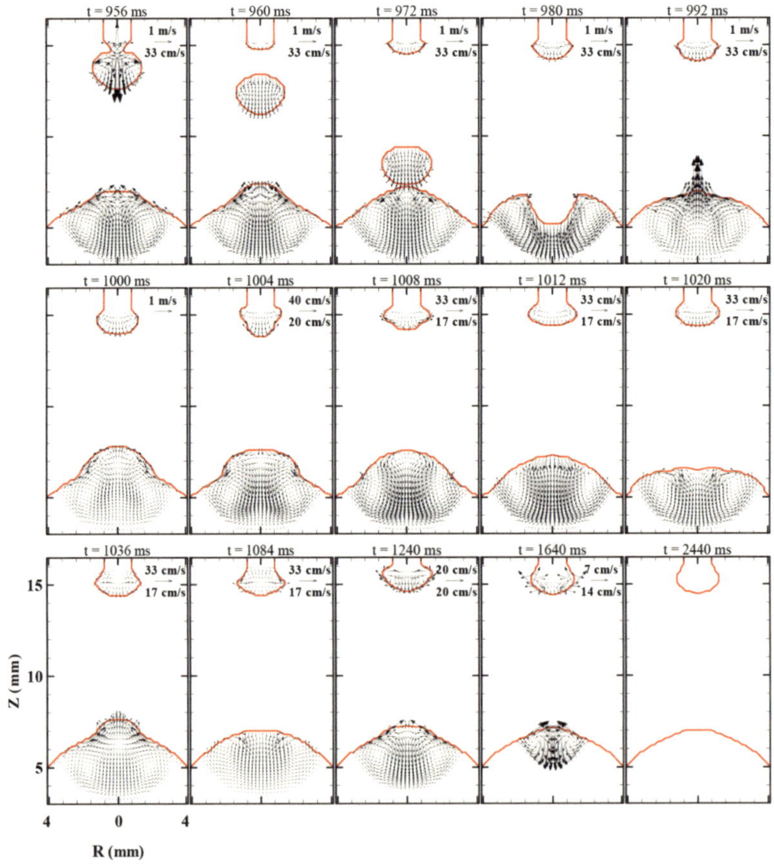

Figure 11. The corresponding velocity distributions of the cases shown in **Figures 10**.

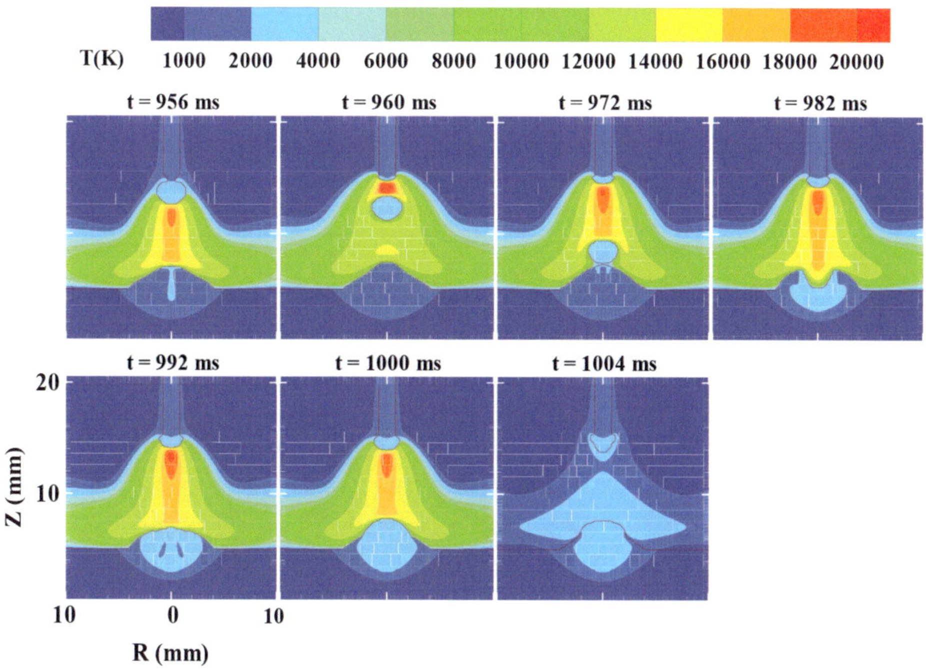

Figure 12. Temperature distributions in the arc plasma during the last droplet impingement and weld-pool dynamics.

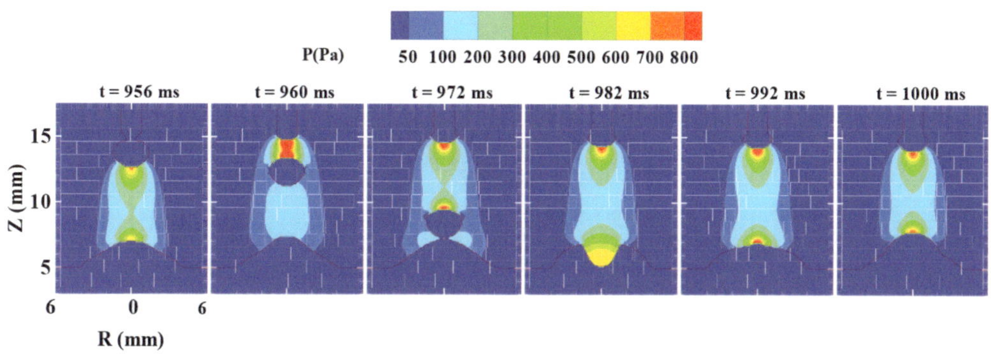

Figure 13. Pressure distributions in the arc plasma during the last droplet impingement and weld-pool dynamics.

At $t = 1000$ ms, the current is turned off and the temperature in the arc plasma decreases rapidly due to the high radiation loss and low heat capacity of the arc plasma. At $t = 1004$ ms, the high-temperature arc plasma is replaced by the nonionized shielding gas, which is continued to protect the solidifying weld pool. After a sudden removal of arc pressure and plasma shear stress at the electrode and weld-pool surfaces, the remnant droplet at the electrode and the weld pool oscillates and the oscillation is balanced by the surface tension. The sizes of the molten droplet and the weld pool become smaller with the heat loss to the solid metal by conduction and to the surroundings by radiation and convection. As steel is a good thermal conductor, the heat loss occurs mainly through conduction to the solid metal. Therefore, the liquid metal adjacent to the solid and liquid interface solidifies first and the solidus line moves outward toward the electrode and weld-pool surfaces. The solidification completes at $t = 2600$ ms in the electrode and at $t = 2440$ ms in the weld pool. **Figure 14** shows the final shape of the weld bead including the weld penetration, which is similar to the reported experimental results [7, 8, 43].

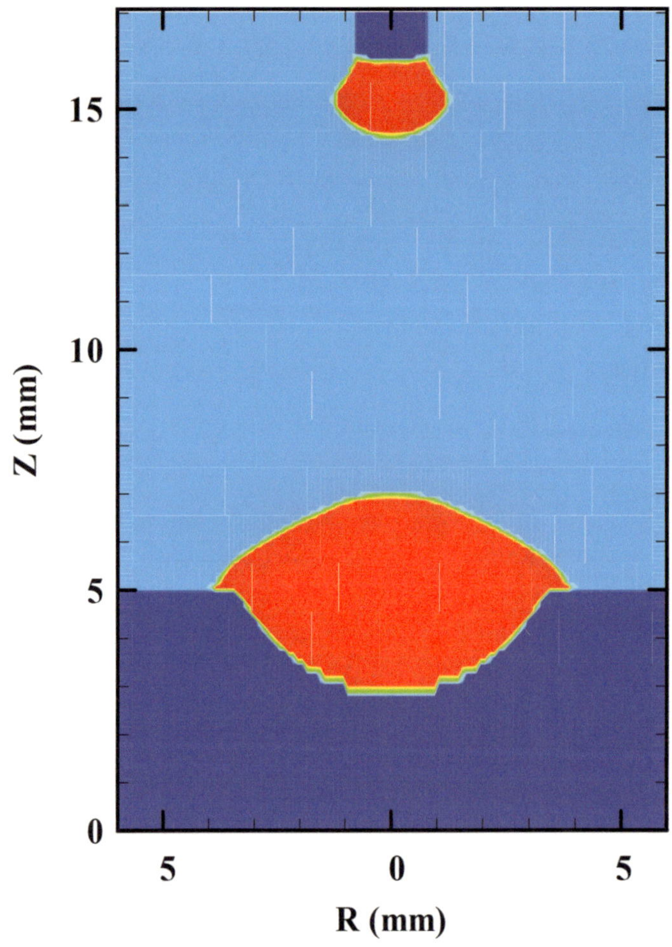

Figure 14. The solidified weld-bead shape.

5. Conclusions

A comprehensive model has been presented to simulate the transport phenomena in a gas metal arc-welding process, including the arc plasma evolution, the melting of the electrode and the droplet generation, detachment, transfer, and impingement onto the workpiece, and the weld-pool dynamics and solidification. This model included all the three regions—the electrode, the arc plasma, and the weld pool—in the computational domain and modeled the interactive coupling between these three regions. The distributions of arc pressure, current density, and heat flux at the weld-pool surface are found to vary in a wide range, and thus cannot be represented by a fixed distribution in many published GMAW models. The simulation results have revealed physical insights which cannot be found with those isolated single-region models in the literature. The transient evolution of the arc plasma was found to influence and also to be influenced by the droplet formation, detachment, transfer in the arc, and weld-pool dynamics. Therefore, a comprehensive model is required to accurately take into account the coupling events in both the arc domain and metal domain. The comprehensive model can be used to study the effects of process parameters on the welding process and the final weld formation, such as droplet generation with pulse currents to achieve one droplet per pulse (ODPP) and the effects of shielding gas and wire feed rate on the welding process.

Author details

Junling Hu[1*], Zhenghua Rao[2] and Hai-Lung Tsai[3]

*Address all correspondence to: jjhu@bridgeport.edu

1 Department of Mechanical Engineering, University of Bridgeport, Bridgeport, CT, USA

2 School of Energy Science and Engineering, Central South University, Changsha, China

3 Department of Mechanical and Aerospace Engineering, Missouri University of Science and Technology, Rolla, MO, USA

References

[1] Choi S.K., Yoo C.D., Kim Y.-S. The dynamic analysis of metal transfer in pulsed current gas metal arc welding. J. Phys. D: Appl. Phys. 1998; 31: 207–215.

[2] Choi S.K., Yoo C.D., Kim Y.-S. Dynamic simulation of metal transfer in GMAW, Part 1: globular and spray transfer mode. Weld. J. 1998: 38–44s.

[3] Choi S.K., Yoo C.D., Kim Y.-S. Dynamic simulation of metal transfer in GMAW, Part 2: short-circuit transfer mode. Weld. J. 1998: 45–51s.

[4] Wang G., Huang P.G., Zhang Y.M. Numerical analysis of metal transfer in gas metal arc welding. Metall. Trans. 2003; 34B: 345–353.

[5] Wang G., Huang P.G., Zhang Y.M. Numerical analysis of metal transfer in gas metal arc welding under modified pulsed current conditions. Metall. Trans. 2004; 35B(5): 857–866.

[6] Wang F., Hou W.K., Hu S.J., Kannatey-Asibu E., Schultz W.W., Wang P.C. Modelling and analysis of metal transfer in gas metal arc welding. J. Phys. D: Appl. Phys. 2003; 36: 1143–1152.

[7] Fan H.G., Kovacevic R. Droplet formation, detachment, and impingement on the molten pool in gas metal arc welding. Metall. Trans. 1999; 30B: 791–801.

[8] Fan H.G., Kovacevic R. Dynamic analysis of globular metal transfer in gas metal arc welding – a comparison of numerical and experimental results. J. Phys. D: Appl. Phys. 1998; 31: 2929–2941.

[9] Tsao K.C., Wu C.S. Fluid flow and heat transfer in GMA weld pools. Weld. J. 1988; 70–75s.

[10] Kumar S., Bhaduri S.C. Three-dimensional finite element modeling of gas metal-arc welding. Metall. Trans. 1994; 25B: 435–441.

[11] Kumar S., Bhaduri S.C. Theoretical investigation of penetration characteristics in gas metal-arc welding using finite element method. Metall. Trans. 1994; 26B: 611–624.

[12] Jaidi J., Dutta P. Modeling of transport phenomena in a gas metal arc welding process. Numer. Heat Transf. A. 2001; 40: 543–562.

[13] Kim J.-W., Na S.-J. A study on the effect of constant tube-to-workpiece distance on weld pool shape in gas metal arc welding. Weld. J. 1995; 74(5): 141–152s.

[14] Ohring S., Lugt H.J. Numerical simulation of a time-dependent 3-D GMA weld pool due to a moving arc. Weld. J. 1999: 416–424s.

[15] Kim C.-H., Zhang W., DebRoy T. Modeling of temperature field and solidified surface profile during gas-metal arc fillet welding. J. Appl. Phys. 2003; 94: 2667–2679.

[16] Kim W.H., Fan H.G., Na S.J. Effect of various driving forces on heat and mass transfer in arc welding. Numer. Heat Transf. A. 1997; 32(6): 633–652.

[17] Chen J., Schwenk C., Wu C. S., Rethmeier M. Predicting the influence of groove angle on heat transfer and fluid flow for new gas metal arc welding processes. Int. J. Heat Mass Transf. 2012; 55(1–3): 102–111.

[18] Cheon J., Kiran D.V., Na S.-J. CFD based visualization of the finger shaped evolution in the gas metal arc welding process. Int. J. Heat Mass Transf. 2016; 97: 1–14.

[19] Jia X., Xu J., Liu Z., Huang, S., Fan Y., Sun Z. A new method to estimate heat source parameters in gas metal arc welding simulation process. Fusion Eng. Des. 2014; 89(1): 40–48.

[20] Cao Z., Yang Z., Chen X.L. Three-dimensional simulation of transient GMA weld pool with free surface. Weld. J. 2004: 169–176s.

[21] Wang Y., Tsai H.L. Impingement of filler droplets and weld pool dynamics during gas metal arc welding process. Int. J. Heat Mass Transf. 2001; 44: 2067–2080.

[22] Wang Y., Shi Q., Tsai H.L. Modeling of the effects of surface-active elements on flow patterns and weld penetration. Metall. Trans. 2001; 32B: 145–161.

[23] Wang Y., Tsai H.L. Effects of surface active elements on weld pool fluid flow and weld penetration in gas metal arc welding. Metall. Trans. 2001; 32B: 501–515.

[24] Guo H., Hu J., Tsai H.L. Three-dimensional modeling of gas metal arc welding of aluminum alloys. J. Manuf. Sci. Eng. 2010; 132: 021011.

[25] Guo H., Hu J., Tsai H.L. Numerical modeling of cold weld formation and improvement in GMAW of aluminum alloys. Numer. Heat Transf. Part A: Appl. 2010; 57: 392–414.

[26] Guo H., Hu J., Tsai H.L. Formation of weld crater in GMAW of aluminum alloys. Int. J. Heat Mass Transf. 2009; 52: 5533–5546.

[27] Hu J., Tsai H.L. Modeling of transport phenomena in 3D GMAW of thick metal with V-groove. J. Phys. D: Appl. Phys. 2008; 41: 065202.

[28] Hu J., Guo H., Tsai H.L. Weld pool dynamics and the formation of ripples in 3D gas metal arc welding. Int. J. Heat Mass Transf. 2008; 51: 2537–2552.

[29] Wu L., Cheon J., Kiran D.V., Na S.-J. CFD simulations of GMA welding of horizontal fillet joints based on coordinate rotation of arc models. J. Mater. Process. Technol. 2016; 231: 221–238.

[30] Jones L.A., Eagar T.W., Lang J.H. Images of steel electrode in Ar-2%O_2 shielding during constant current gas metal arc welding. Weld. J. 1998: 135–141s.

[31] Jones L.A., Eagar T.W., Lang J.H. Magnetic forces acting on molten drops in gas metal arc welding. J. Phys. D: Appl. Phys. 1998; 31: 93–106.

[32] Jones L.A., Eagar T.W., Lang J.H. A dynamic model of drops detaching from a gas metal arc welding electrode. J. Phys. D: Appl. Phys. 1998; 31: 107–123.

[33] Haidar J. A theoretical model for gas metal arc welding and gas tungsten arc welding. I. J. Appl. Phys. 1998; 84(7): 3518–3529.

[34] Haidar J., Lowke J.J. Predictions of metal droplet formation in arc welding. J. Appl. Phys. D: Appl. Phys. 1996; 29: 2951–2960.

[35] Haidar J. An analysis of the formation of metal droplets in arc welding. J. Phys. D: Appl. Phys. 1998; 31: 1233–1244.

[36] Haidar J. Prediction of metal droplet formation in gas metal arc welding. II. J. Appl. Phys. 1998; 84(7): 3530–3540.

[37] Haidar J. An analysis of heat transfer and fume production in gas metal arc welding. III. J. Appl. Phys. 1998; 85(7): 3448–3459.

[38] Hertel, M., Spille-Kohoff A., Füssel U., Schnicket M. Numerical simulation of droplet detachment in pulsed gas–metal arc welding including the influence of metal vapour. J. Phys. D: Appl. Phys. 2013; 46(22): 224003.

[39] Schnick, M., Fuessel U., Hertel M., Haessler M., Spille-Kohoff A., Murphy A.B. Modelling of gas–metal arc welding taking into account metal vapour. J. Phys. D: Appl. Phys. 2010; 43(43): 434008.

[40] Murphy A.B. Why the arc, and its interactions with the electrodes, are important in predictive modelling of arc welding. Plasma Phys. Technol. 2015; 2: 233–240.

[41] Murphy A.B. A self-consistent three-dimensional model of the arc, electrode and weld pool in gas–metal arc welding. J. Phys. D: Appl. Phys. 2011; 44(19): 194009.

[42] Lu F., Wang H.-P., Murphy A.B., Carlson B.E. Analysis of energy flow in gas metal arc welding processes through self-consistent three-dimensional process simulation. Int. J. Heat Mass Transf. 2014; 68: 215–223.

[43] Fan H.G., Kovacevic R. A unified model of transport phenomena in gas metal arc welding including electrode, arc plasma and molten pool. J. Phys. D: Appl. Phys. 2004; 37: 2531–2544.

[44] Hu J., Tsai H.L. Heat and mass transfer in gas metal arc welding, Part I: the arc. Int. J. Heat Mass Transf. 2007; 50: 833–846.

[45] Hu J., Tsai H.L. Heat and mass transfer in gas metal arc welding, Part II: the metal. Int. J. Heat Mass Transf. 2007; 50: 808–820.

[46] Hu J., Tsai H.L. Effects of welding current on droplet generation and arc plasma in gas metal arc welding. J. Appl. Phys. 2006; 100: 053304.

[47] Hu J., Tsai H.L. Metal transfer and arc plasma in gas metal arc welding. J. Heat Transf. 2007; 129: 1025–1035.

[48] Xu G., Hu J., Tsai H.L. Three-dimensional modeling of arc plasma and metal transfer in gas metal arc welding. Int. J. Heat Mass Transf. 2009; 52: 1709–1724.

[49] Rao Z.H., Hu J., Liao S.M., Tsai H.L. Modeling of the transport phenomena in GMAW using argon-helium mixtures, Part I – the arc. Int. J. Heat Mass Transf. 2010; 53: 5722–5732.

[50] Rao Z.H., Hu J., Liao S.M., Tsai H.L. Modeling of the transport phenomena in GMAW using argon-helium mixtures, Part II – the metal. Int. J. Heat Mass Transf. 2010; 53: 5707–5721.

[51] Rao Z.H., Liao S.M., Tsai H.L. Effects of shielding gas compositions on arc plasma and metal transfer in gas metal arc welding. J. Appl. Phys. 2010; 107(4): 044902.

[52] Rao Z.H., Zhou J., Tsai H.L. Determination of equilibrium wire-feed-speeds for stable gas metal arc welding. Int. J. Heat Mass Transf. 2012; 55(23–24): 6651–6664.

[53] Diao Q.Z., Tsai H.L. Modeling of solute redistribution in the mushy zone during solidification of aluminum-copper alloys. Metall. Trans. 1993; 24A: 963–973.

[54] Carman P.C. Fluid flow through granular beds. Trans. Inst. Chem. Eng. 1937; 15: 150–166.

[55] Kubo K., Pehlke R.D. Mathematical modeling of porosity formation in solidification. Metall. Trans. 1985; 16A: 823–829.

[56] Beavers G.S., Sparrow E.M. Non-Darcy flow through fibrous porous media. J. Appl. Mech. 1969; 36: 711–714.

[57] Lowke J.J., Kovitya P., Schmidt H.P. Theory of free-burning arc columns including the influence of the cathode. J. Phys. D: Appl. Phys. 1992; 25:1600–1606.

[58] Lancaster J.F. The physics of welding. 2nd ed. Oxford: Pergamon Press; 1986.

[59] Torrey M.D., Cloutman L.D., Mjolsness R.C., Hirt C.W. NASA-VOF2D: a computer program for incompressible flows with free surfaces. LA-10612-MS. Los Alamos National Laboratory; 1985.

[60] Brackbill J.U., Kothe D.B., Zemach C. A continuum method for modeling surface tension. J. Comput. Phys. 1992; 100: 335–354.

[61] Celic A., Zilliac G.G. Computational study of surface tension and wall adhesion effects on an oil film flow underneath an air boundary layer. NASA Ames Research Center; August 1997.

[62] Finkelnburg W., Segal S.M. The potential field in and around a gas discharge, and its influence on the discharge mechanism. Phys. Rev. Lett. 1951; 83: 582–585.

[63] Zacharia T., David S.A., Vitek J.M. Effect of evaporation and temperature-dependent material properties on weld pool development. Metall. Trans. 1992; 22B: 233–241.

[64] Granger R.A. Fluid mechanics. New York: CBS College; 1985.

[65] Patanka S.V. Numerical heat transfer and fluid flow. New York: McGraw-Hill; 1980.

Chapter 6

The Analysis of Temporary Temperature Field and Phase Transformations in One-Side Butt-Welded Steel Flats

Jerzy Winczek

Additional information is available at the end of the chapter

http://dx.doi.org/10.5772/63994

Abstract

In this chapter, the welding method applied for modelling the temperature field and phase transformations is presented. Three-dimensional and temporary temperature field for butt welding with thorough penetration was determined on the basis of analytical methods of an integral transformation and Green's function. Structural changes of heating and cooling, proceeding in a weld (in the heat-affected zone), were described using the existing formulations of phase transformations. Considerations were illustrated by an example, for which analysis of temperature fields, developed by a moving heat source, and calculations of the distribution of particular phases (structures) were carried out. Metallographic studies of the butt joints, which were arc welded under a flux, were carried out in the empirical part of this work. Their results enabled the verification of the numerical simulation results of the phase transformations.

Keywords: butt-welded joint, temperature field, HAZ, phase transformations, numerical modelling, metallographic examination

1. Introduction

Welding is characterised by many specific features associated with variable temperatures and variable physical and mechanical properties of the welding material. The moving heat source, characteristic of welding, partial melts the joint surface and fuses an electrode. The electrode fills a joint space with liquid metal. Hence, welding elements are subjected to varying temperature ranges, that is, from ambient to that of a liquid metal.

Crystallisation and solidification, segregation of alloy elements and solutes and structural changes caused by intensive cooling occur extensively. Thermal and mechanical states and microstructure directly state about the quality of the welding joint.

Modelling the temperature field during welding was first initiated by Rosenthal [1] and Rykalin [2], who supposed the point and linear models of heat source, respectively. The adoption of a point heat source, as in the above-mentioned studies, yields results with respect to the points located near the centre of the weld, which are significantly different from the actual temperature values. Therefore, Eagar and Tsai [3] proposed a two-dimensional (2D) Gaussian-distributed heat source model and developed a solution of temperature field in a semi-infinite steel plate. Subsequently, Goldak et al. [4] introduced a double ellipsoidal three-dimensional heat source model. There are two ways of modelling the temperature field during welding: analytical [5–14] and numerical (the finite difference methods, infinitesimal heat balances and finite element method) [15–30]. The welding methods and types of joints can be studied through these approaches [6, 20, 21, 31–33]. The construction of numerical models with heightening complexity allows more essential factors for the exact description of the structural changes in the welded steel.

2. Temperature field in the butt-welded joint with thorough penetration

Welding is characterised by an application of the movable, concentrated heat source, which in turn makes the temperature field movable in time and space:

$$T = T(r,t) = T(x,y,z,t) \tag{1}$$

Studies are being conducted to develop models of temperature field. Such models should have a real-time shape and temperature gradients based on the geometrical dimension of the welding element and also time. Referring to the formulated problem, the solution of heat equation for isotropic medium is essential to determine a temporary temperature field:

$$\nabla^2 T(r,t) = \frac{1}{a}\frac{\partial}{\partial t}T(r,t) - \frac{f(r,t)}{\lambda} \tag{2}$$

where $T(r,t)$ is temperature at r position at t time, a is the coefficient of temperature compensation, λ thermal conductivity and $f(r,t)$ supplied energy per volume and time unit.

Analytical method, proposed by Geissler and Bergmann [34, 35], was chosen to solve this differential equation. A short description of the method, described in detail in the above-mentioned studies, is presented below.

The following assumptions were accepted in the calculations:

- quantities characterising the material properties, such as thermal conductivity, temperature compensation and thermal capacity, are constant (independent from temperature),

- heat waste by convection and radiation is negligible,
- reciprocal interaction of temperature field and phase changes is not taken into account,
- heat of fusion is not taken into consideration.

A sample with thickness of D and width of $B_1 + B_2$, heated by the movable welding heat source, which is displaced at velocity v along the x-axis (**Figure 1**), is an illustration of the below consideration.

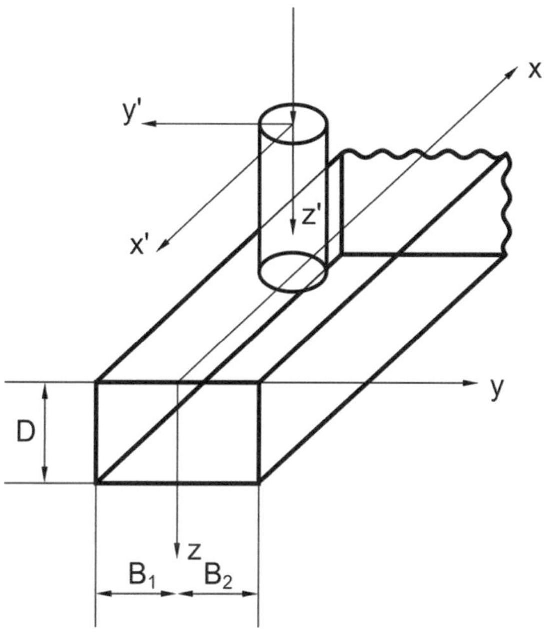

Figure 1. Schematic of heating of a steel sample using welding heat source.

According to Geissler and Bergmann [34, 35], the solution of Eq. (2) can be written as a superposition of Green's function. This leads to the following convolution of integrals as a general expression of temperature

$$T(r,t) = \frac{a}{\lambda} \int_0^t \iiint f(r',t') G(|r - r'|, t - t') dx'dy'dz'dt' \quad (3)$$

where $r(x,y,z)$ is a vector pointing the place on the sample, while $r' = (x',y',z')$ determines the source position. Green's function G describes the temperature field in the point of material defined by r in time t. It is caused by a point heat source acting in the r' position and at $t' < t$ time. G depends on the geometry of the sample and can be determined by transformation method, while f defines the cross-section of the welding heat source.

A three-dimensional temperature field with the possibility of acceptance of different geometries of samples as well as the shape of the heat source can be determined from Eq. (3).

In the case of the Gauss model of heat source, we have

$$f(x,y)=\frac{P}{2\pi R^2}\exp\left(-\frac{x^2+y^2}{2R^2}\right), -\infty < x < \infty, -\infty < y < \infty, \tag{4}$$

where the power of source is denoted by P and determined for R radius and corresponds to 1/e of its peak value (**Figure 2**).

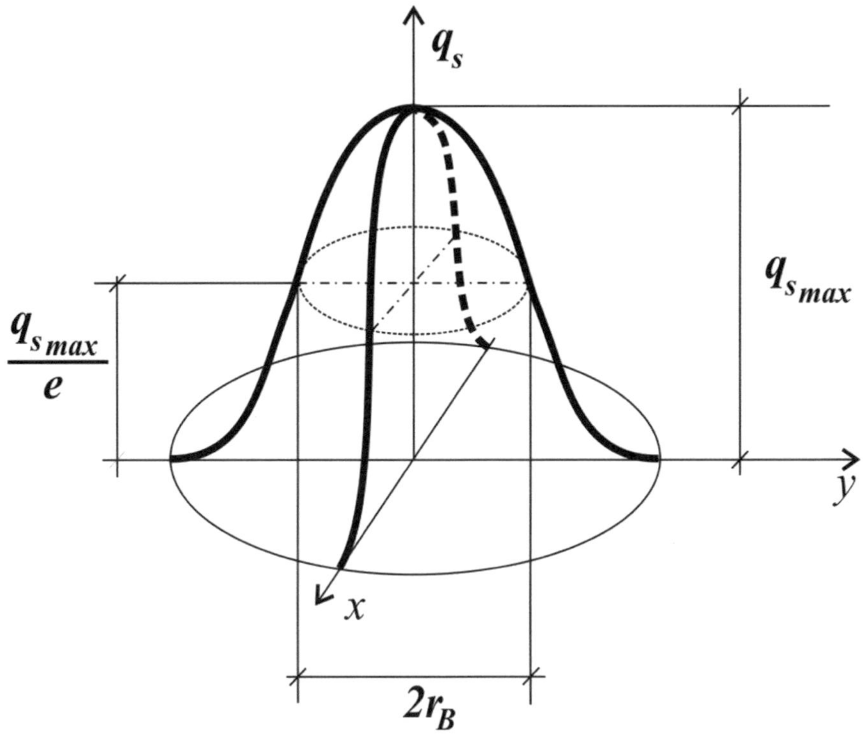

Figure 2. Gauss distribution of power of the heat source.

An infinitely long bar with the above-mentioned dimensions of cross-section was accepted in the considered example (**Figure 1**).

This can be written in the Cartesian coordinate system as follows:

$$-\infty < x < \infty, -B_1 \le y \le B_2, 0 \le z \le D \tag{5}$$

Boundary conditions defining the surface and Green's function were taken from the study by Carslaw and Jaeger [36]. Green's function takes the following form:

$$G(|r-r'|,t-t') = G_{(x)}(x-x',t-t')G_{(y)}(y-y',t-t')G_{(z)}(z,t-t') \tag{6}$$

$$G_{(x)} = \left[4\pi a(t-t')\right]^{-1/2} \exp\left(-\frac{\left(x+v(t_0-t')-x'\right)^2}{4a(t-t')}\right) \tag{7}$$

$$\begin{aligned} G_{(y)} = \left[4\pi a(t-t')\right]^{-1/2} \times \Bigg[& \sum_{n=-\infty}^{\infty} \exp\left(-\frac{\left[y-y'-2n(B_2+B_1)\right]^2}{4a(t-t')}\right) \\ & + \sum_{n=1}^{\infty} \exp\left(-\frac{\left[y-y'-2nB_2-2(n-1)B_1\right]^2}{4a(t-t')}\right) \\ & + \sum_{n=1}^{\infty} \exp\left(-\frac{\left[y-y'+2nB_1+2(n-1)B_2\right]^2}{4a(t-t')}\right) \Bigg] \end{aligned} \tag{8}$$

$$G_{(z)} = \left[4\pi a(t-t')\right]^{-1/2} \sum_{n=-\infty}^{\infty} \exp\left(-\frac{(z-2nD)^2}{4a(t-t')}\right) \tag{9}$$

where n is the transformation number of the source.

The relationship of movement of the welding heat source to the welding element is included in $G(x)$ function. Because a flat model of heat source was assumed, the z' coordinate is not present in $G(z)$ function. The coordinate systems connected with the heat source and welding material coincide with t_0 time. Considering that the shape of the heat source is independent of time, z' is dependent on f and G is eliminated; a modified integral to count temperature profile is thus obtained:

$$T(r,t) = \frac{a}{\lambda} \int_0^t \iint f(x',y') G(|r-r'|,t-t') \, dx' dy' dt' \tag{10}$$

z' integral is removed in comparison to Eq. (3).

Temperature distribution can be calculated after substituting Green's function and Eq. (4) into Eq. (10). The integral over a range of variables can be evaluated, which yields the following result:

$$T(x,t) = \frac{\eta P}{2\pi\rho Cp}\int_0^t u(t,t') \sum_{n=-\infty}^{\infty} \exp\left(-\frac{(z-2nD)^2}{4a(t-t')}\right)$$
$$\times\left\{\sum_{n=-\infty}^{\infty} \exp\left(-\frac{(x+v(t_0-t'))^2+(y-4nB)^2}{2R^2+4a(t-t')}\right)F_1(y)\right.$$
$$+\sum_{n=-\infty}^{\infty} \exp\left(-\frac{(x+v(t_0-t'))^2+(y-4nB)^2}{2R^2+4a(t-t')}\right)F_2(y)$$
$$\left.+\sum_{n=-\infty}^{\infty} \exp\left(-\frac{(x+v(t_0-t'))^2+(y-4nB)^2}{2R^2+4a(t-t')}\right)F_3(y)\right\}dt' + T_0 \quad (11)$$

where

$$u(t,t') = \frac{1}{2\sqrt{\pi a(t-t')}\left(R^2+2a(t-t')\right)} \quad (12)$$

$$F_{1(y)} = erf\left(\sqrt{\frac{R^2+2a(t-t')}{4aR^2(t-t')}}\left(B_2 - \frac{R^2(y-4nB)}{R^2+2a(t-t')}\right)\right) + $$
$$-erf\left(\sqrt{\frac{R^2+2a(t-t')}{4aR^2(t-t')}}\left(-B_1 - \frac{R^2(y-4nB)}{R^2+2a(t-t')}\right)\right) \quad (13)$$

$$F_{2(y)} = erf\left(\sqrt{\frac{R^2+2a(t-t')}{4aR^2(t-t')}}\left(B_2 - \frac{R^2(y-2(2n-1)B)}{R^2+2a(t-t')}\right)\right) + $$
$$-erf\left(\sqrt{\frac{R^2+2a(t-t')}{4aR^2(t-t')}}\left(-B_1 - \frac{R^2(y-2(2n-1)B)}{R^2+2a(t-t')}\right)\right) \quad (14)$$

$$F_{3(y)} = erf\left(\sqrt{\frac{R^2+2a(t-t')}{4aR^2(t-t')}}\left(B_2 - \frac{R^2(y+2(2n-1)B)}{R^2+2a(t-t')}\right)\right) + $$
$$-erf\left(\sqrt{\frac{R^2+2a(t-t')}{4aR^2(t-t')}}\left(-B_1 - \frac{R^2(y+2(2n-1)B)}{R^2+2a(t-t')}\right)\right) \quad (15)$$

http://dx.doi.org/10.5772/63994

3. Kinetics of phase transformations in a solid state

Heating processes of steel lead to the transformation of a primary structure into austenite, while cooling leads to the transformation of austenite into ferrite, pearlite, bainite and martensite. Structural changes of a welded joint, connected with its cooling (also with hardening), develop heterogeneous image of material structure, which influences the state of stress after welding. The zone with a yield point lesser or greater than that of an indigenous material can occur in the welded joint.

Mechanical properties of the joint mostly depend on the type of welding material (its primary structure and chemical constitution of steel) and the characteristics of heat cycles accompanying welding. Temperature levels attained during heating, the hold time at a particular temperature and velocity of cooling in the 800–500°C range determine the type of structure present in the joint during and after welding.

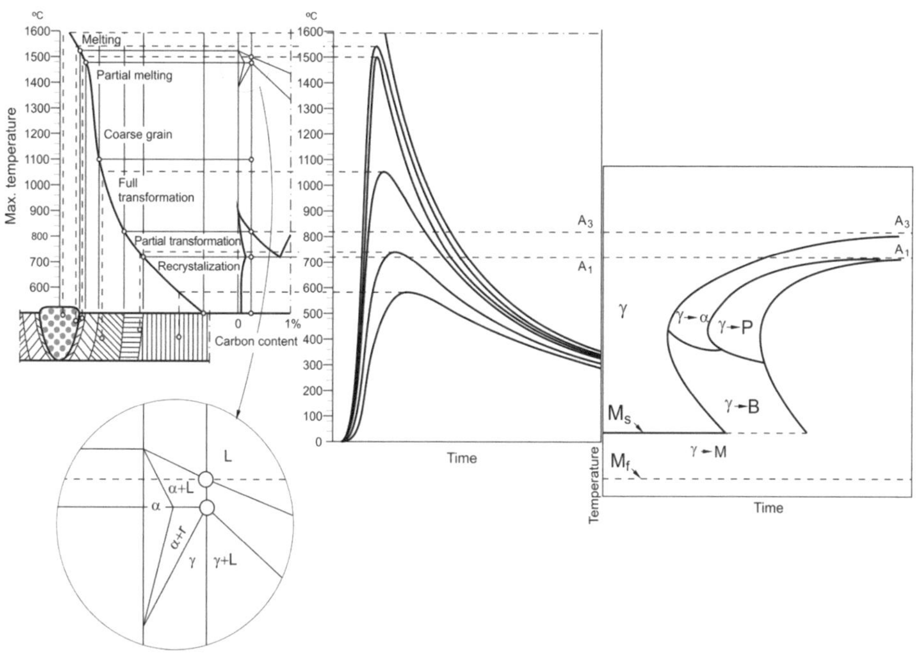

Figure 3. Characteristic structural areas of a welded joint, depending on the temperature and the share of the carbon in the steel.

Figure 3 shows the distribution characteristics of the weld joint zone of structural carbon steel [37] with a schematic of the fragment of an iron-carbon system and fragment of the TTT-welding diagram. Together, it is categorised into the following zones:

– fusion zone, which undergoes a thorough penetration and is characterised by the dendritic structure of solidification,

- partial joint penetration, where material is in a semi-fluid state and creates the border between the melted material and the material being converted into austenite,
- the course-grained structure, the so-called overheating zone,
- proper transformation, where perfect conversion of primary structure into austenite occurs,
- partial transformation between temperature A_1 at the beginning of austenisation and A_3 at the end of austenisation, where only a part of the structure changes into austenite,
- recrystallisation.

Several studies have focused on the description and numerical modelling of steel phase transformations. These studies have been reviewed by Rhode and Jeppson [38].

The type of a newly created phase depends heavily on the kinetics of heating and cooling processes. Kinetics of those processes is described by Johnson-Mehl-Avrami's and Kolomogorov's (JMAK) rules [39]. The amount of austenite ϕ_A created while heating the ferrite-pearlitic steel is therefore defined according to the following formula:

$$\varphi_A(T) = \sum_j \varphi_j^0 \left(1 - \exp\left(-b_j(T) t^{n_j(T)}\right)\right) \tag{16}$$

where ϕ_j^0 constitutes an initial share of ferrite ($j \equiv F$), pearlite ($j \equiv P$) and bainite ($j \equiv B$), while constants b_j and n_j are determined using conditions of the beginning and the end of transformation:

$$n_j = \frac{\ln(\ln(0.99))}{\ln(A_1 / A_3)}, \; b_j = \frac{0.01 n_i}{A_1} \tag{17}$$

In welding processes, the volume fractions of particular phases during cooling depend on the temperature, cooling rate, and the share of austenite (in the zone of incomplete conversion $0 \le \phi_A \le 1$). In a quantitative perspective, the progress of phase transformation during cooling is estimated using additivity rule by voluminal fraction ϕ_j of the created phase, which can be expressed analogically in Avrami's formula [40] by equation:

$$\varphi_j(T,t) = \varphi_A \varphi_j^{\max} \left\{1 - \exp\left[b_j\left(T(v_{8/5})\right) t(T)^{n(T(v_{8/5}))}\right]\right\} \tag{18}$$

where φ_j^{max} is the maximum volumetric fraction of phase j for the determined cooling rate estimated on the basis of the continuous cooling diagram (**Figure 4**), while the integral volumetric fraction equals to:

$$\sum_{j=1}^{k} \varphi_j = 1 \tag{19}$$

and k denotes the number of structural participations.

The quantitative description of dependence of the material structure and quality on temperature and transformation time of overcooled austenite during surfacing is made using the time-temperature-transformation diagram during continuous cooling, which combines the time of cooling $t_{8/5}$ (time when material stays within the range of temperature between 500 and 800°C, or the velocity of cooling ($v_{8/5}$ – (800 – 500)/$t_{8/5}$) and the temperature with the progress of phase transformation (**Figure 4**). Those diagrams are called TTT–welding diagrams.

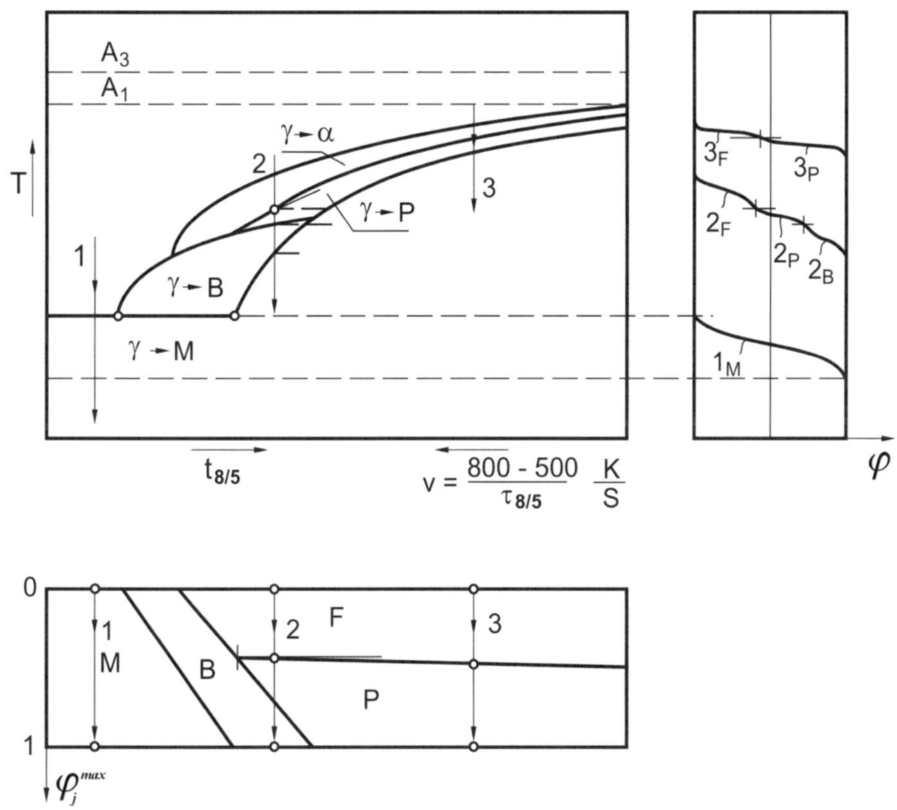

Figure 4. Scheme of phase changes of overcooled austenite depending on cooling velocity within temperature range 800–500°C.

Quantitatively, the progress of phase transformation is estimated by volumetric fraction ϕ_j of the created phase, where i denotes ferrite (j≡F), pearlite (j≡P), bainite (j≡B) or martensite (j≡M). The volumetric fraction ϕ_j of the created phase can be expressed using a formula given in Eq. (27), wherein time t is replaced with a new independent variable, temperature T [41]:

$$\varphi_j = \varphi_A \varphi_j^{\max}\left(1-\exp\left(-b_j T^{n_j}\right)\right)+\varphi_j^0 \tag{20}$$

where

$$n_j = \frac{\ln\left(\ln\left(1-\varphi_j^s\right)/\ln\left(1-\varphi_j^f\right)\right)}{\ln\left(T_j^s/T_j^f\right)}, b_j = \frac{n_j\left(1-\varphi_j^f\right)}{T_j^s} \tag{21}$$

$$\frac{\varphi_j^s}{\varphi_j^{\max}} = 0{,}01, \frac{\varphi_j^f}{\varphi_j^{\max}} = 0{,}99 \tag{22}$$

ϕ_j^0 is the volumetric participation of j-th structural component, which has not been converted during austenitisation; $T_j^s = T_j^s(v_{8/5})$ and $T_jf = T_j^f(v_{8/5})$ are, respectively, the initial and final temperature of phase transformation of this component.

The fraction of martensite formed below the temperature M_s is calculated using the Koistinen-Marburger formula [42, 43]:

$$\varphi_M\left(T\right) = \varphi_A / \varphi_M^{\max}\left\{1-\exp\left[-\mu\left(M_s - T\right)\right]\right\}, \mu = -\frac{\ln\left(\varphi_M^{\min} = 0.1\right)\right)}{M_s - M_f} \tag{23}$$

where ϕ_m denotes the volumetric fraction of martensite; M_s and M_f denote the initial and final temperature of martensite transformation, respectively; T is the current temperature of the process.

4. Thermal and phase transformation strains

Changes in temperature during welding cause deformations associated with the thermal expansion and deformation of the material resulting from the structural phase transformation. Deformation during the whole thermal cycle is the total deformation created during heating and cooling [44]:

$$\varepsilon\left(x,y,z,t\right) = \varepsilon^H + \varepsilon^C \tag{24}$$

where ε^H and ε^C denote the thermal and phase transformation strains during heating and cooling, respectively.

Heating leads to an increase in the material volume, while transformation of the initial structure (ferritic, pearlitic or bainitic) in austenite causes shrinkage which is associated with different densities of the given structures. Then, the strain caused during heating is calculated as follows:

$$\varepsilon^{H} = \varepsilon^{Th} - \varepsilon^{Trh} \tag{25}$$

where ε^{Th} is the strain caused by thermal expansion of the material:

$$\begin{aligned}\varepsilon^{Th} = \sum_{i=A,P,F,B,M} &(\alpha_i\varphi_{i0}\left(T-T_0\right)H\left(T_{A_1}-T\right)+\alpha_i\varphi_i\left(T-T_{A_1}\right)H\left(T_{A_3}-T\right)H\left(T-T_{A_1}\right)+\\ &+\alpha_A\left(T-T_{A_3}\right)H\left(T-T_{A_3}\right))\end{aligned} \tag{26}$$

while ε^{Trh} is the phase transformation strain during heating:

$$\varepsilon^{Trh} = \sum_{i=P,F,B,M} \varphi_i\gamma_{iA} \tag{27}$$

where γ_{iA} is the structural strain of the i-th structure in austenite, T_0 is the initial temperature, α_i is the linear thermal expansion coefficient of the i-th structure and $H(x)$ is the function defined as follows:

$$H\left(x\right)=\begin{cases}1 & for\ x>0\\ 0{,}5 & for\ x=0\\ 0 & for\ x<0\end{cases} \tag{28}$$

During cooling, the total strain (similarly as during heating) is the sum of strains associated with thermal expansion (in this case, the shrinkage of the material) as well as structural strains. Volumetric increase can be attributed to the high density of austenite (highest among the hardening structures such as martensite, bainite, ferrite and pearlite). The strain caused during cooling can be described by the following relation [44]:

$$\varepsilon^{C} = \varepsilon^{Tc} + \varepsilon^{Trc} \tag{29}$$

where ε^{Tc} is the strain caused by thermal shrinkage of material:

$$\begin{aligned}\varepsilon^{Tc} &= \alpha_A\left(T-T_{SOL}\right)H\left(T-T_s\right)+\alpha_A\left(T_s-T_{SOL}\right)H\left(T_s-T\right)+\\ &+\sum_{i=A,P,F,B,M}\alpha_i\varphi_i\left(T-T_{si}\right)H\left(T_{si}-T\right)\end{aligned} \tag{30}$$

while ε^{Trc} is the strain caused by phase transformation during cooling:

$$\varepsilon^{Trc} = \sum_{i=P,F,B,M} \varphi_i \gamma_{Ai} \tag{31}$$

where T_{SOL} denotes solidus temperature, T_s the initial temperature of phase transformation, T_{si} the initial temperature of austenite transformation in the i-th structure and γ_{Ai} the structural strain of austenite in the i-th structure. In addition, due to the limitation of the existence of solid state material:

$$\varepsilon(x,y,z,t) = 0 \quad for \quad T > T_{SOL} \tag{32}$$

5. The example of calculation of temperature field and phase transformations in welded flats

In the considered example, it is assumed that the welded material is steel S235 with the following material constants: specific heat C_p = 670 J/(kg·K), density ρ = 7800 kg/m^3, thermal diffusivity $a = 1.2 \times 10^{-5}$ m^2/s, the temperatures at the beginning and end of austenite transformation, respectively, A_1 = 996 K (723°C) and A_3 = 1108K (835°C), solidus temperature T_S = 1763K (1490°C) and the liquidus temperature T_L = 1793K (1520°C). The model of joining two butt-welded flats with a thickness of 0.012 m and a width of 0.1 m is shown in **Figure 5**.

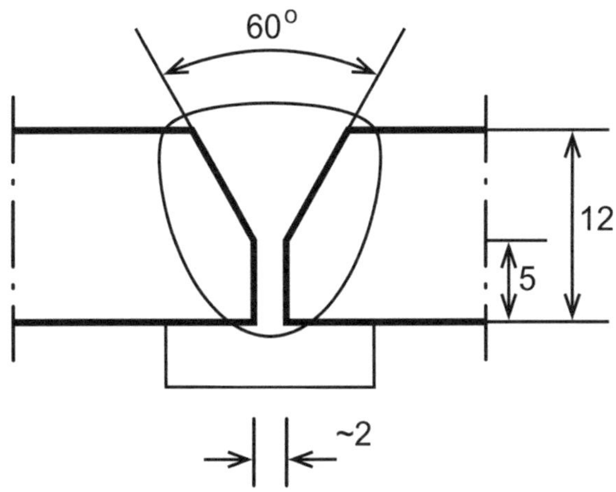

Figure 5. Scheme of the welded joint with technology chamfering.

The speed and power of the movement source are assumed to be v = 0.005 m/s and P = 18000 W, respectively. Time-varying temperature field was determined according to the formulas (Eqs. (11)–(15)). The temperature field on the upper surface of the welded flats and in the longitudinal section determined by the trace of the source transition is shown in **Figures 6** and **7**.

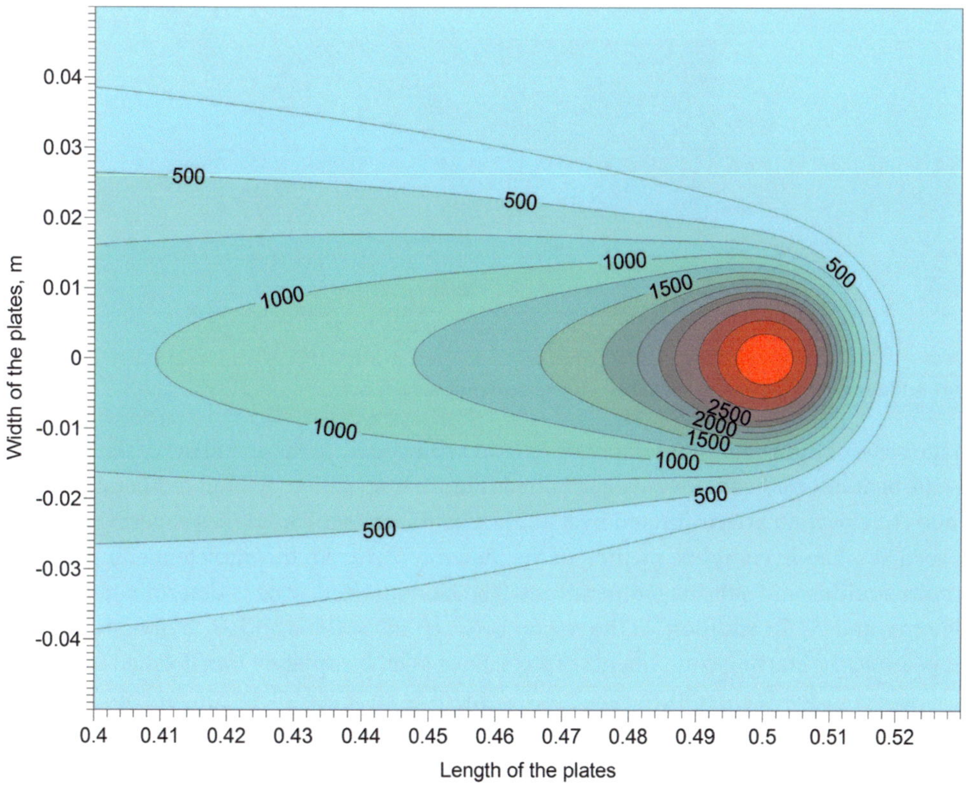

Figure 6. Temperature distribution (°C) on the surface of flats at time t = 102 s from beginning of welding.

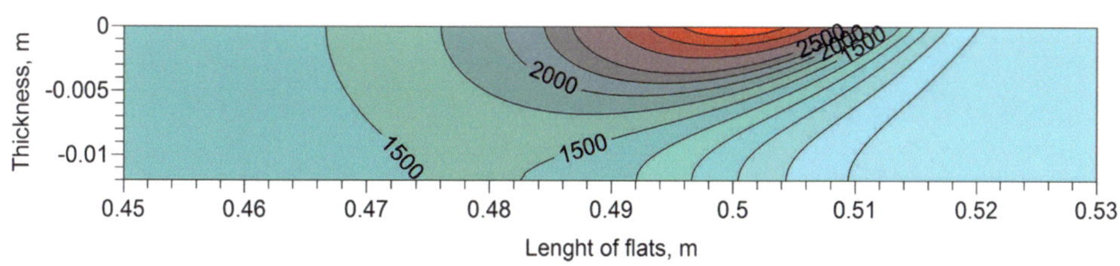

Figure 7. Temperature distribution (°C) in longitudinal section at time t = 102 s from beginning of welding.

The quantities $B_1 = B_2 = B$ are considered in calculations, which are equal to the width of one flat, thereby obtaining a temperature field symmetrical to the plane defined by vertical axis of the source and the direction of its movement. Thus, both this and the further calculation illustrations are shown on the right symmetrical cross-section which is perpendicular to a moving heat source. The image of maximum temperature isotherms in a cross-section of the joint (on the front contact of flats) is shown in **Figure 8**.

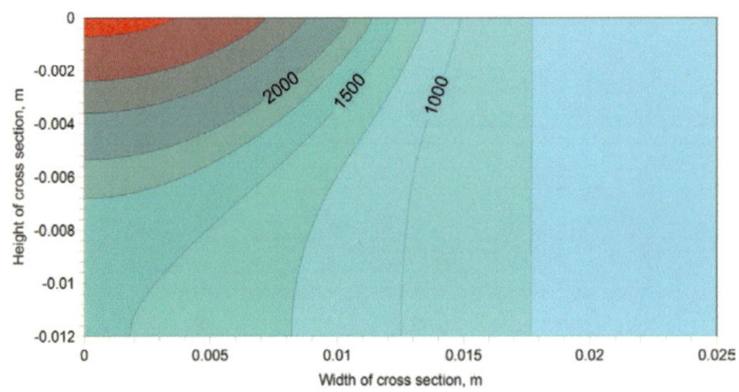

Figure 8. Maximum temperatures (°C) in half of cross-section of welded flats.

On the basis of the maximum temperature field achieved in particular areas of the weld joint, specific heat-affected zones were determined (**Figure 9**) specified by limit temperatures T_L, T_S, A_3 and A_1. The weld area is limited by a liquidus isotherm, which implies covering the part of the section wherein complete melting of the material occurred. Incomplete melting is limited by both liquidus and solidus temperatures. The heat-affected zone is determined by solidus isotherms and A_1. In addition, in the region between the solidus and A_3 temperatures, ferrite and pearlite, the starting components of steel, have been completely transformed into austenite, whereas in the temperature range A_1-A_3, there is an incomplete conversion. In the area where the temperature did not exceed A_1, certainly, phase transitions did not occur and the structure of the parent material (ferritic-pearlitic) was thus preserved. The participation of the heating structures was determined on the basis of the Fe-C diagram. The participation of the cooling structure was determined on the basis of knowledge of the TTT-welding diagram for welding steel S235 and the variations in temperature during cooling (**Figure 10**) [45].

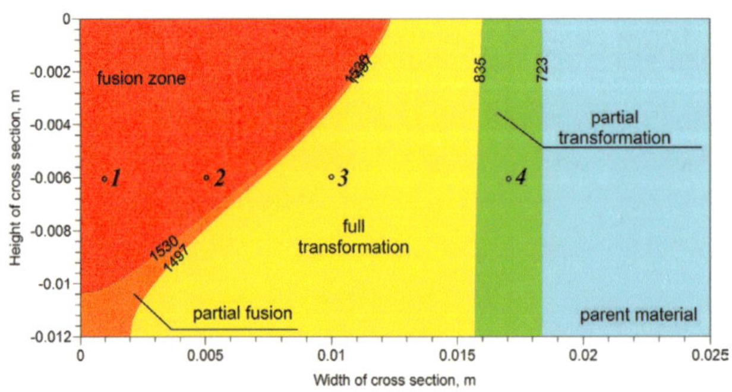

Figure 9. Heat-affected zones (HAZ) with selected points.

http://dx.doi.org/10.5772/63994

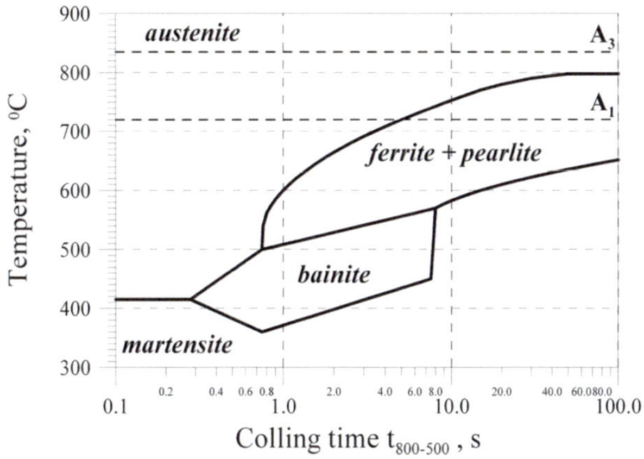

Figure 10. TTT-welding diagram for S235 steel.

The analysis of the cooling speed $v_{8/5}$ showed that after complete cooling in both weld and heat-affected zone, wherein there has been complete and partial transformation of ferrite and pearlite into austenite, supercooled austenite resulted in a ferrite-pearlite structure. Changes in the temperature and structure in the considered cross-section for the selected time of the welding cycle are shown in **Figures 11–19**. The analysis of these changes was investigated on a symmetrical half of the cross-section of the flats' connection, which is perpendicular to the direction of source movement and x_0=0.5 m from the beginning of the flats (in half their length), that is, from the beginning of the coordinate system associated with flats. For t = 0, the beginning of the coordinate systems associated with the source and the welded object overlaps. **Figures 11–13** illustrate the condition where the joint area was completely filled with the welded material, there were no cooling phase transitions and the largest participation of austenite is observed. By contrast, **Figures 14–19** illustrate the gradual decrease in the share of austenite which forms a ferritic-pearlitic structure following the temperature drop.

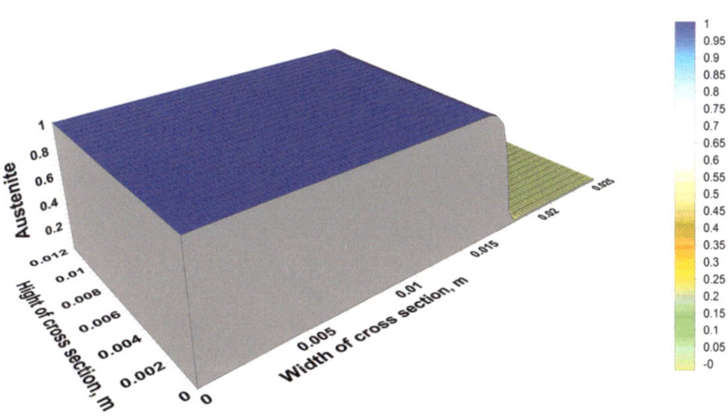

Figure 11. Volume fraction of austenite in cross-section at time t = 109 s from process beginning.

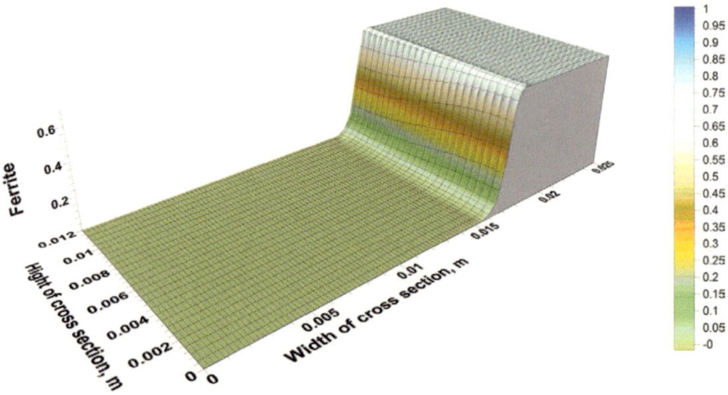

Figure 12. Volume fraction of ferrite in cross-section at time $t = 109$ s from process beginning.

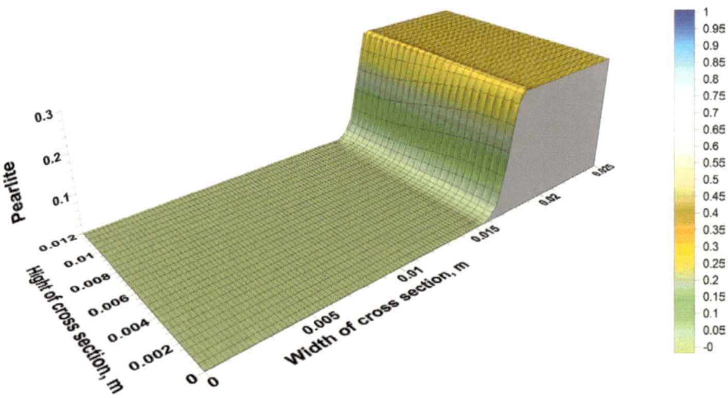

Figure 13. Volume fraction of pearlite in cross-section at time $t = 109$ s from process beginning.

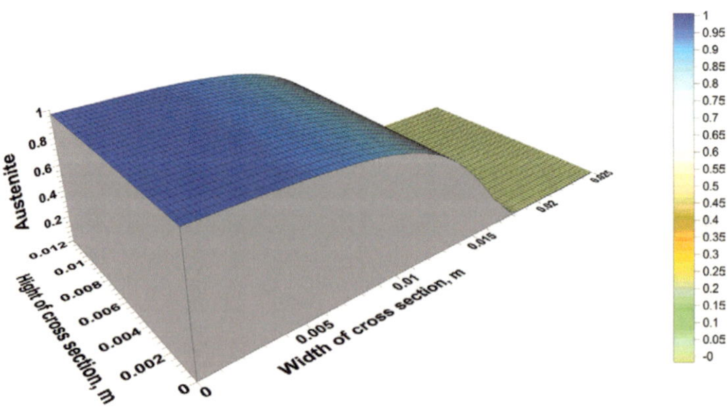

Figure 14. Volume fraction of austenite in cross-section at time $t = 134$ s from process beginning.

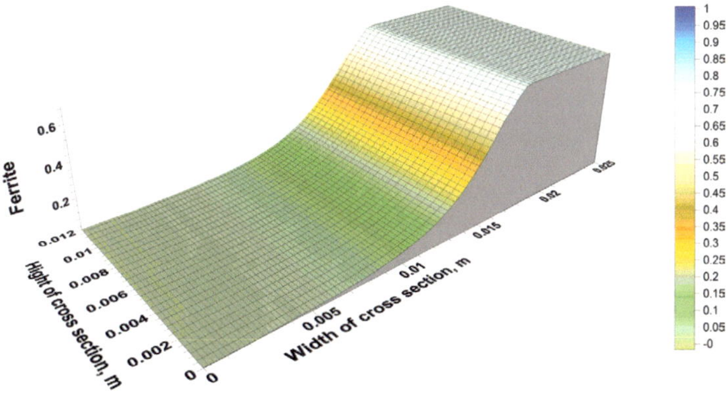

Figure 15. Volume fraction of ferrite in cross-section at time t = 134 s from process beginning.

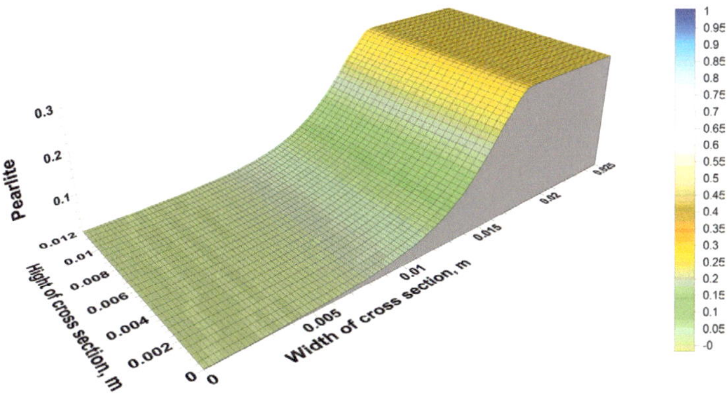

Figure 16. Volume fraction of pearlite in cross-section at time t = 134 s from process beginning.

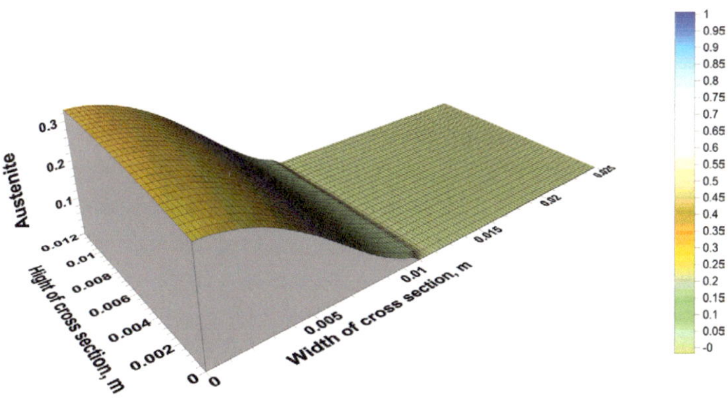

Figure 17. Volume fraction of austenite in cross-section at time t = 146 s from process beginning.

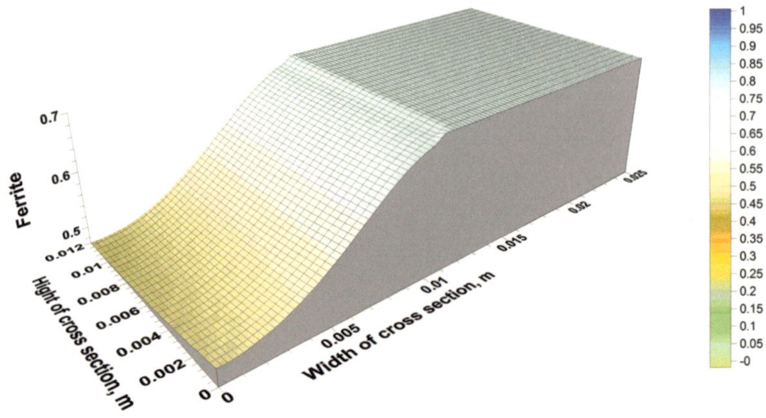

Figure 18. Volume fraction of ferrite in cross-section at time t = 146 s from process beginning.

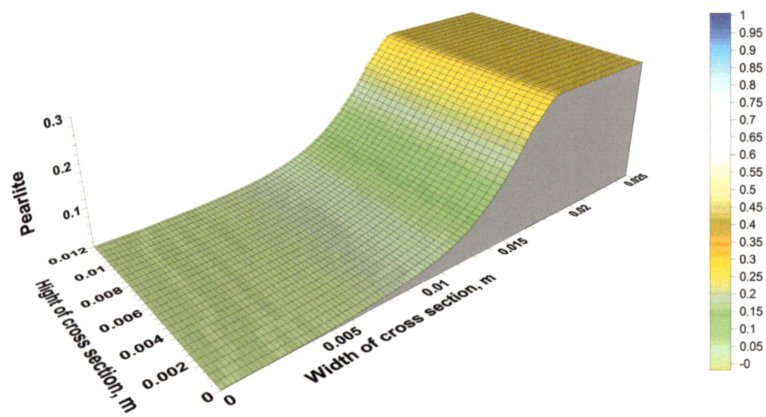

Figure 19. Volume fraction of pearlite in cross-section at time t = 146 s from process beginning.

Changes in temperature and volume fraction of the individual structural components at the selected points in the cross-section (comp. **Figure 9**) are shown in **Figure 20**.

Point *1* is located in the area of chamfering of flats; hence, the graph of temperature and austenite volume fraction starts at the moment of the joint execution. At points *2* (fusion zone) and *3* (full-transformation zone), a complete austenite transformation occurs. But at point *2*, apart from phase transformations in the solid state, melting and solidification phase transformations occur, resulting in a diverse dendritic structure shown during metallographic testing. At point 4, a partial transformation of the primary structure into austenite occurs, which is visible on the graph.

In strain calculations, linear expansion coefficients of the particular structural elements were assumed and structural stresses (**Table 1**) were determined on the basis of the author's own dilatometric research [46].

http://dx.doi.org/10.5772/63994

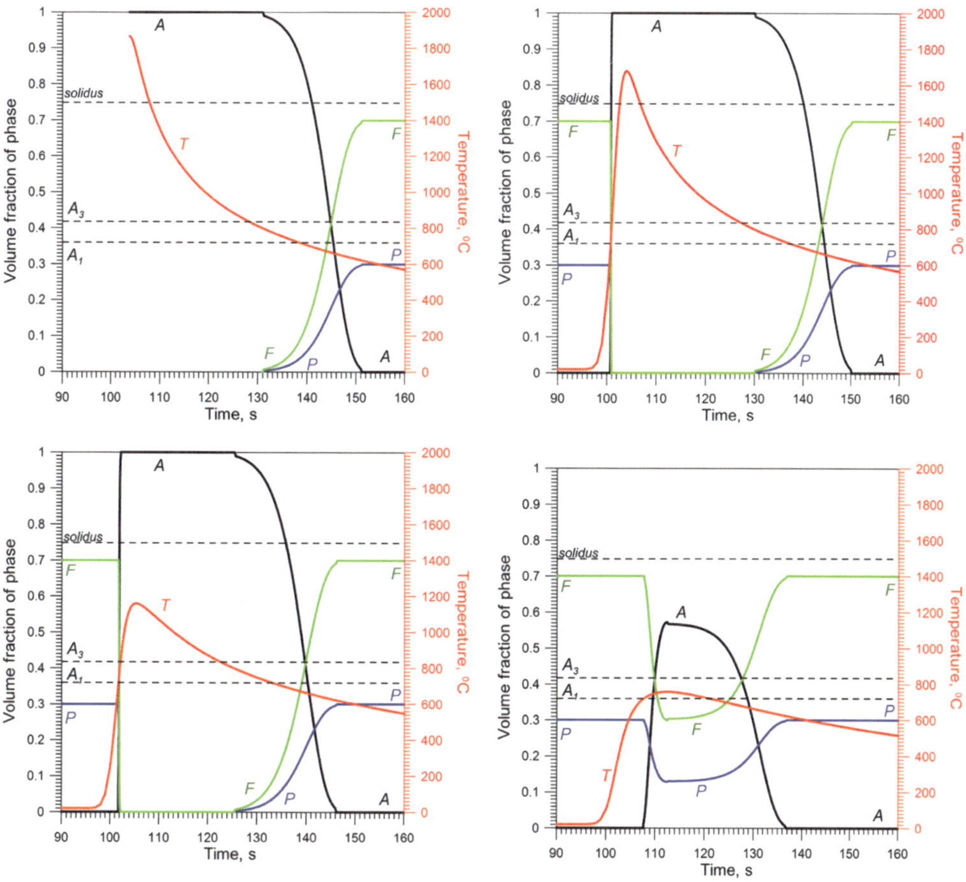

Figure 20. Welding thermal cycles and volume fractions of phases at selected points: T—temperature, A—austenite, P—pearlite, F—ferrite.

	α [1/°C]		γ
Austenite	2.178×10^{-5}	$\gamma_{F,P,S \to A}$	1.986×10^{-3}
Ferrite	1.534×10^{-5}	$\gamma_{B \to A}$	1.440×10^{-3}
Pearlite	1.534×10^{-5}	$\gamma_{A \to F,P}$	3.055×10^{-3}
Bainite	1.171×10^{-5}	$\gamma_{A \to B}$	4.0×10^{-3}
Martensite	1.36×10^{-5}		

Table 1. Structural (γ) and thermal (α) expansion coefficients of phases.

Dilatometric graphs (thermal and structural strains as a function of temperature) for selected points of the section (comp. **Figure 9**) are shown in **Figure 21**. In **Figure 21a**, a dilatometric graph for point *1* of the weld area is presented, where the material is deposited in the liquid phase as the molten material of the electrode, flux and partially melted edges of the flats; therefore, we observe only shrinkage "negative" strains of the cooling metal with a clear fault reflecting the strain of austenite transformation into ferrite and pearlite. Point *2* (**Figure 21b**) is located in the area of melting of the starting material, and the dashed line in the figure reflects the change of state from solid to liquid. The solid line (bottom) is initiated by weld solidification.

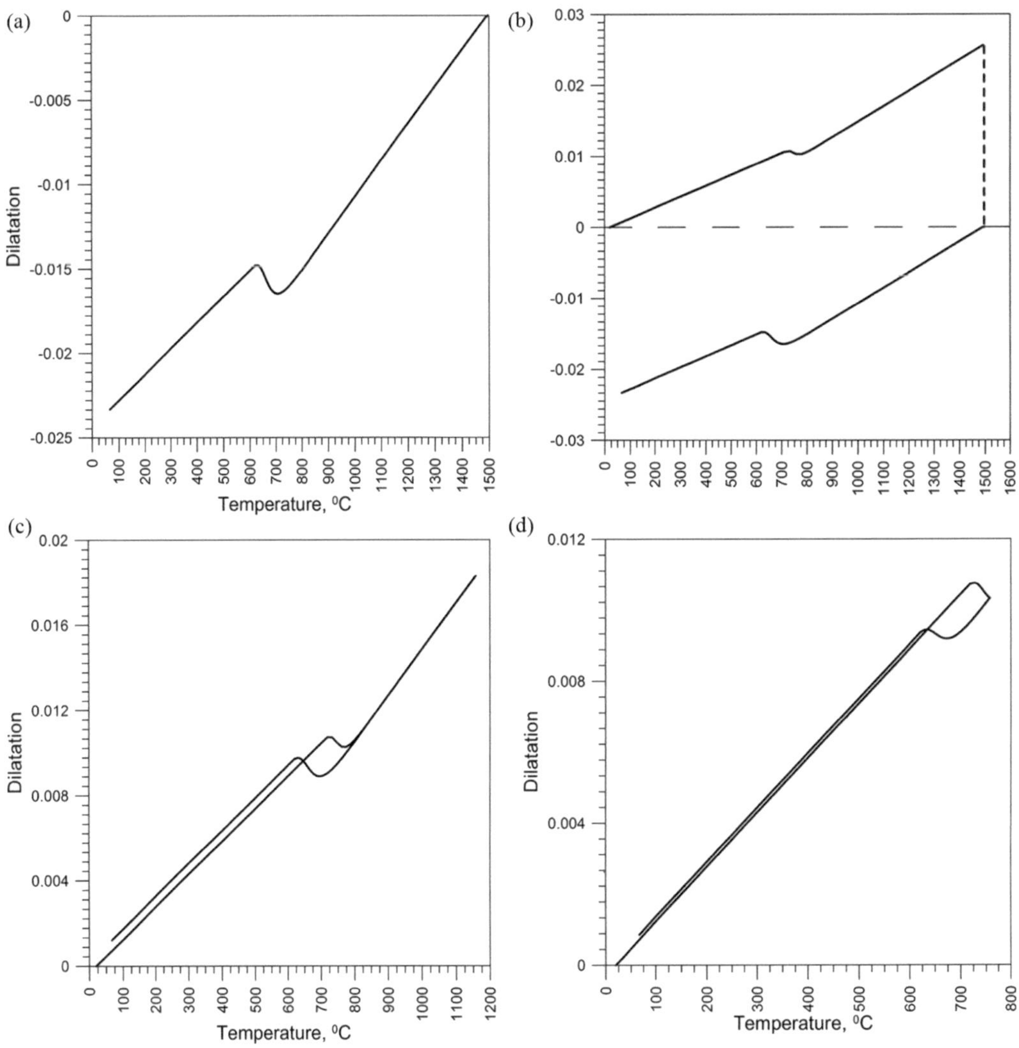

Figure 21. Dilatometric curves for selected points of cross-section.

Dilatometric graph in point *3* (**Figure 21c**) illustrates the history of strain changes for a full thermal cycle. In the considered point, the material with a ferritic-pearlitic structure is completely transformed into austenite during heating (lower line), which is shown with a fault in the diagram in the temperature range 723–835°C. At point 4 (**Figure 21d**) partial conversion

http://dx.doi.org/10.5772/63994

of ferrite and pearlite in austenite was observed, that is, at that point, the temperature during heating has exceeded the temperature of the beginning of austenitizing A_1, but has not reached the final temperature of austenitizing A_3. The stage of the austenite transformation is determined linearly in a numerical model by making the austenite transformation progress conditionally on the ratio $(T_{max} - A_1)/(A_3 - A_1)$, where T_{max} is the maximum temperature reached at a particular point of the weld joint.

6. Verification of the results of numerical simulation of phase transitions with the results of metallographic research

In order to verify the results of the numerical calculations, metallographic tests of the butt-welded joint were carried out. For this purpose, two flats with identical geometry as calculated, that is, two flats with a thickness of 0.012 m and width of 0.1 m, were welded. The material of the flats was steel S235. Before making the joint, sheet chamfering was conducted. Then, welding was carried out with the submerged arc welding method, under welding flux Taste-3 and with SPGΦ15 wire. Welding parameters were voltage U = 30 V, current I = 600 A and welding speed 20 m/h. The diagram of the elements prepared for joining is presented in **Figure 5** (identical to the welded joint adopted for numerical analysis). Cross-section of the weld joint (the image of the sample taken for metallographic examinations) is presented in **Figure 22**. Metallographic analysis was performed for specific zones of a welded joint, that is, for the area of weld, heat-affected zone and parent material. **Figure 23** shows an image of the middle part of the welded joint (at the junction of welded flats) with a clearly visible dendritic structure which is characteristic of solidification. **Figure 24** shows an image of the structure in the right symmetrical part of the welded joint in the area from the weld to the ferrite-pearlite structure of the parent material. On the border of the weld, dendrites are visible, which change in the heat-affected zone into a structure with the Widmannstatten structure elements.

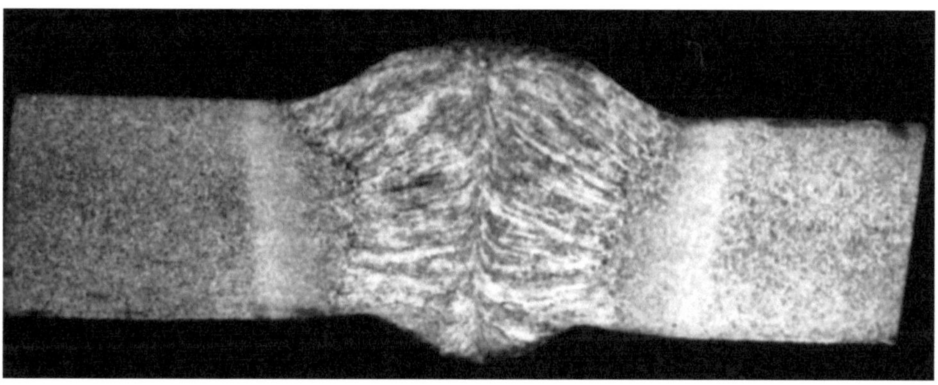

Figure 22. Cross-section of a welded joint: sample taken for metallographic analysis, magnification 2×.

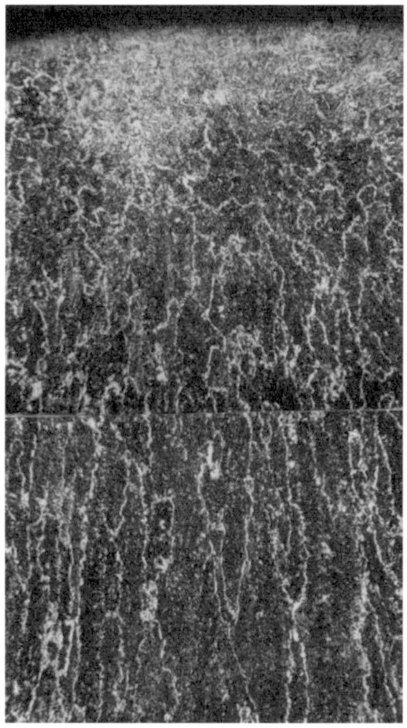

Figure 23. Weld, magnification 140×.

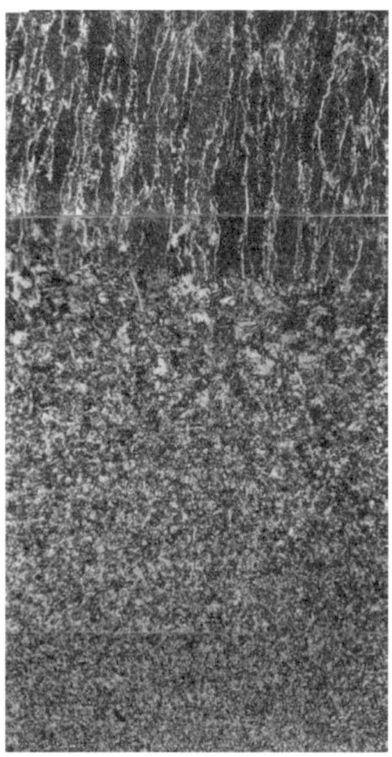

Figure 24. Transition zone, magnification 140×.

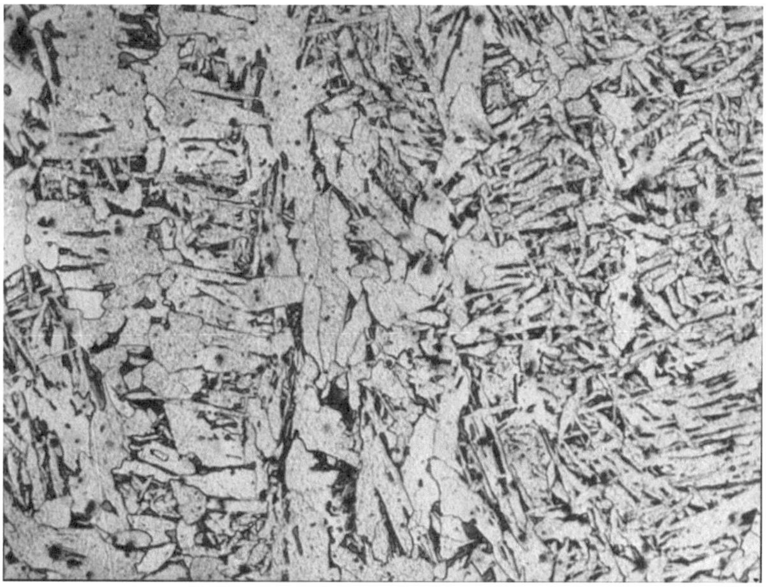

Figure 25. Weld, Nital Etch, magnification 250×, ferritic-pearlitic dendritic structure.

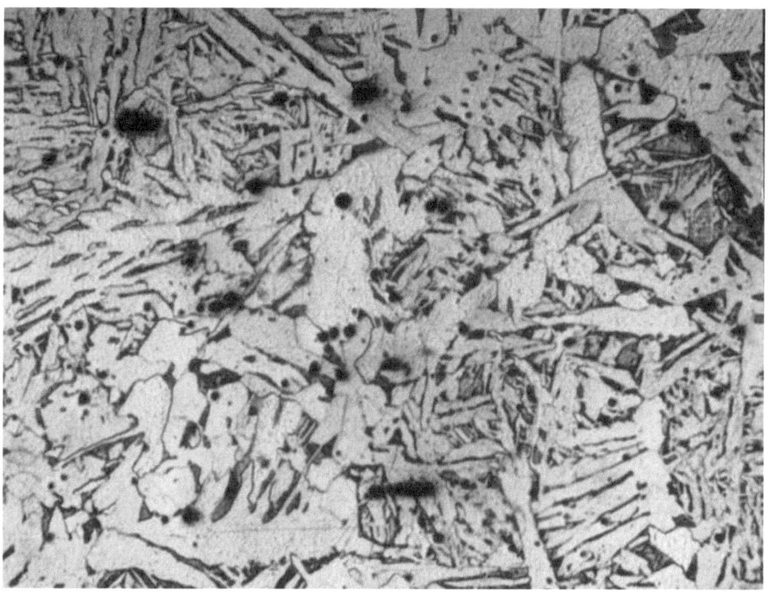

Figure 26. Heat affected zone, Nital Etch, magnification 250×, ferritic-pearlitic structure.

Next, the images of structures of individual areas are presented. In the area of the weld (**Figure 25**), we observe the ferritic-pearlitic structure with a small amount of supercooled pearlite and island bainite in the dendritic system characteristic of solidified structures. In the heat-affected zone (**Figure 26**), ferrite, pearlite and pearlite balls supercooled with a number of non-metallic inclusions are visible. Primary structure of the parent material (**Figure 27**) consists of ferrite and pearlite in the band system (hardly visible).

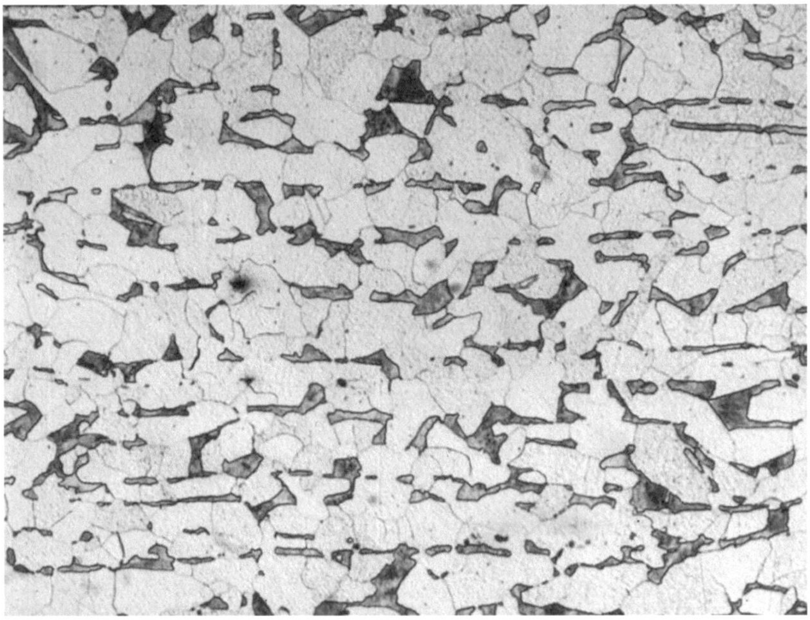

Figure 27. Parent material, Nital Etch, magnification 250×, ferritic-pearlitic banded structure.

The results of metallographic tests show high conformity with the results of numerical simulation and testify to the correctness of the developed numerical model.

7. Conclusion

The proposed model for determining temperature field allows to obtain characteristic zones of the weld joint with shapes and dimensions similar to the real ones. Metallographic verification of the results of numerical simulation of phase transitions also provided satisfactory results.

Analysis of phase transformations occurring during the welding process allowed determination of the quantitative composition of the cooling structures in the weld area (in the heat-affected zone).

This approach allows for an accurate tracking of changes in phase participations and the course of thermal and phase strains, which enable to determine, within the framework of thermoplasticity theory, welding stresses temporary and residual welding stresses.

http://dx.doi.org/10.5772/63994

Author details

Jerzy Winczek

Address all correspondence to: winczek@gmail.com

Czestochowa University of Technology, Czestochowa, Poland

References

[1] Rosenthal D. Mathematical theory of heat distribution during welding and cutting. Welding J. 1941;20:220s–234s

[2] Rykalin N N. Fundamentals of heat flow in welding. AN SSSR: Moskva;1947. 272 p.

[3] Eagar T W, Tsai N S. Temperature fields produced by traveling distributed heat sources. Welding J. 1983;62:346s–355s

[4] Goldak J, Chakravarti A, Bibby M. A new finite element model for welding heat sources. Metall Trans. 1984;15B:299–305

[5] Bo K S, Cho H S. Transient temperature distribution in arc welding of finite thickness plates. Proc Inst Mech Eng. 1990;204B3:175–183

[6] Radaj D. Heat effects of welding. Temperature field, residual stress, distortion, Springer-Verlag; Berlin Heidelberg. 1992; 348 p.

[7] Jeong S K, Cho H S. An analytical solution to predict the transient temperature distribution in fillet arc welds. Welding J. 1997;76:223s–232s

[8] Nguyen N T, Mai Y W, Simpson S, Ohta A. Analytical approximate solution for double ellipsoidal heat source in finite thick plate. Welding J. 2004;84:82s–93s

[9] Kwon Y, Weckman D C. Analytical thermal model of conduction mode double sided arc welding. Sci Technol Weld Join. 2008;13:539–549

[10] Fachinotti V D, Anca A A, Cardona A. Analytical solutions of the thermal field induced by moving double-ellipsoidal and double elliptical heat sources in a semi-infinite body. Int J Num Meth Biomech Eng. 2011;27:595–607

[11] Antonakakis T, Maglioni C, Vlachoudis V. Closed form solutions of the heat diffusion equation with Gaussian source. Int J Heat Mass Transf. 2013;62:314–322

[12] Ghosh A, Barman N, Chattopadhyay H, Hloch S. A study of thermal behaviour during submerged arc welding. Strojniški vestnik – Journal of Mechanical Engineering. 2013;59:333–338

[13] Franko A, Romoli L, Musacchio A. Modelling for predicting seam geometry in laser beam welding of stainless steel. Int J Thermal Sci. 2014;79:194–205

[14] Salimi S, Bahemmat P, Haghpanahi M. A 3D transient analytical solution to the temperature field during dissimilar welding processes. Int J Mechn Sci, 2014;79:66–74

[15] Na S J, Lee S Y. A study on the tree- dimensional analysis of the transient temperature distribution in gas tungsten arc welding. Proc Inst Mech Eng. 1987;201B3:149–156

[16] Radaj D. Finite element analysis of welding residual stress. International Conference on Residual Stresses 2, Nancy 1988, Nov. 23–25, Elsevier Applied Science, London. 1989; p. 510–516

[17] Karlsson R I, Josefson B L. Three dimensional finite element analysis of temperatures and stresses in a single – pass butt-welded pipe. Trans ASME. 1990;112:76–84

[18] Mundra K, DebRoy T, Kelkar K M. Numerical prediction of fluid flow and heat transfer in welding with a moving heat source. Num Heat Transf. 1996;29A:115–129

[19] Komanduri R, Hou Z B. Thermal analysis of the arc welding process: part I. General solutions. Metall Mater Trans. 2000;31B:1353–1370

[20] Mahapatra M M, Datta G L, Pradhan B. Three-dimensional finite element analysis to predict the effects of shielded metal arc welding process parameters on temperature distributions and weldment zones in butt and one-sided fillet welds. Proc I Mech – J Eng Manuf. 2006;220:837–845

[21] Kumar A, DebRoy T. Heat transfer and fluid flow during gas-metal-arc fillet welding for various joint configurations and welding positions. Metall Mater Trans. 2007;38A: 506–519

[22] Wang S, Goldak J, Zhou J, Tchernov S, Downey D. Simulation on the thermal cycle of a welding process by space-time convection-diffusion finite element analysis. Int J Thermal Sci. 2009;48:936–947

[23] Lindgren L E. Finite element modeling and simulation of welding. Part 1: increased complexity. J Thermal Stresses. 2001;24:141–192

[24] Hongyuan F, Qingguo M, Wenli X, Shude J. New general double ellipsoid heat source model. Sci Tech Weld Join. 2005;10:361–368

[25] Deng D. FEM prediction of welding residual stress and distortion in carbon steel considering phase transformation effects. Mater Dsgn. 2009;30:359–366

[26] Jiang W, Liu Z, Gong J M, Tu S T. Numerical simulation to study the effect of repair width on residual stresses of stainless steel clad plate. Int J Pres Ves Pip. 2010;87:457–463

[27] Lee H T, Chen C T, Wu J L. 3D numerical study of effects of temperature field on sensitisation of Alloy 690 butt welds fabricated by gas tungsten arc welding and laser beam welding. Sci Tech Weld Join. 2010;15:605–612

[28] Joshi S, Hildebrand J, Aloraier A S, Rabczuk T. Characterization of material properties and heat source parameters in welding simulation of two overlapping beads on a substrate plate. Comp Mater Sci. 2013;69:559–565

[29] Fu G, Lourenco M I, Duan M, Estefen S F. Effect of boundary conditions on residual stress and distortion in T-joint welds. J Constr Steel Res. 2014;02:121–135

[30] Chen Y, He Y, Chen H, Zhang H, Chen S. Effect of weave frequency and amplitude on temperature field in weaving welding process. Int J Adv Manuf Techn. 2014;75:803–813

[31] Tsai N S, Eagar T W. Distribution of the heat and current fluxes in gas tungsten arc. Metall Trans. 1985;16B:841–846

[32] Piekarska W, Kubiak M. Three-dimensional model for numerical analysis of thermal phenomena in laser-arc hybrid welding process. Int J Heat Mass Transf. 2011;54:4966–4974

[33] Chen J, Schwenk C, Wu C S, Rethmeier M. Predicting the influence angle on heat transfer and fluid flow for new gas metal arc processes. Int J Heat Mass Transf. 2012;55:102–111

[34] Geissler E., Bergmann H W. Calculation of temperature profiles heating and quenching rates during laser processing. Optoelektronik Magazin. 1987;3:430–434

[35] Geissler E., Bergmann H W. 3D Temperature fields in laser transformation hardening, part I: quasi-stationary fields. Optoelektronik Magazin. 1988;4:396–403

[36] Carslaw H S, Jaeger J C. Conduction of heat in solids. Clarendon Press: London 1996.

[37] Ruge J. Handbook of welding technology, vol. 1: Materials. Springer Verlag;Berlin:1980

[38] Rhode J, Jeppson A. Literature review of heat treatment simulations with respect to phase transformation, residual stresses and distortion. Scand J Metall. 2000;29:47–62

[39] Piekarska W, Kubiak M, Bokota A. Numerical simulation of thermal phenomena and phase transformations in laser-arc hybrid welded joint. Archiv Metall Materials. 2011;56:409–421

[40] Avrami M. Kinetics of phase change. I. General theory. J Chem Physics. 1939;7:1103–1112

[41] Parkitny R, Winczek J. Modelling of phase transformations during multipass surfacing, In: Conf. Proc. XXXVIII Sympozjon Modelling in Mechanics, Silesian University of Technology, Gliwice. 1999; p. 219–224

[42] Koistinen D P, Marburger R E. A general equation prescribing the extent of the austenite-martensite transformation in pure iron-carbon alloys and plain carbon steels. Acta Metall. 1959;7:59–60

[43] Domański T, Bokota A. Numerical models of hardening phenomena of tools steel base on the TTT and CCT diagrams. Archiv Metall Mater. 2011;56:325–344

[44] Winczek J. A simplified method of predicting stresses in surfaced steel rods, J Mater Proces Technol, 2012;212:1080–1088

[45] Brózda J, Pilarczyk J, Zeman M. TTT-welding diagrams transformation of austenite. Śląsk; Katowice. 1983; 139 p.

[46] Winczek J, Kulawik A. Dilatometric and hardness analysis of C45 steel tempering with different heating-up rates. Metalurgija. 2012;51(1):9–12

Chapter 7

Laser and Hybrid Laser-Arc Welding

G. A. Turichin

Additional information is available at the end of the chapter

http://dx.doi.org/10.5772/64522

Abstract

Laser and hybrid laser-arc welding are used at present in modern industry, having many advantages over traditional welding technology. Sectors such as the automotive industry, shipbuilding, aviation and space industry, chemical machinery, defense industry, and so on cannot be imagined without these technologies. Possibility of dramatic increase of weld joint properties, robustness, and high level of process automation makes the technology of laser and hybrid material processing a prospective part of the industry. At the same time, physical complexity of these processes, their cross-science nature, and necessity in high-level skilled stuff require many efforts for wide and successful industrial implementation. Present manuscript, devoted to discussion of physical peculiarity of laser and hybrid laser-arc welding of metals, approaches to physical-based design of technological equipment, as well as examples of industrial applications of laser and hybrid welding concerning the possibility to control welded metal structure and properties, is one of the steps on this way.

Keywords: laser, welding, laser material processing, hybrid welding, laser-induced plasma, technological equipment, process control, process monitoring, kinetics of phase transformations

1. Introduction

There is a tendency of modern industry to decrease construction weight that is connected to the necessity of increasing fuel efficiency. For this purpose, new high-strength, two- and three-phase steels are applied, as well as new Al- and Ti-based alloys, and their properties are defined by parameters of inclusions ensembles. New technologies for automotive bodywork use tailored blanks when the welded blanks are exposed by stamping. Weight reduction blanks with the same mechanical characteristics are possible by using new high-strength steels. Nevertheless, there is the problem of providing required characteristics of welds. It is also important to minimize welding stress and distortion, provide quality assessment, and process automation.

For these problems to be solved, intensive development and wide industrial implementation of laser and hybrid laser-arc welding (HLAW) becomes necessary in near future. Having a lot of evident advantages, beam welding due to complicity of technological processes needs to be successfully used with a deep understanding of process peculiarities, new CAE-based approaches of technology design, design of technological equipment on the basis of high brightness fiber lasers, technical vision, and process monitoring, as well as creation of new classes of welding materials.

Use of laser radiation and electric arc together for welding of metals and alloys so that both sources of heating influence on a material within just one heating zone was born 30–35 years ago [1]. Until recently, CO_2 lasers with radiation in far-infrared region were used. Metal interaction with the laser radiation of 10.6 and 1.06 μm is principally different. Lasers with such wavelength radiation are of poor quality and low accuracy. Only in recent years continuous fiber power lasers with good quality has been developed. They possess high beam quality and high accuracy.

Hybrid laser-arc welding is one of the most promising technologies for joining thick and heavy parts for production of gas and oil pipes, shipbuilding industry, building constructions, and bridge sections. The main benefit of hybrid laser-arc welding is the possibility to weld by one path materials with thickness of up to 20 mm and more, including new type of steels and modern alloys. Hybrid welding can also provide high-quality weld seam whose properties are comparable with laser weld seam properties, but the use of this technology in the case of real production is restricted by high complicity of the process [2] and appearance of different defects, such as porosity, cracks, spiking, and humping in the weld seam.

The analysis of results of the carried investigations of hybrid laser-arc welding process enables to determine the series of problems, in which decision is needed to develop reliable welding technology of thick metals and alloys. It is necessary to exclude: an undesirable direction of crystals growth; dramatic increase of the seam width in the top part of its cross-section; existence of hardening structures in deep penetration zone; presence of set of gas pores; and inadequate values of impact strength of the axial zone, notably at negative test temperatures. The overview of modern trends and problem for solution is presented in [3]. The only way to develop a reliable technology of hybrid laser-arc welding is the use of computer engineering-based approach to determine and optimize technological parameters as well as for finding and testing of technological methods.

2. Theoretical base of laser and hybrid welding

For understanding the weld pool formation and behavior during hybrid laser-arc welding it is comfortable to use two process models: first, for steady-state case, and second, for dynamic behavior of melt pool. The steady-state model is described in more detail in [4]. All incomplete models consider main peculiarities of interconnected physical processes. Problem with laser beam absorption and reflection inside the keyhole, heat transfer in solid face, and task about vapor flow in keyhole are solved by the same way as it was done in the model of laser welding [4]. The arc and plasma models, which use boundary layer approximation for mass, momentum, current, and energy equations [5], are very specific for laser-arc welding. Media compressibility, volumetric heating by laser beam and arc current, mixing of metal vapor, shielding and arc gas, and temperature influence on kinetic coefficients have to be considered as well as workpiece surface influence on arc and shielding gas flow. The hybrid electric discharge, which defines values of ionization rate, and spatial distributions of conductivity and thermal diffusivity, is also very specific. Physical nature of listed processes is also important for the formation of melt pool, so it is necessary to look into it more deeply.

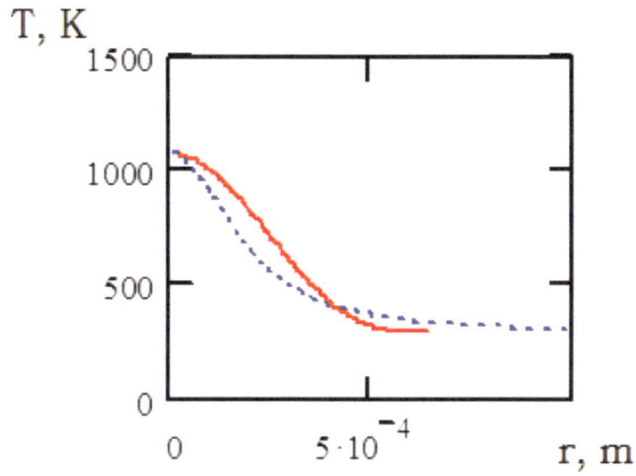

Figure 1. Jet radial temperature distribution 1 cm above surface, solid line – compressible gas, dot line – oncompressible.

The laser-induced plasma plume above the workpiece surface have an influence on welding process due to laser radiation absorption and refraction [6], and also can be used as an information source for online process monitoring [7] and control. As it is well known plasma plume structure and parameters are strongly dependent on radiation wavelength [8] and shielding gas nature and rate [9]. From the gas dynamics point of view, plasma plume is a subsonic submerged jet [10] of metal vapor in shielding gas with volumetric heating due to laser radiation absorption in the plasma. The numerical schemes [11] or well-known analytical solution for noncompressive submerged jet [12] are usually applied for calculation of the parameters of vapor-jet plasma. The volumetric heating, strongly influencing on jet flow, depend on plasma absorption coefficient. The theoretical descriptions of laser-induced plasma are usually supposed as thermodynamic equilibrium. The temperature of equilibrium plasma is defined ionization degree and others plasma parameters [13]. However, because absorption of radiation energy by plasma electrons and energy transfer from electrons to heavy component require an energy gap between light and heavy plasma components, supposition of thermodynamic equilibrium is not correct [14]. So the description of laser-induced plasma and also of plasma combined laser-arc discharge is to be based on solution of a Raiser kinetic equation [15] for electrons energy spectrum, like it was done for laser-induced plasma in the keyhole [14], taking into account the chemistry and gas dynamics of plasma plume.

The analytic description of plasma plume in vapor-jet shielding gas mix with consideration of volumetric heating, heat conductivity, diffusion, viscosity, and compressibility effects for laser and hybrid welding is possible to get by means of application of boundary layer approach for plume gas dynamics and approach of physical kinetic for plasma of laser-arc discharge [5].

Enthalpy of vapor-gas mixture in plasma plume according to [5] in this case is given by the formula:

$$h = \left[\left(\frac{v\beta(1+B(x))}{x} \frac{\chi/2v}{1-\chi/2v} - \frac{\alpha^2 PmC_p}{2\eta kx^2} \frac{r^2}{2} \right) \frac{1-\chi/2v}{\frac{\chi}{2v} \left(\frac{v\beta(1+B(x))}{x} \right)^{\chi/2v}} \right]^{\frac{1}{1-\chi/2v}} \quad (1)$$

where r and x are the polar coordinates, constants α and β are given by initial conditions of momentum and energy fluxes [5], $B(x, y) = \int_0^x \frac{Q(x', y)\gamma}{h_c(x', y=0) \cdot h^*} dx'$, h_c is the initial value of vapor enthalpy, h^* is the vapor enthalpy for evaporation temperature, c is the heat capacity, Q is given by the formula $q = \frac{4m_e c\rho}{\sqrt{\pi} M_{Fe} M_{He}^2} \cdot \frac{v_\infty^2}{N_\infty^2 \alpha_{Fe}} \cdot J_{Fe} \exp\left(-\frac{J_{Fe}}{\langle\varepsilon\rangle}\right) = Q\frac{c}{h}$, $q = \mu I$, I is the laser beam intensity, μ is the plasma absorption factor, η and v are the dynamic and kinematic gas viscosity, and χ is the temperature diffusivity.

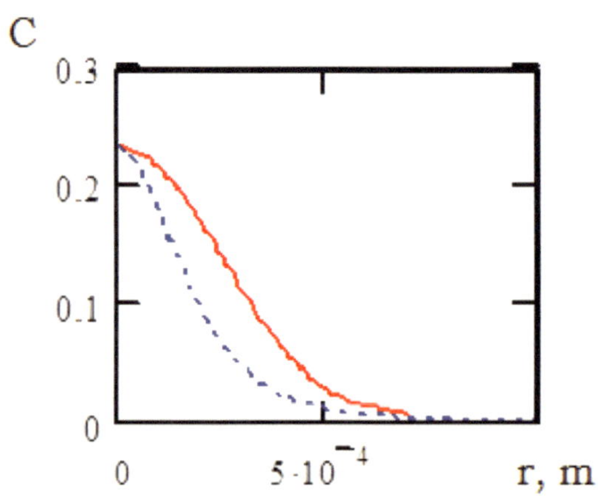

Figure 2. Iron concentration field in jet 1 cm above surface, solid line – compressible gas, dot line – non-compressible.

http://dx.doi.org/10.5772/64522

Figures 1–3 show examples of calculated distributions of temperature, mix density, and concentration in the plasma plume during laser welding of mild steel with initial temperature of 3200 K, flowing out of the keyhole with initial speed of 200 m/s into the helium shielding atmosphere of room temperature.

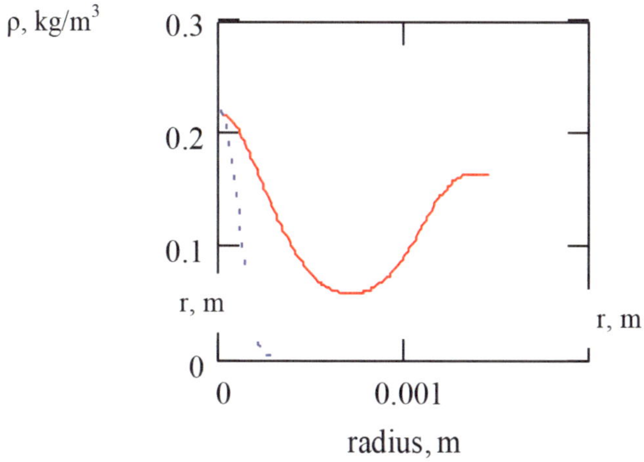

Figure 3. Density distribution on keyhole outlet, solid line – total density, dot line – iron vapor density.

The plasma plume interferometry experiments (**Figure 4**), made with the Michelson scheme, confirm the developed theory.

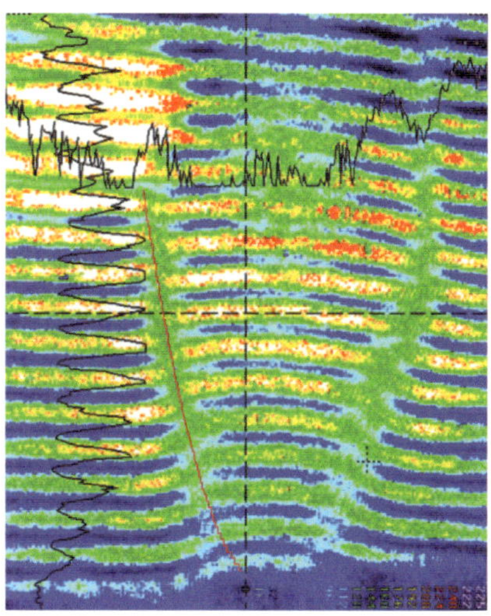

Figure 4. Plasma plume interferogram with calculated boundary of the jet (red solid line) for NdYAG laser welding.

Plasma plume volumetric heating due to laser beam absorption and Joule heating of electric arc make strong effect on parameters of plasma plume. Calculation shows that volumetric heating due to plasma absorption lead to increases of jet thickness (**Figure 5**). The same behavior is typical for velocity and concentration (**Figure 6**) distributions. In calculation, the beam parameters which are typical for welding with deep penetration by CO_2 laser: beam power $W = 6$ kW, wavelength is 10.6 μm, focal radius $rf = 0.23$ mm, target material = mild steel, and shielding gas = He. The keyhole outlet radius (0.35 mm), initial jet velocity (200 m/s), and temperature (3200 K) were calculated by the simulation software LaserCAD [16].

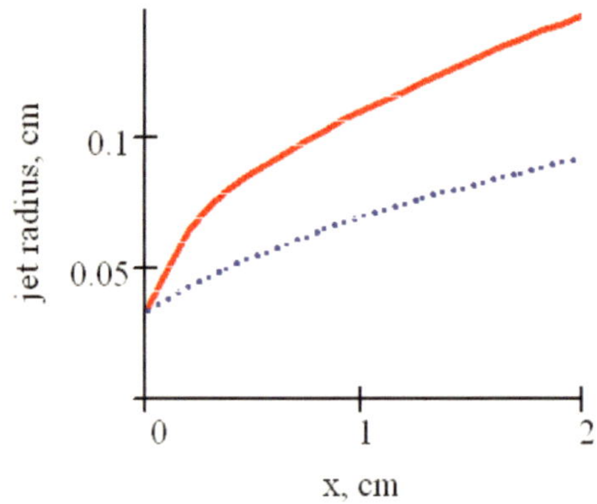

Figure 5. Boundary of the jet, solid line – with plasma absorption, dot line – without.

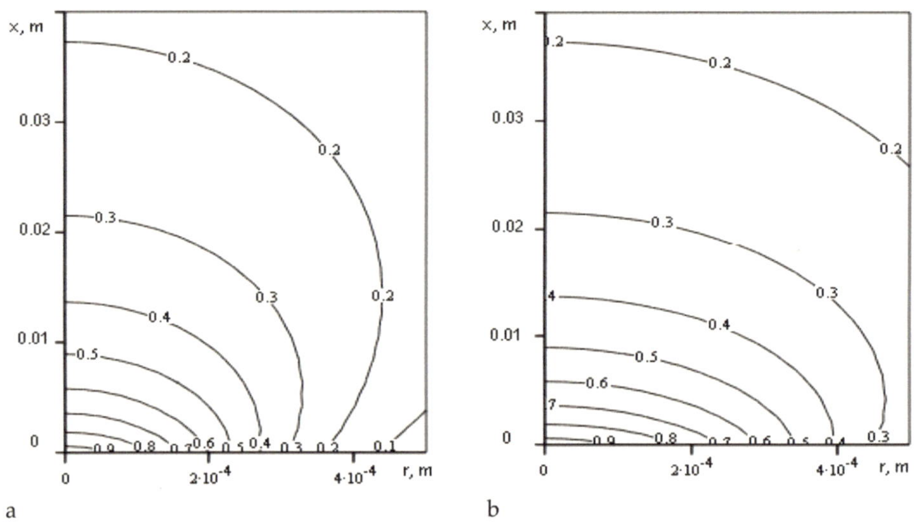

Figure 6. Distribution of metal vapor concentration in plasma plume, (a) without plasma absorption, (b) with consideration of plasma absorption.

Influence of plasma absorption with concurrence of heat convection with jet velocity and conduction in radial direction in plasma plume can shift a temperature maximum from the workpiece surface, even in the surface focusing (**Figures 7** and **8**). Also the difference between the temperatures of electron and heavy plasma components, which characterize the plasma nonequlibrity, changes along the plume.

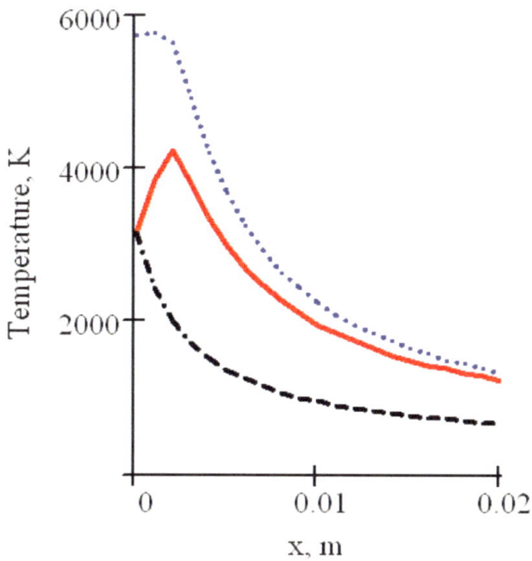

Figure 7. Temperature distribution along the plume axis, dot line – electron temperature, vapor-gas mix temperature – solid line – with, double dot – without plasma absorption.

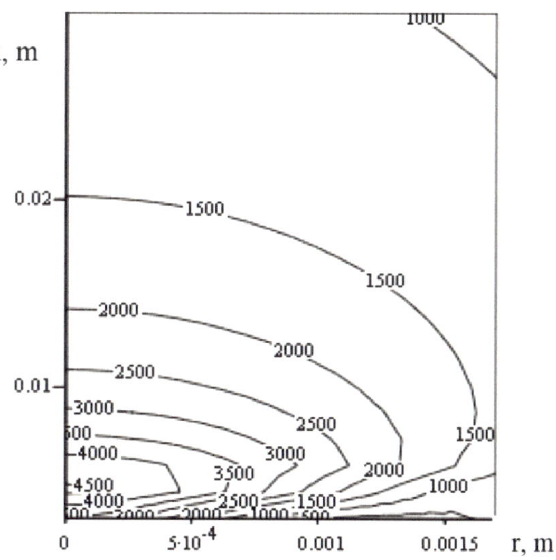

Figure 8. Temperature distribution in plasma plume.

Fast mixing of metal jet with surrounded gas limit the conductive kern formation by the region near the workpiece surface, as shown in **Figure 9**.

So, it is possible to mark that boundary layer gas dynamic model with consideration of discharge kinetics allow to explain an electric arc compression during hybrid welding by conductive kern formation in the near-surface region. The results of calculation also allowed to explain a shift of plume maximum temperature from the workpiece surface, which is often visible experimentally, not by the beam over focusing, but by the concurrence of heating with convection and conduction heat transfer.

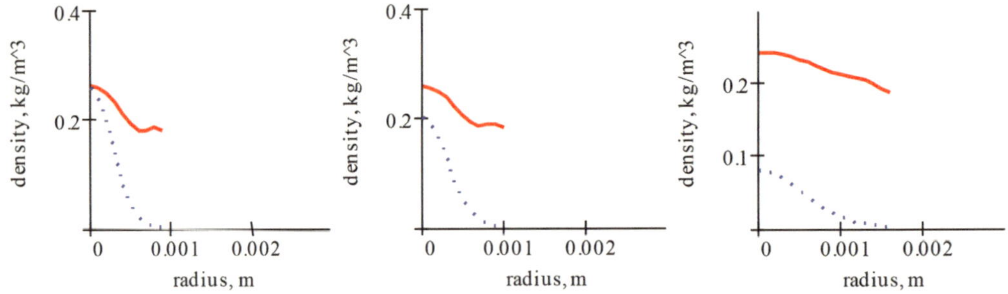

Figure 9. Radial distribution of mix density (solid) and metal density (dot) on the keyhole outlet and on the distances, correspondingly 0.3 and 2 cm.

Another specific problem for laser-arc welding in the case of laser-MAG and laser-MIG technology is a problem of filler wire melting. This problem can be described on the basis of one-dimensional approach with consideration of Stephan conditions on solid-liquid interface and the action of the electric force for drops transferring [17]. For solution of melt flow and heat transfer task in the melt pool, the approximation of potential flow of ideal liquid with viscous boundary layers on the melting front and keyhole surface has been used [4]. Because hybrid welding is often used for the welding of large and heavy parts, influence of gap between parts becomes especially important for this case.

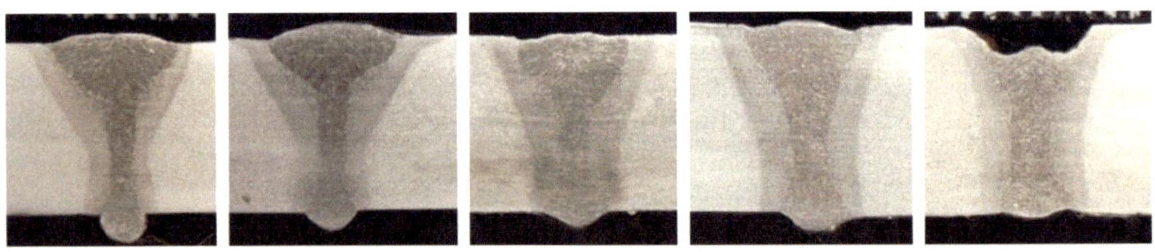

Figure 10. Weld macrosections with different gap width: a – 0 mm; b – 0.3 mm; c – 0.6 mm; d – 0.9 mm; e – 1.2 mm.

For most tasks of welding sheets and pipes, gap width influence on the quality of welding is important. In this area a large amount of research was carried out [18]. The authors found a reduction of the tensile strength and the destruction of the HLAW sample in the HAZ [19]. Experiments show that the highest efficiency and deepest penetration of alloy elements of the welding wire at the HLAW also depend on gap width [20]. Good weld appearance was created at the HLAW of the low alloy high-strength steel with thickness of 16 mm with a various gap widths from 0 to 0.7 mm [21].

Experiments show that gap width makes an effect not only on weld formation stability but also on welding process heat efficiency, as illustrated in **Figure 10**.

It can be seen that the width of the top of the weld decreases with the increase of the gap width, subsequent increase in the gap width did not affect. The next increasing gap width did not have a strong influence on the weld metal width on the top plate. The width at the root of the weld and in the middle is increased by increasing the gap. As a result, the amount of filler metal is increased. Influence of slot width to the width of the weld is shown in **Figure 11**.

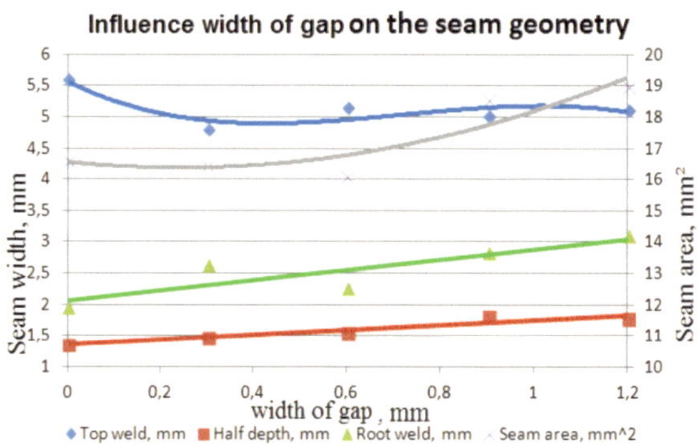

Figure 11. Influence of the gap width on the weld metal geometry.

It is also visible (**Figure 12**) that the optimum size of the gap is 1.2 mm. It can be explained that with increasing gap the welding pool width increases too, and, therefore, the volume of the dropping melting metal is higher. The reason for this is gravity.

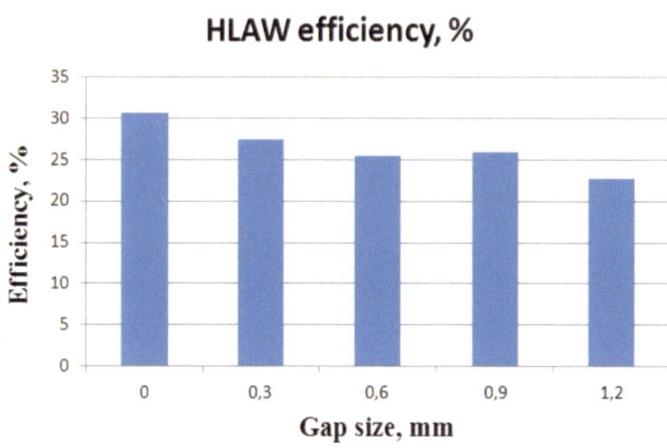

Figure 12. Influence of the gap width on the HLAW efficiency.

Analysis of the influence of the gap between the workpieces showed that its optimum size varies in the range of 0–0.3 mm. If a gap of 0.6–0.9 mm undercuts were observed, a gap width of 1.2 m lack of filler material was also observed. The volume of the welding wire in the root and half depth of the weld increased simultaneously with increasing gap width. Also, increasing the gap width from 0 to 1.2 mm decreased HLAW efficiency from 30.6 to 22.7%. The results of the experiments described above allow to create a mathematical model for the prediction of the welded joint geometry at HLAW with gap. The integration of the model with the CAE system, LaserCAD [22], allows to predict the welded joint geometry at the HLAW with different gap widths.

Schematically the interaction of different physical processes that are important for processing HLAW is shown in **Figure 13**. Different tasks are connected through boundary conditions; when solution of one task determines boundary condition for another, as through direct influence of equation coefficients.

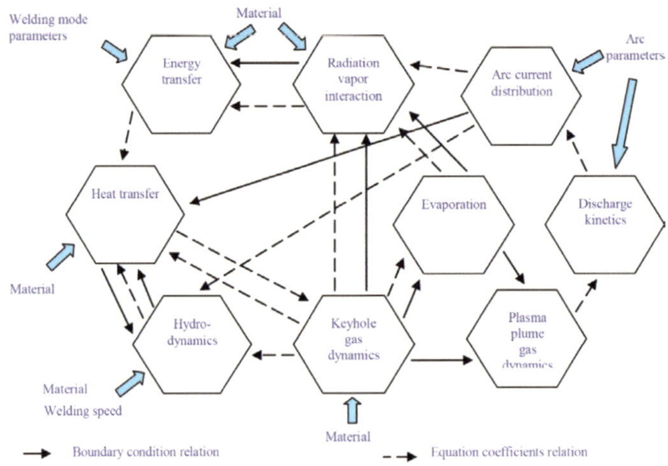

Figure 13. Structure of physical model of HLAW process.

Mathematical formulation of this physical model have been put on the basis of a number of steady-state process models, such as [23, 24], or [25], which allow to simulate a shape and size of melt pool as well as temperature distribution in weld bath and HAZ during hybrid HLAW welding.

For successful technology development it is also necessary to have a physical adequate description of melt pool dynamic behavior, which is responsible of formation of such welding defects as humping, porosity, spiking, and undercuts. This description has been designed on the basis of dynamic model of laser welding process [26]. The model is based on the formalism of Lagrange mechanics, which allow consider phenomena—wave motion of the cavity surface, change of the shape and sizes of the weld pool in time, and influence of the cavity motion as the whole on oscillations of its depth and radius.

The system of Lagrange equations for description of melt pool dynamic behavior can be represented as [27]:

$$\frac{\mathrm{d}}{\mathrm{dt}}\frac{\partial L}{\partial \dot{q}_i}-\frac{\partial L}{\partial q_i}=Q_i+R_i\ , \tag{2}$$

where q_i assumes H (depth of penetration, s_0 (keyhole cross-section area), $s_1,\ldots s_n$ (amplitudes of waves of different length), consecutively, Q_i is the generalized forces, and R_i is the dissipative function [27].

The mathematical formalism of solution of this system allows to carry out the dynamic analysis of occurrence of porosity and spiking, as shown in **Figure 14**.

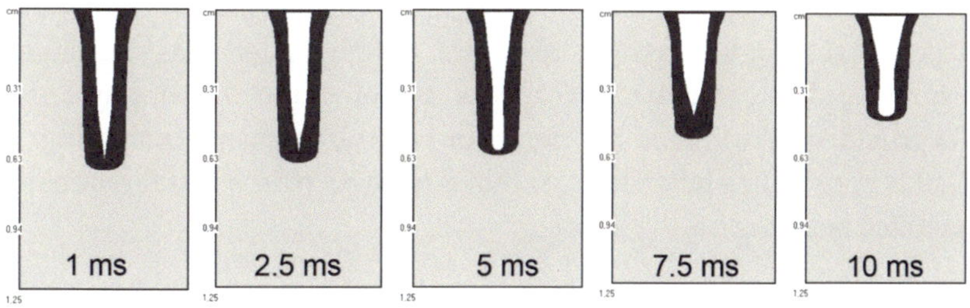

Figure 14. Simulation of the melt pool dynamics at HLAW, material – mild steel, laser beam power – 4.5 kW, welding speed – 12 mm/s, focal radius – 0.2 mm, focal distance – 30 cm, arc power – 2.5 kW.

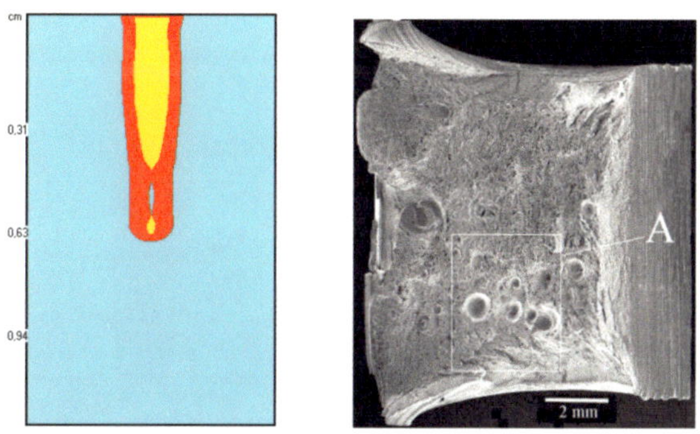

Figure 15. Simulation of porosity appearance due to cavity collapse (left) and results of experiment.

Analysis of the obtained results shows that dense occupation of the confined areas on the phase portraits by the phase trajectories explain the turbulent character of the cavity oscillations. It explains the calculation results independence for initial conditions. Sizes attractor is determined by the welding regimes.

The simulation results showed that hybrid welding of different generalized coordinates have different spectra of oscillations. The low frequencies below 100 Hz are equal to the radius of the cavity and the depth of the oscillation. Increasing the depth of penetration leads to a shear rate range toward lower frequencies. Long (S_1) and short (S_2) waves have the highest frequency range to 10 kHz. These spectra also depend on the cavity depth. The increase of feeding velocity also decreases the low-frequency oscillations. The same approach can be used for analysis of porosity and spiking appearance (**Figure 15**).

3. Design of technological equipment for hybrid processes

The peculiarities of HLAW process lead to specific requirements for technological equipment for this process. Development of technological hybrid welding equipment is easier using the block-module approach. It allows inheritance in design and interoperability and also decreases development time. According to this approach, each installation includes laser power source, arc source, special welding tool, system of control, cooling system, gas distribution system, welding parts manipulator, seam tracking system, and process monitoring system. Joining of subsystem is provided by common interfaces for control, mechanic connecting, gases, and water. Using this approach, a number of machines have been designed for industry; some of these are described below.

First, technological installation, which we describe below in details, has been developed for hybrid laser-MAG welding of oil and gas pipes of large diameter (**Figure 16**). It provides welding of pipe steel with thickness of more than 12 mm together by one pass with a speed of up to 3 m/min. Simulation with LaserCAD allowed to determine parameters of installation for this purpose: laser source power is not less than 15 kW; beam diameter in focus is 0.3–0.4 mm; welding current is not less than 250 A; and diameter of the electrode wire is in range of 1–2 mm.

Figure 16. Hybrid welding equipment.

This hybrid laser-arc complex contains: IPG fiber laser LS-15, arc power source with current of up to 1500 A, and numerical control filler wire feeding equipment, special working tools, CNC module of preparation, and distribution of used gases, monitoring system of the welded joint, tracking system with scanner laser sensor, process monitoring system, and control system.

This module is used in complex welding of metallic workpieces with a possible gap of up to 2 mm, MAG torch is installed in front of the laser. This complex mainly includes positions already mentioned above.

Stabilization of the hybrid module position and control is described in more detail in [2]. Precision of keeping the focus point position of the laser head relative to welded blanks in the vertical direction is ±0.2 mm, and in the cross-section direction is ±0.5 mm.

The control system of hybrid laser-arc complex is developed as a hardware-software complex. Present system controls the components of the complex. The system includes a control subsystem forming welds subsystem monitoring and automatic control system.

It provides reading profile blanks geometry control, tracking the welding process at speeds of up to 6 m/min, the welding head positioning, control of the laser source, the control operation of the arc, and gas control system. The system also has a built-in protection against harmful conditions and operational control using the monitoring system.

Development of the control and monitoring systems for process of laser-arc welding is still a relevant task. Using of such systems is essential for the adoption of these technologies into the industry.

To solve this problem, it is necessary to carry out researches of the weld pool dynamics and also define basic mechanisms of defects formation specified for welding with deep penetration. For developing this system, secondary emission signals coming from laser-arc action zone were also analyzed and sensors for their registration were included.

The monitoring system, designed for HLAW, is based on registration of optical emission in different spectral ranges, depending on the range [5]. Sensors are installed in two directions and are equipped with a video camera. It simplifies the guidance process and allows tracing the spatial variation of the active zone. Synchronous registration, processing, and recording of signals are realized by using developed software.

Series of experiments on welding of model samples was carried out for verifying the monitoring system. Test results confirm the possibility of weld formation monitoring using the multisensor monitoring system (**Figure 17**). The presented results confirm the ability to monitor porosity level in weld using the developed system.

However, to use the monitoring system in real production it requires further research aimed at understanding the characteristics of particular process [28], identifying typical process defects, and adaptation monitoring system for this technological process.

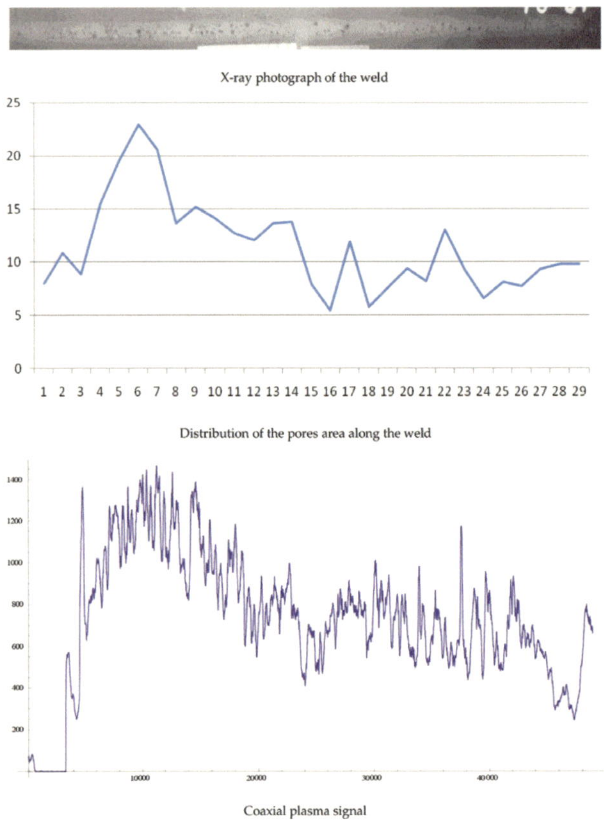

Figure 17. Signal response of the coaxial plasma sensor to change the level of porosity. X-ray photograph of the weld. Distribution of the pores area along the weld. Coaxial plasma signal.

For solving problems, the control system consists of several subsystems: laser control; arc equipment control; gas equipment control; welding head stationing; determination of the metal joint geometry; the control of parameters and protection of the laser welding head; the central controller module; and operating computer. For communication with other modules of the complex a card of CAN interface is put in the computer.

Other HLAW system [29], shown in **Figure 18**, based on soft direction belt, on which a motion unit with hybrid module, seam tracking, and filler wire feeding are mounted. Another subsystem is located in the stationary unit. In the equipment, 20 kW fiber laser with two direction switches is used. It allows to weld two joints in one time. This system uses a new inverter arc source. A used subsystem has a number of design features, for example, tracking sensor operating at high power in welding, and with a high degree of reflected radiation. However, they require special processing techniques of received signals [30]. Monitoring system for providing quality control process should capture the emergence of various defects (porosity, humping) [28]. The control system of this complex is implemented on the same base as the previous one.

Hybrid technology permits to create mobile technological equipment, which is able to perform all spectra of process advantages.

Control and stabilization of position of the hybrid laser-arc module relative to the joint is fulfilled by linear drives system installed on moving platform. It operates the same way as the stationary machine.

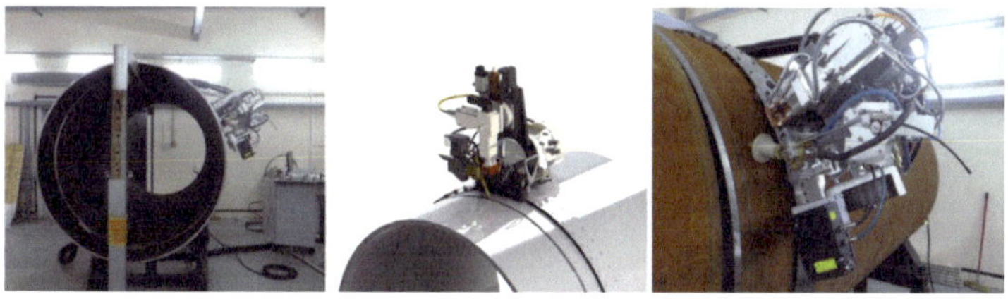

Figure 18. Mobile hybrid orbital system.

Another example of HLAW machine, based on the same design approach, is robot-based installation (**Figure 19**).

Figure 19. Robot-based HLAW machine on the base of 25 kW fiber laser.

Robot-based scheme of machine allows to weld parts of up to 4 m in length in different space positions.

4. Technology of HLAW

There are several fields of applications [31], in which usage of HLAW is especially prospective. First of them is production of large diameter pipes for oil and gas. The technology for different thickness of pipe walls was designed and the examples are shown in **Figure 20**.

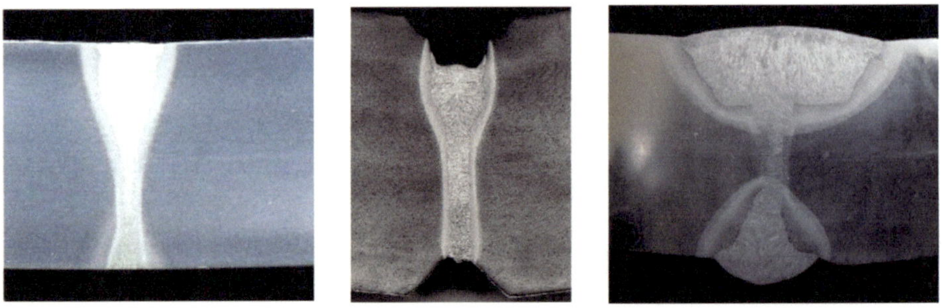

Figure 20. Macrosection of the one pass weld (left), three passes weld (center – first technological pass), right – technological pass and two filled passes by submerged arc welding. Depth is 15 mm left and 24 mm center and right. Material is steel X80.

Optimization of arc torch position relative to laser beam and optimization of composition of arc gas mixture together with usage of specially designed filler wire with ultrafine admixtures allow to avoid appearance of different defects, such as porosity, hot cracks, and humping. Also, it provides value of impact energy on temperature −40°C in the bounds of 140–200 J for pipe steel X80. HLAW also could be used for welding of nonturnable joints during pipeline building. In this case, hybrid scheme allows to increase not only the process productivity and joint quality but also process tolerance. Technology allows appropriate weld formation even in the case of more than 3 mm of vertical displacement of pipe joint edges (**Figure 21**).

Figure 21. Root pass side. Orbital pipeline hybrid welding with 2 mm gap and 3 mm vertical displacement.

Next prospective field of HLAW is shipbuilding [32]. Usage of this technology allows to increase dramatically the productivity of creation of flat sections for ship body. Because HLAW allows to get a weld with parallel walls displacement (**Figure 22**), it is possible to minimize welding stress and distortions, as shown in **Figure 23**.

http://dx.doi.org/10.5772/64522

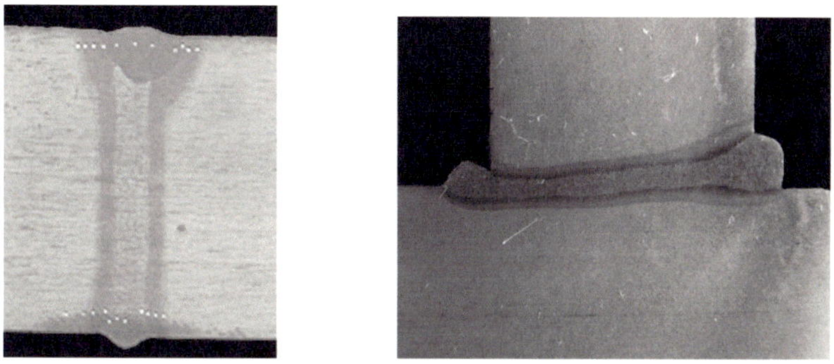

Figure 22. Butt HLAW joint in shipbuilding section (Material – steel PCE36, 20 mm thickness) – left, angle joint in shipbuilding section (Material – steel 09Г2C 15 mm thickness) – right.

Figure 23. Comparison of MAG welded ship body section (left) with HLAW case (right).

5. Peculiarity of metal structure formation in hybrid welding

To reduce the constructions weight new high-strength steels are applied. Characteristics of these steels are determined by parameters of ensembles of ultrafine inclusions. New technologies for car bodies are based on tailored blanks. Weight reduction without decreasing the strength is achieved by using high-strength steels and alloys in the production blanks. However, it is necessary to ensure satisfactory plastic weld characteristics.

This requires the use of welding technique, which provides the required level of ductility of weld seams and the same quality as laser welding process provides. Using a high-speed thermal cycling is the one of most promising techniques for getting good microstructure and mechanical properties [33]. This may be accomplished by welding two heat sources following one after another, for example, tandem laser welding [34]. However, from an economic standpoint it is advisable to use cheap local sources of heating as the second source, such as a powerful lamp.

To select treatment regimes, which provide the requisite microstructure and properties of the alloy, it is necessary to have a quantitative definition of the effect of temperature cycling on phase and structural transformations in the material. Modern concepts of phase transformations mechanisms under the laser exposure on Fe-based alloys [35], which became the basis of thermodynamics of phase transitions, do not take into account the nonequilibrium characteristics of fast transformation processes that result in the appearance of nonequilibrium microstructures in laser processing treatment. Thermodynamic characteristics do not allow to predict the temperature mutations for both transformations depending on the rate of heating and do not provide results for the nonstationary leak diffusion process. A quantitative description of the structural components is impossible without these factors. They can be formulated using only the kinetic theory of phase transformations, which has been made for the case of welding steels [36], and will be discussed in more detail in the example.

Phase transformations in Fe alloys under the beam treatment are defined as the metal structure consisting of two components. The first one is connected to decay (or formation) of solid solution of carbon in iron and formation (decay) of ferric carbide. The second process involves the conversion of FCC-BCC upon cooling after heating. A kinetic model was developed to determine the parameters of the formation and growth of the carbide inclusions.

The ratio between the surface area and the volume of the carbide inclusions depends on its shape. For the growth process, it can be considered as a point source. Concentration field around inclusion has a spherical symmetry. The inclusion can be considered as a sphere with an effective radius a. The nonequilibrium growth of new phase carbide inclusions is described using kinetic equation of the chemical reaction:

$$\frac{da}{dt} = K_1(T) \cdot C - K_2(T) \tag{3}$$

where $K_1(T)$ and $K_2(T)$ is a constant forward and reverse reactions, C is the carbon concentration on the surface of the inclusions, and T and t are the local temperature and time. The reaction speed constants are given by the Arrhenius formula:

$$K_{1,2}(T) = K_{1,2}{}^{(0)} \exp\left(U_{f,s} / kT \right), \tag{4}$$

where U_f, U_s are activation energy of the direct and reverse reaction, $K_1{}^{(0)}$, $K_2{}^{(0)}$ are frequency factors.

The superficial concentration is determined by a solution of the diffusion equation:

$$\frac{dC}{dt} - D\Delta C = D\frac{1}{r^2}\frac{\partial}{\partial r}\left(r^2 \frac{\partial C}{\partial r} \right) \tag{5}$$

The boundary condition on the surface of the growing inclusion (at $r = a$) is the condition of the solute flux continuity:

$$-D\frac{\partial C}{\partial r}\bigg|_{r=a}=K_1(T)\cdot C\big|_{r=a}-K_2(T) \tag{6}$$

The admixture concentration far from the growing inclusion is determined by its average value C_0:

$$C\big|_{r\to\infty}\to C_0 \tag{7}$$

The solution of this problem has been obtained by standard methods of mathematical physics:

$$C=C_0-\frac{1.56}{D\sqrt{\pi}}\frac{(K_1C_0-K_2)}{\left(\frac{K_1}{D}-\frac{1}{a}\right)}. \tag{8}$$

Pay attention to the mutual influence in the carbide inclusions ensemble and having denoted the carbon concentration in the cementite as C′ (for Fe_3C $C' \cong 0.25$) and inclusion number density as n, the ensemble of spherical inclusions can be obtained as:

$$C_0=C_0-nC'\frac{4}{3}\pi a^3. \tag{9}$$

To determine a value of the parameter n, it is possible to connect the n value with a solubility limit, which can be determined from the thermodynamic phase diagram. Having denoted the solubility limit as $C_{\lim}(T)$, one can get from a substance conservation condition:

$$n=\frac{C_0-C_{\lim}(293)}{\frac{4}{3}\pi a_{\lim}^3C'} \tag{10}$$

where $a_{\lim}$ is a limit size of the growing inclusion. Then the equation for inclusion radius can be rewritten as:

$$\frac{da}{dt}=K_1\left(C_0-nC'\frac{4}{3}\pi a^3-\frac{1.56}{\sqrt{\pi}D}\frac{K_1C_0-K_2}{\frac{K_1}{D}-\frac{1}{a}}\right)-K_2 \tag{11}$$

Euler's method is used for solving the equations of growth of carbides. According to the results, $a_{\lim}$ should be included in the equation. Preliminary value of $a_{\lim}$ should be introduced, which provides the same final value of the inclusion radius a. The values of parameters, used for calculation in this article for low carbon steel, are given by the next list: $K_1^{(0)} = 45$ m/s, $K_2^{(0)} = 2.2$ m/s, $U_f = 1.26 \cdot 10–19$ J/atom, $U_s = 1.60 \cdot 10^{-19}$ J/atom, $D_0 = 2 \cdot 10^{-6}$ m²/s, $U_d = 1.38 \cdot 10^{-19}$ J/atom.

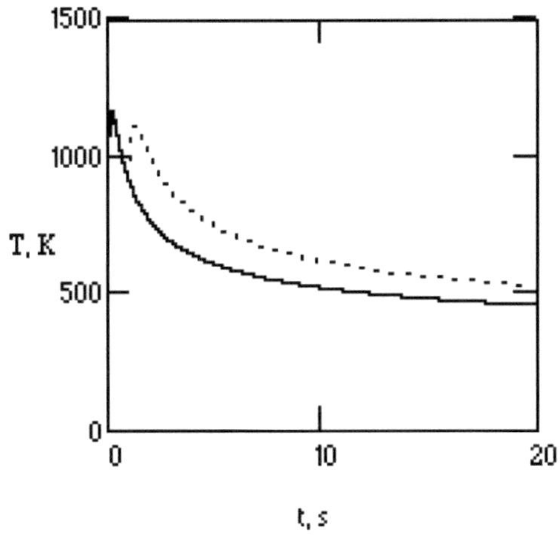

Figure 24. Temperature cycle for steel 08U (solid line – laser processing, dotted line – hybrid processing). Delay between temperature peaks 1 s.

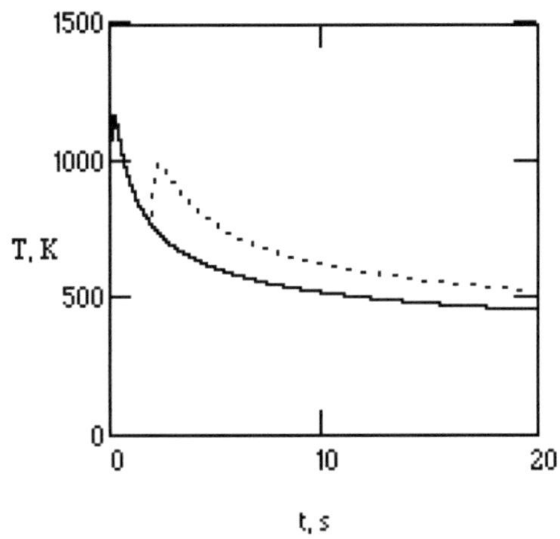

Figure 25. Temperature cycles for steel 08U (solid line – laser processing, dotted line – hybrid processing). Delay between temperature peaks 2 s.

Eventually, the kinetic model for the transformation $\alpha - \gamma$ should be formulated on the same principles as the model of nucleation and growth of inclusions. The interphase border movement rate in this case is that the diffusion Peclet number is not small, the diffusion equation, as against a problem about carbides growth, has been solved considering a convective term. On the other hand, since growing grain sizes are significantly greater than the thickness of the diffusion layer, the problem is seen as univariate. Simultaneous solution of the related tasks about kinetics of new phase grains growth and the diffusion of carbon allow us to calculate the amount of a new phase at any point in the cycle. The result is a self-consistent system of equations, which describe material microstructure formation at the high-speed heating and cooling. The original structure adopted by the initial size of the carbide inclusions and initial grain sizes affect the diffusion coefficient. The thermal cycle is an input parameter, the parameters of which are determined by the technological cycle. The solution allows to calculate the evolution of the phase composition of the steel in the treatment zone.

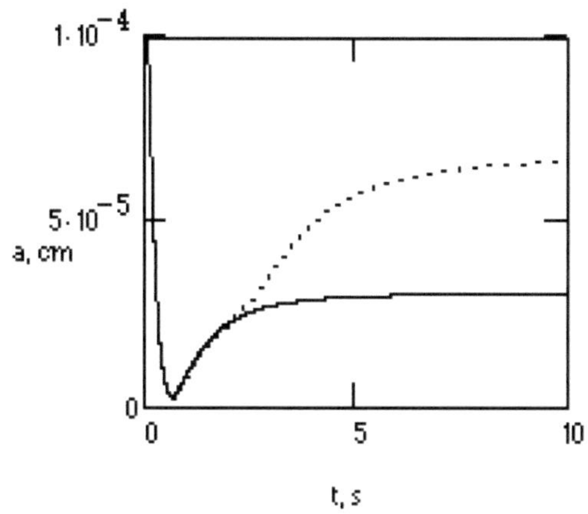

Figure 26. Calculation results for carbide nanoinclusions sizes for steel 08U (solid line – laser processing, dotted line – hybrid processing). Delay between temperature peaks 1 s.

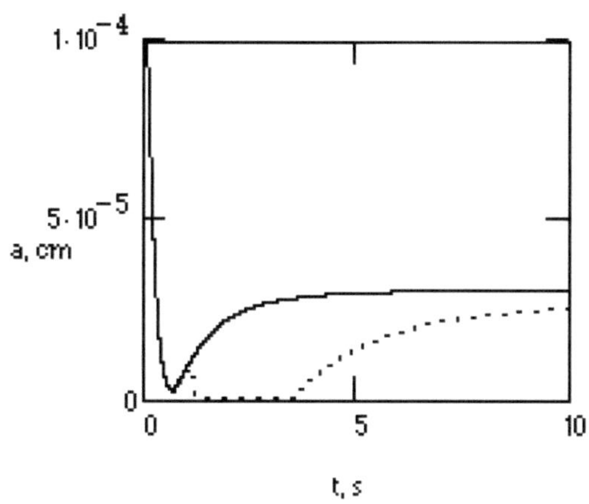

Figure 27. Calculation results for carbide nanoinclusions sizes for steel 08U (solid line – laser processing, dotted line – hybrid processing). Delay between temperature peaks 2 s.

Changing the form of a temperature cycle, which can be realized not only by HLAW, but also by another hybrid technologies, such as laser-light, dual beam, and others, it is possible to achieve both size reduction and enlargement of the carbide inclusions in comparison with a temperature cycle, which is typical for laser action without an additional heat source (**Figures 24–29**).

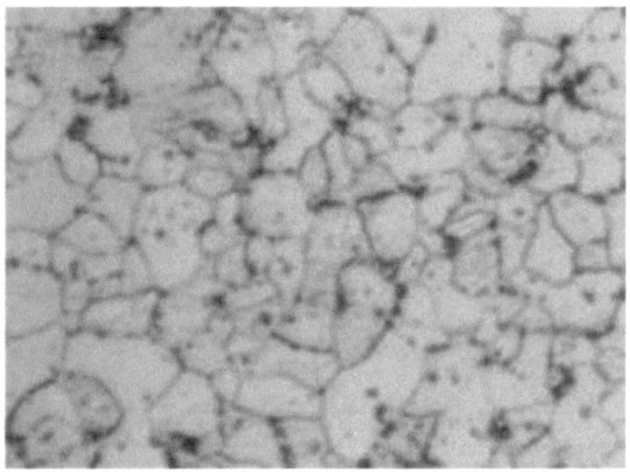

Figure 28. Microstructure after hybrid action, the maximal heating temperature is 1200 K, field width 50 μm. Delay between temperature peaks 1 s.

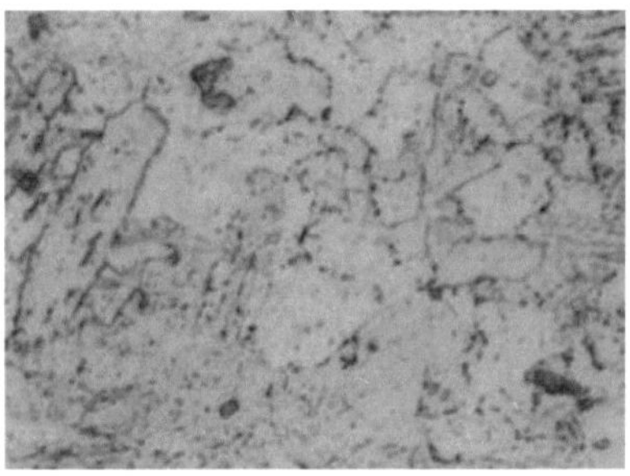

Figure 29. Microstructure after hybrid action, the maximal heating temperature is 1200 K, field width 50 μm. Delay between temperature peaks 2 s.

Influence of processing parameters for the delay between the first and second peaks of the cycle is shown in **Figures 24–29**. As obtained by theoretical studies, decrease in the time delay between the peaks results in a significant reduction in the size of carbide inclusions.

Microhardness distribution analysis showed that the welds obtained by hybrid welding have significantly less variation in the values of microhardness versus laser welding samples. Decrease of martensite part in weld metal allows raising its plasticity by more than 40%. So it can be concluded that hybrid welding technology suggests an additional possibility to control phase structure parameters in comparison with laser welding as it has been shown in the example of hybrid laser-light process.

As stated above, the carbides precipitation and nature of thermal cycles influence the process of phase composition formation. For estimation of steels phase composition, the time-temperature transformation diagrams of austenite decay together with curves corresponding to thermal cycles are usually used. The method of phase composition quantitative assessment based on Quench factor analysis was described in [37] and successfully tested on aluminum

http://dx.doi.org/10.5772/64522

alloys. Later on Quench factor analysis was applied for investigation possibility of predicting steel's mechanical properties during heat treatment [38], but in these works, processes of carbide formation and their influence on actual position of austenite decay C-curves were not considered. As was shown in [36], the nonequilibrium kinetics of new phase nucleus formation and nonstationary diffusion are influence by phase condition formation in practical welding conditions. Such point of view was used for logical design of quantity precipitate calculation for low-carbon steel welding, and realized as a computer program for the estimation of thermal field, which was used in the model described above. The calculation was carried out for steel 08U, the comparison of results and experimental data revealed their acceptable coincidence (**Figures 30** and **31**).

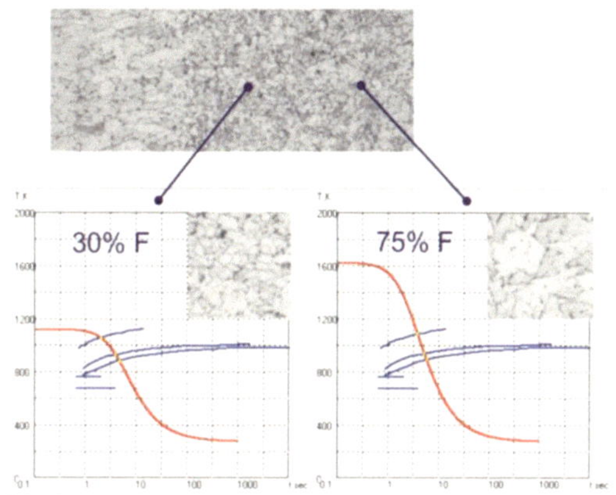

Figure 30. Comparison of calculation results of phase combination with experiment, low carbon steel 08U – laser frequency 5 Hz, pulse duration 7 ms, radiation power – 2.25 kW, lamp electric power – 5.16 kW, welding speed 2 m/s.

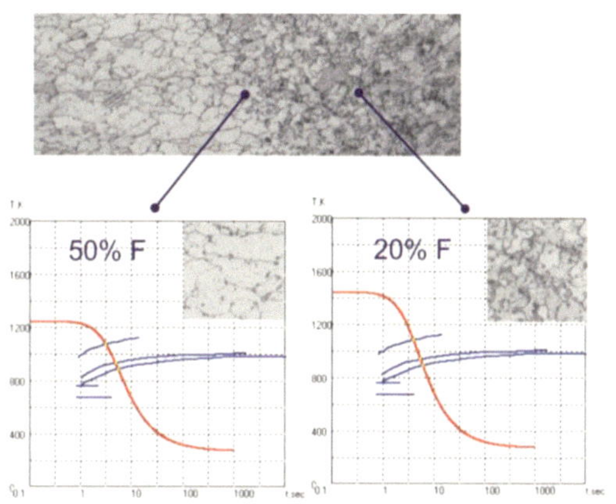

Figure 31. Comparison of calculation results for phase combination with experiment, low carbon steel 08U – laser frequency 10 Hz, pulse duration 3.5 ms, radiation power – 4.5 kW, lamp electric power – 5.726 kW, welding speed 4 mm/s.

Finally, it is possible to conclude that the use of hybrid welding instead of laser one allows to increase the melting efficiency and to provide stability of weld formation at welding metals in large variety of thickness. Also, hybrid welding technology provides additional possibility to control phase structure parameters, allowing to decrease the level of welding stress and distortion, and increasing welding productivity. Present level of technology development allows to design reliable and cost-effective technological equipment for hybrid laser-arc welding on the basis of high-power fiber lasers.

Author details

G. A. Turichin

Address all correspondence to: olgaklimova.ti@yandex.ru

Institute of Laser and Welding Technologies, Saint-Petersburg, Russian Federation

References

[1] W. M. Steen, Methods and apparatus for cutting, welding, drilling and surface treating. Patent 1547172 Great Britain, MKI V23K 26/00, 9/00, Publ. 06.06.79.

[2] F. O. Olsen, Hybrid laser arc welding, Woodhead Publishing, Denmark, 2009, 336 p.

[3] B. Ribic, T. A. Palmer and T. DebRoy, Problems and issues in laser-arc hybrid welding, International Materials Reviews, 2009, 54(4), 223–244.

[4] V. A. Lopota, Y. T. Sukhov and G. A. Turichin, Computer simulation of laser beam welding for technological applications, Izvestija Akademii Nauk, Ser. Phys, 1997, 61(8), 1613.

[5] G. A. Turichin, A. M. Grigor'ev, E. V. Zemlyakov, E. A. Valdaitseva, U. Dilthey and A. Gumeniuk, Special features of formation of plasma torch under conditions of hybrid laser-arc welding, High Temperature, 2006, 44(5), 647–655.

[6] M. Capitelli, A. Casavola, G. Colonna and A. De Giacomo, Laser-induced plasma expansion: theoretical and experimental aspects, Spectrochimica Acta Part B, 2004, 59, 271–289.

[7] D. F. Farson and K. R. Kim, Generation of optical and acoustic emissions in laser weld plumes, Journal of Applied Physics, 1999, 85(3), 1329–1336.

[8] L. J. Radziemski and D. A. Cremers, Laser-induced plasmas and applications, Dekker, New York, 1989, p. 336.

[9] T. X. Phuoc, An experimental and numerical study of laser-induced spark in air, Optics and Lasers in Engineering, 2005, 43, 113–129.

[10] L. G. Loitzynski, Mechanics of liquid and gas, Moscow, Drofa, 2003.

[11] K. Ramachandran et al., Structural analysis of converging jets in a triple torch plasma system, Journal of Physics D: Applied Physics, 2003, 36, 1198–1203.

[12] A. Revuelta et al., Confined axisymmetric laminar jets with large expansion ratios, Journal of Fluid Mechanics, 2002, 456, 319.

[13] V. Golant et al., Basis of plasma physics, Moscow, Atomizdat, 1977, 348 p.

[14] U. Diltey, A. Gumeniuk, V. Lopota and G. Turichin, Kinetic description of keyhole plasma in laser welding, Journal of Physics D, 2000, 33(21), 2747–2753.

[15] Y. Raizer, Laser spark and discharge propagation, Moscow, Science, 1974, 308 p.

[16] V. Lopota, G. Turichin, E. Valdaitseva, E. Pozdeeva and A. Gumeniuk, Computer system of electron beam and laser welding modeling, The Paton Welding Journal, 2006, 4, 29.

[17] G. Turichin, E. Pozdeeva and E. Zemlyakov, Numerical-analytical model of electrode wire melting in laser-arc welding, Physics and Chemistry of Material Processing, 2007, 4, 41-45./rus/.

[18] G. Turichin, I. Tzibulsky and M. Kuznetsov, Technology of laser-arc welding, Saint-Petersburg, Russia, Polytechnic University, 2015, 48 p.

[19] F. Kong, W. Liu, J. Ma and E. Levert, Feasibility study of laser welding assisted by filled wire for narrow-gap butt-jointed plates of high-strength steel. Welding in the World, 2013, 57, 693–699.

[20] 20. I. Tsibulskiy, M. Kuznetsov and A. Akhmetov, Effect of welding position and gap between samples on hybrid laser-arc welding efficiency, Applied Mechanics and Materials, 2014, 682, 35–40.

[21] M. Rethmeier, S. Gook and A. Gumenyuk, Perspectives of application of laser-GMA-hybrid girth welding for pipeline construction. Proc. of the 6 Int. Conf. "Beam Technology and Laser Aplication", Russia, Saint-Petersburg, 2009. pp. 278–288.

[22] G. Turichin, E. Valdaytseva and I. Tzibulsky, Computer analysis of beam material processing: LaserCAD simulation system, Photonics, 2008, 6, 18–21.

[23] E. Beyer, M. Dahmen, B. Fuerst, E. W. Kreutz, H. Nitchs, G. Turichin and W. Schulz, A tool for efficient laser processing, Proceedings of 14 Int. Congress on application of lasers – ICALEO-95, San Diego, USA.

[24] A. Kaplan, Influence of the beam profile formulation when modeling fiber-guided laser welding, Journal of Laser Applications, 2011, 23, 042005.

[25] G. Turichin, E. Valdaytseva, I. Tzibulsky, A. Lopota and O. Velichko, Simulation and technology of hybrid welding of thick steel parts with high power fiber laser, Physics Procedia, Proceedings of the 6th International WLT Conference on Lasers in Manufacturing, LiM" Munich, 2011. pp. 646–655.

[26] G. Turichin, E. Valdaitseva, E. Pozdeeva, U. Dilthey and A. Gumeniuk, Theoretical investigation of dynamic behavior of molten pool in laser and hybrid welding with deep penetration, Paton Welding Journal, 2008, 7, 11–15.

[27] V. A. Lopota, G. A. Turichin, I. A. Tzibulsky, E. A. Valdaytzeva, E.-W. Kreutz and W. Schulz, Theoretical description of dynamic phenomena in laser welding with deep penetration, Proceedings of SPIE – Proceedings of the 1998 6th International Conference on Industrial Lasers and Laser Applications, ILLA-98. Shatura, RUS, 1999. pp. 98–107.

[28] G. Turichin, E. Zemlyakov, K. Babkin and A. Kuznetsov, Monitoring of laser and hybrid welding of steels and Al-alloys, Physics Procedia, 2014, 56, 1232–1241.

[29] G. Turichin, O. Velichko, A. Kuznetsov, J. Pevzner, O. Grinin and M. Kuznetsov, Design of mobile hybrid laser-arc welding system on the base of 20 kw fiber laser, Proceedings – 2014 International Conference Laser Optics, LO, Saint-Petersburg, Russia, 2014. C. 6886481

[30] K. Babkin, E. Zemlyakov, G. Turichin and A. Kuznetsov, Use a technical vision in in automatic control of scanning laser welding, Nauchno-Technicheskie vedomosti SPbPU, 2015, 3(226), 142–149.

[31] G. Turichin, I. Tzibulsky, E. Valdaytseva and A. Lopota, Hybrid laser arc welding of metals of large thickness, Welding and Control (RUS), 2009, 3, 16–23.

[32] G. Turichin, I. Tzibulsky, V. Levshakov, N. Steshenkova and E. Valdaytseva, Laser and laser-arc welding of shipbuilding steels. Beam Technologies & Laser Application: Proc. of VII International scientific and technical Conference. Saint-Petersburg, 2013. pp. 90–103.

[33] E. L. Gyulikhandanov and A. D. Khaidorov, Effect of thermocycling on the structure and properties of high-speed steel obtained by electroslag remelting, Journal of Metal Science and Heat Treatment, 2002, 44(9–10), 426.

[34] C. Glumann, J. Rapp, F. Dausinger and H. Hugel, Welding with combination of two CO2 lasers — Advantages in processing and quality. ICALEO' 93, Orlando, FL, 1993, pp. 672–678.

[35] D. M. Gureev et al., Mechanisms of phase transformations in iron and steels at laser heating. Samara, Publishing House, Samara University, 1999.

[36] U. Dilthey, A. Gumenyuk and G. Turichin, Calculation of the kinetics of diffusion phase transformations in low-alloyed steels in beam welding, Paton Welding Journal, 2006, 2, 11–16.

[37] J. T. Staley, Quench factor analysis of aluminium alloys. Material Science and Technology, 1987, 3(11), 923–935.

[38] G. E. Totten, Y. H. Sun and C. E. Bates, Simplified property predictions AISI 4140 based on quench factor analysis. Third International Conference on Quenching and Control of Distortion, Prague, Czech Republic, 24–26 Mar. 1999. pp. 219–225.

Chapter 8

Current Issues and Problems in the Joining of Ceramic to Metal

Uday M.B., Ahmad-Fauzi M.N.,
Alias Mohd Noor and Srithar Rajoo

Additional information is available at the end of the chapter

http://dx.doi.org/10.5772/64524

Abstract

Ceramics and metals are two of the oldest established classes of technologically useful materials. While metals dominate engineering applications, ceramics have some attractive properties compared to metals, which make them useful for specific applications. The properties of individual ceramics and metals can vary widely; however, the characteristics of most materials in the two classes differ significantly. Joints between a metal and ceramic are becoming increasingly important in the manufacturing of a wide variety of technological product. But joining ceramics to metallic materials often remains an unresolved or unsatisfactorily resolved problem. This chapter deals with problems of various studies in recent years on the joining between two materials.

Keywords: ceramic, metal, joining, problems, reliability

1. Introduction

The successful application of ceramics in many devices and structures requires some type of joining with metal [1]. Therefore, the ceramic-metal joints are used widely in the different applications such as vacuum tubes, high voltage feed through, transistor packages, sapphire-metal windows, rocket igniter's bodies and many others [2]. The new joining purposes for these materials involve automobile engine components, such as the silicon carbide, silicon nitride and yettria-stabilized zirconia. The ceramic rotor was joined to metal shaft by new method which compensated problems in both shrink fitting and active brazing methods. The designing of ceramic rotor was carried out in order to ensure the strength and durability of the component as well as to obtain the same aerodynamic characteristics as in the metal rotor. All applications have depended upon improved mechanical and thermal properties, such as strength, and resistance to fatigue, creep and oxidation.

Achieving high integrity joints between ceramics and metals, however, is a challenge. The properties of ceramics that make them attractive may pose major handicaps for joint fabrication. Due to the chemical inertness of ceramics, conventional joining methods for metals cannot be used. To obtain adequate bond quality, high temperature and pressure are often required [3] and bonding media with reactive elements have been used [4]. The chemical phenomena occurring at interfaces determine the structure of the interface and hence, its properties. The chemical reaction between the ceramic and the metal may easily initiate bond formation; however, thick brittle reaction layers or intermetallics formed at the interface often cause premature failure at very low stresses [5].

Even successful joint formation does not guarantee mechanical soundness of the joint. The inherent differences in physical properties between the ceramic and the metal make it very difficult to find an effective process to join that keeps detailed and comprehensive strength and flexibility. There are two primary factors that cause the reliability issue of joint such as the coefficient of thermal expansion (CTE) mismatch and the difference in the nature of the interface bond. The thermal residual stresses are induced in a joint during cooling due to the CTE mismatch and differing mechanical responses of ceramic and metal. This may lead to a detrimental influence on joint strength [5, 6].

The aim of this chapter deals with problems of various studies in recent years on the joining between two dissimilar materials. The focus is on the general problems, solutions and factors influencing reliability with different ceramic-metal joining processes.

2. General problems in ceramic-metal joint

There exist many problems between ceramic and metal materials, such as the atom bond configuration, chemical and physical properties, etc. These problems make the joining of ceramics to metals difficult. The following main problems such as ionic bonds and covalent bonds are characteristic atomic bond configurations of ceramic materials. The peripheral electrons are extremely stable. Using the general joining method of fusion welding to join ceramics with metals is almost impossible, and the molten metal does not generally wet on ceramic surfaces [7].

When joining ceramics to metals with the brazing method, for example, metallization on the ceramic surface is necessary with general inactive brazing filler metal or the use of active brazing alloys in order to get a reliable joint. The thermal expansion coefficients of ceramics are generally much lower than metals. Stress will be generated in the ceramic/metal joint due to the thermal expansion mismatch and will degrade the mechanical properties of the joint and can cause joint cracking immediate after the joining process. The thermal stress in the joint due to the thermal expansion mismatch should be carefully considered when joining ceramic with metal. Many ceramics have low thermal conductivity and susceptibility to thermal shock. Using the fusion welding method to join ceramics by concentration heating or with a high energy density heat source, cracking in the ceramic easily occurs. It is necessary to reduce the temperature gradient in and around the fusion zone as much as possible and to carefully control the heating and cooling speed during the joining process.

3. Factors influencing reliability of ceramic-metal joint

Joining ceramics to metallic materials is not so easy to be carried out without considerations of several problems originating from the differences in physical and chemical natures between ceramics and metals to be joined [8, 9]. **Figure 1** summarizes the several points, which may cause large scatter in the strength directly. From the microscopic view, interface contact formed by wetting, chemical and physical reaction at interfaces should be of concern in the first place [10]. The cracking in the layer frequently reduces joint strength. Thermal or residual stress in a joint becomes the other important factor. Large thermal stress both in joining process and in services induces flaws into joints. These factors will reflect the distribution of unbonded or weakly bonded is a land like defects on interfaces resulting in substantial reduction in joint strength [11, 12].

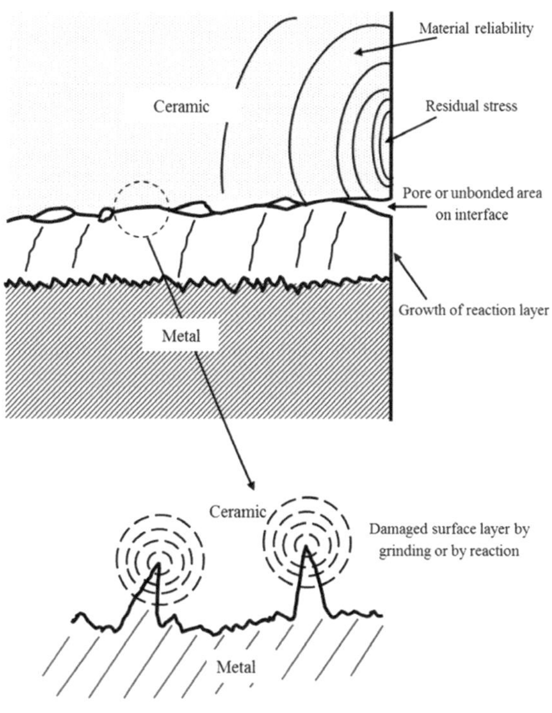

Figure 1. Schematics of factors influencing on reliability of ceramic/metal joint [1, 12].

The development of residual stresses is one of the major problems in the ceramic/metal joining at the interface when the material is cooled down from the bonding temperature to room temperature [13]. These residual stresses reduce the strength of the bonded material and in some cases lead to catastrophic failure at or near an interface, during the joining process. The mechanical analysis of a joint metal to ceramic is a very complex problem. There are many different characteristics to look at ceramic/metal joints. Depending on the detailed application, some characteristics are more important than others [14]. Therefore, in the following sections, we will focus in the joining problems researches for factors influencing on reliability of ceramic/ metal joint.

3.1. Material reliability

The ceramic, because of its inherent brittleness, is the most critical material for obtaining reliable joints [15]. The base properties of the bulk ceramic member are essential. When the properties of the bulk ceramic are not sufficient, the thermal stress simply fractures the ceramic member. Furthermore, the surface condition of the ceramic is also very important for the joint reliability. The ceramics are produced by the different forming methods and a subsequent densification during sintering at high temperatures. Due to high hardness and brittleness of ceramic, any shaping complicated treatment often needs diamond cutting tools and abrasives. Whereas it should avoid sharp edges and corners that may cause the concentration of tensile stress [16]. Moreover, when the ceramic material is ground by a metal bonded diamond wheel, microcracks are introduced at the surface of the ceramic. The size of the microcracks depends on the diamond grit size of the wheel and also on the rate of material removal. The surface damage can initiate major cracks in the ceramic by the thermal stress and, hence, result in an unreliable joint. Therefore, the ceramic surface should be free of damage to obtain high reliability joints. This condition can be met simply by using sintered ceramic materials. However, nearly all sintered ceramic parts over about 2 cm in size should be grounded, because distortion of the parts during the sintering requires grinding for dimensional control. Ground ceramic materials should be treated further to obtain a defect free surface condition. This can be performed by a resintering or lapping process. In the resintering process, the damaged layer is healed through sintering. In the case of the lapping process, the damaged layer is physically removed. It should be mentioned that the thickness removed by the lapping must completely eliminate the surface damages [15].

3.2. Thermal expansion and residual stress

Residual stresses are stresses that remain in the materials joining after the original cause of the stresses have been removed. Thermal residual stresses play the key role in the mechanical behaviour of various joint materials. Thermal stresses may occur in a heated structure which is rigidly constrained, and also in a structure with temperature gradients. Thermal residual stresses in the ceramic/metal joints can be classified into three groups in accordance with the mechanism that produces them. First, thermal stresses caused by a volumetric change, either expansion or shrinkage, associated with phase transformation. For these stresses arise from a phase change, the temperature must change to cause the phase change. Second, thermal

http://dx.doi.org/10.5772/64524

stresses caused by a difference in CTE mismatch between two materials joined together. For these stresses to arise from a difference in coefficients of thermal expansion, the temperature may be changing or it may have stabilized. Third, thermal stresses caused by a thermal stresses caused by a temperature gradient resulting in the thermal differential rates within the volume of the material or within the structure and potentially lead to cracking. For these stresses to arise from differential rates of expansion or contraction, the temperature must change and produce a gradient, which may or may not persist. Whether the temperature gradient persists or not, the thermally induced stresses from this source persist [17].

Ceramic-metal joints represent an important class of components because of their applications in hostile environments. Examples can be found in different application such as automotive, microelectronics, the aerospace industry or biomedical applications. Generally, a ceramic-metal joint develops a residual stress field, which has its origin in the thermomechanical fabrication process and is due to the difference in CTE between the ceramic and metal (**Figure 2**). Residual stresses have significant effects on the mechanical stability of the interface, since they may cause plastic deformations on the metal side and cracking in the ceramic, thus compromising the adhesion or even inducing failure of the joint.

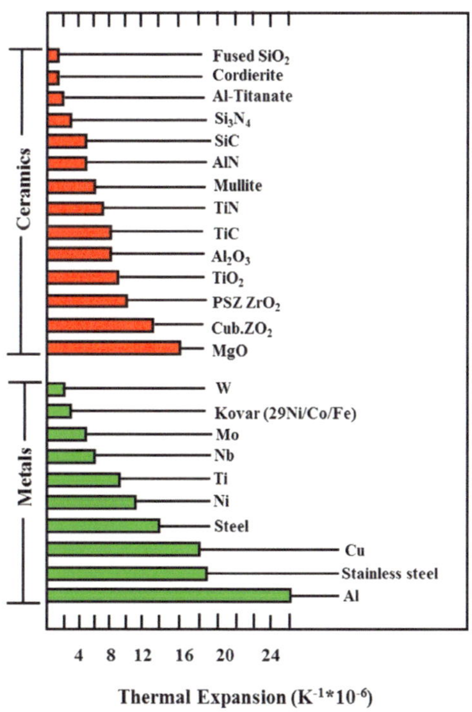

Figure 2. Comparison of thermal expansion coefficients of metals and ceramics [18].

The residual stresses produced in the ceramic metal joint could be estimated for full elastic conditions according to this equation [7]:

$$\sigma_C = \frac{\Delta\alpha \times \Delta T \times E_m \times E_C}{(E_m + E_C)} \tag{1}$$

where σ_C is the residual stress after the joint cools to room temperature, $\Delta\alpha$ is the difference of thermal expansion coefficient between materials, ΔT is the difference between joining temperature and room temperature, E_m is a Young's model of metal, E_C is a Young's model of ceramic. If the thermal stresses in the metal exceed its yield strength, the residual stresses in the joint could be determined by [7]:

$$\sigma_C = \sigma_{my} + \Delta\alpha.\Delta T.E_{mp} \tag{2}$$

where E_{mp} is the linear strain hardening coefficient and σ_{my} is the yield strength of the metal (linear elastic-linear plastic conditions are assumed).

The distribution of thermal residual stress is not uniform in the joint and even along the interface between these different materials. The concentration of thermal stress becomes more intense with the proximity of the interface [19]. The most harmful effect of thermal stress is caused by the tensile stress at the interface or in the ceramic. The direction of the maximum tensile stresses is mainly perpendicular to the interface and the free surface direction, causing the crack opening and failure occurs. The breadth of thermal residual stress depends on the shape and dimension of the ceramic/metal interface [20]. For example, the diameter dependence of the thermal stress of the Si_3N_4/invar alloy joint measured on the surface near the interface as shown in **Figure 3**. The larger diameter leads to generate more thermal residual stress. It is also noteworthy that stress concentration at the corner of the rectangular bond face joint is more serious. The joint strength tends to decrease with increasing thermal expansion

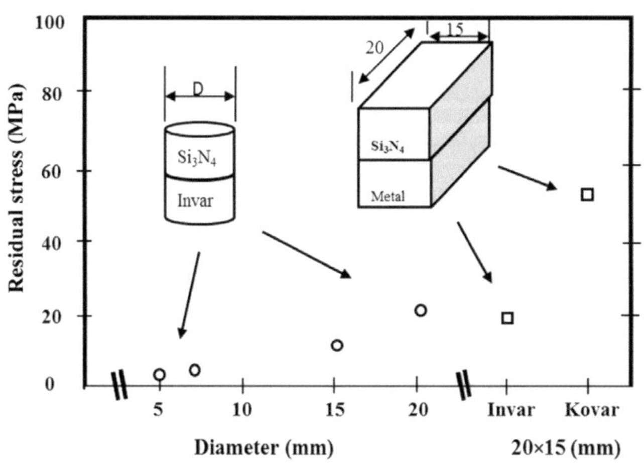

Figure 3. Effect of size and shape of bond face of residual stress on Si_3N_4/invar alloy joints. The residual stress was vertical to the interface on the Si_3N_4 surface [22].

http://dx.doi.org/10.5772/64524

mismatch. However, it occasionally happens that some specimen will be strong but the other will be weak even if they are the same kind [20]. This depends on the presence and distribution of internal flaws induced by residual stress during joining. The strengths of the Si_3N_4/invar (iron-nickel alloy) and Si_3N_4/kovar (iron-nickel-cobalt alloy) joints, which are differing in the amplitude of thermal stress, were examined statistically [21].

The thermal stress may be relieved by two different methods according to Lemus-Ruiz's thesis [23]. One method inserts a metal with approximately the same thermal expansion coefficient as that of the ceramic to decrease the magnitude of thermal stress generated, while the other method involves thermal stress relief by using a ductile metal that easily develops plastic deformation under thermal stress. These two methods may also be employed in combination. **Figure 4** shows a schematic illustration of thermal stress at a joint interface and the mode of cracking due to difference of thermal expansion coefficient [24]. When the thermal expansion coefficient, α_C of the ceramic is smaller than that of the metal, α_M, the ceramic is subjected to tension stress and cracks at the edges, as schematically illustrated in **Figure 4a**, on the other hand, when the thermal expansion coefficient, α_M, of the metal is smaller than that of the ceramic, α_C; tensile stress acts on the core of the ceramic and cracks the ceramic, not at the edges, but transversely at the core, as shown in **Figure 4b**.

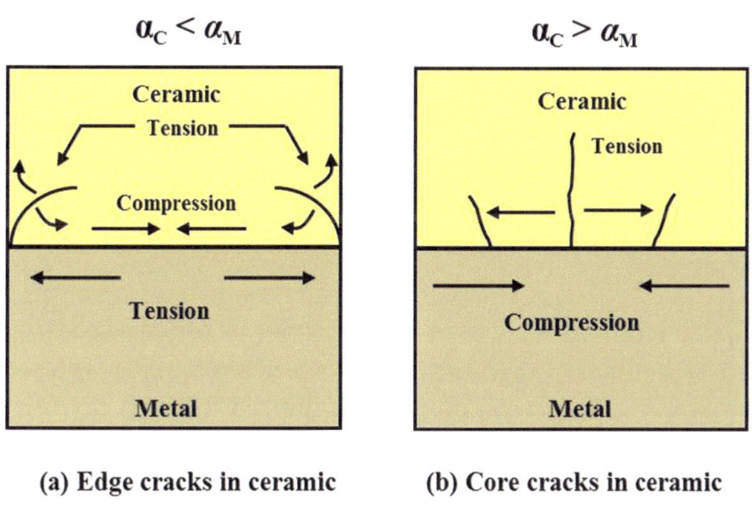

Figure 4. Schematic illustration of thermal stress in joint interface and mode of cracking due to difference of thermal expansion coefficient [23, 24].

To overcome for reducing the residual stress mentioned above, induced by the mismatch of the thermal expansion coefficient between the materials to be joined, the following methods can be used as reported by Zhou [7]: (1) Using soft filler metals, the soft filler metals have low yield strength and could release the residual stress. (2) Using soft interlayer, the residual stress could be reduced by the elastic and plastic deformation of an interlayer, e.g. when using Al or Cu as interlayer, the residual stress is decreased. According to Eq. (1), the residual stress will decrease with Young's model E_m decreasing. (3) Using hard metals of which the thermal expansion coefficient is close to ceramics as the interlayer. Using hard metals such as W, Mo or invar as the interlayer, could reduce the residual stress. Their validity is not obvious when hard metals with high yield strength are the interlayer. (4) Using composite interlayer where the composite interlayers often constitute hard metals and soft metals, like Cu/Mo-Cu/Nb, have a noticeable effect on reducing residual stress, with a combination of merits of those two kinds of metals. (5) Joining under low temperature where the joining ceramic to metal at a low temperature is good for reducing the joint deformation and effectively decreasing the residual stresses. (6) Heat treatment after joining because the proper heat treatment post joining sometimes releases the stress and the strength will vary based on the heat treatment. (7) Appropriate configuration of the joint could decrease the stress concentration extent and reduce the residual stress.

3.3. Interface reliability of the joints

Interfaces play critical roles in properties of many material systems such as composites, coatings and joints. Particularly in ceramic to metal joints, the properties of interfaces have a significant effect on the mechanical reliability of the joints. The mechanism of bond formation at the interface determines the interface structure, which depends strongly on processing conditions as well as materials. The bonding mechanisms can be categorized in terms of mass transfer across the interface. When there is only charge transfer without mass transfer across the interface, the bonding is called chemical bonding. In some literature, it is also called physical bonding or adhesive bonding. When there is mass transfer across the interface such as chemical reaction and diffusion, the bonding mechanism is called chemical reaction bonding [5].

3.3.1. Chemical bonding

While atoms are the smallest units for solid-state physicists, interfaces are the smallest building units for material scientists. Heterogeneous interfaces between two different types of materials change the chemical bonding and new properties are formed [25]. Thus, the chemical bonding holds a significant position as a joining technique in this case and includes a chemical bond being created between both parts of the work through utilization of chemical reactions occurring at the ceramic/metal interface. The chemical bonding problem in that joints can be widely produced by chemical bonding at the interface between ceramics (ionic bonding, covalent bonding) and metals (metal bonding), which basically have different bonding modes [26].

The driving force for formation of ceramic-metal interfaces is the decrease in free energy (ΔG) that occurs when intimate contact is established between the ceramic and metal surfaces [27]. The free energy change per unit area of interface formed is given by the Dupré equation [5]:

$$\Delta G = \gamma_{\mathrm{M}} + \gamma_{\mathrm{C}} + \gamma_{\mathrm{MC}} \tag{3}$$

where γ_M and γ_C are the surface energies of the metal and ceramic, respectively, and γ_{MC} is the metal/ceramic interfacial energy (**Figure 5**). When the bonding is chemical bonding and interfacial separation occurs without plastic deformation of the metal and the ceramic, ΔG is identical to the work of adhesion, W_{ad}, which is the work required to separate a unit area of interface into the two original surfaces. Combined with the Young's equation [28], Eq. (3) can be expressed as [26]:

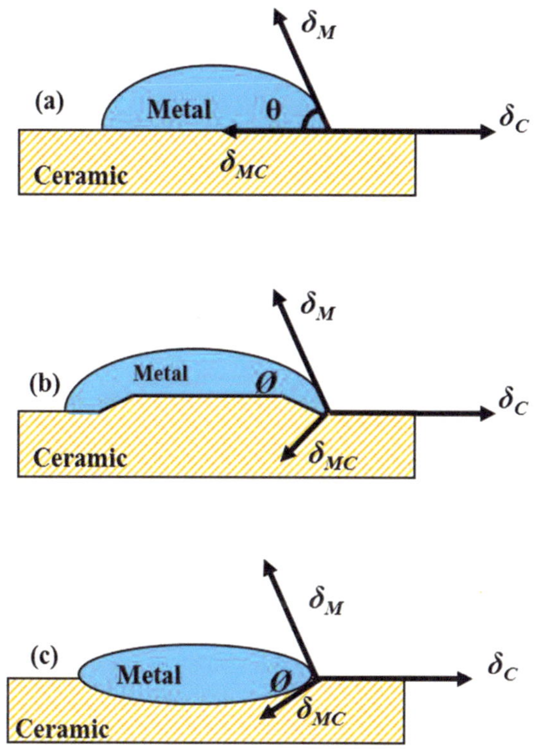

Figure 5. Liquid metal drop shape depending on the contact time: (a) initial contact or rigid solid surface, (b) formation of a ridge (vertical scale exaggerated) and (c) final equilibrium configuration on deformable solid [29].

$$W_{ad} = \gamma_M + \gamma_C - \gamma_{MC} = \gamma_M \left(1 + \cos\theta\right) \tag{4}$$

where θ is the measured contact angle between the liquid or the solid and in equilibrium with a solid substrate. From Eq. (4), it is clear that the interfacial energy of the ceramic/metal, γ_{MC}, decreases as W_{ad} increases.

From **Table 1** it can be seen that, in general, γ_{MC} for Al_2O_3-metal systems tends to increase with the cohesive energy of the metal, which is directly related to its melting temperature (T_m). On the other hand, if the ceramic-metal is a solid-solid system, W_{ad} can be estimated by measuring the dihedral angle, Ø, associated with residual voids on diffusion bonded interfaces [23].

System	γ_{MC} (J/m²)	T_m of metal (°C)
Al_2O_3-Ag	1.57 at 700°C	960
Al_2O_3-Au	1.80 at 1000°C	1063
Al_2O_3-Cu	2.21 at 900°C	1083
Al_2O_3-Ni	2.20 at 1000°C	1453
Al_2O_3-Fe	2.73 at 1000°C	1536

Table 1. Interfacial energies of solid-solid Al_2O_3-metal systems [23].

If the interface ruptures in a brittle fashion, Ø can be measured using an atomic force microscope and W_{ad} is then obtained from [23]:

$$W_{ad} = \gamma_m (1 - \cos \varnothing) \tag{5}$$

Another important consequence of Eq. (3) is that a stable interface requires a positive ΔG (or W_{ad}). For a number of ceramic-metal systems, W_{ad} varies with the temperature, which provides an explanation for the minimum temperature requirements to achieve bonding.

3.3.2. Chemical reaction bonding

When there is mass transfer across the interface, bonding is formed by diffusion or chemical reactions. Chemical reactions at the interface lead to the formation of interfacial reaction layers with properties that differ from both the ceramic and the metal [5]. This can have favourable effects on joint quality by increasing the initial wettability of the metal on ceramic surfaces; however, thick reaction layers increase volume mismatch stresses and thermal residual stresses that detrimental to joint strength. **Figure 6** shows a schematic illustration of the chemical bonding methods and processes. Brazing for instance is a joining technique including the anomalies and gaps which happen on the surfaces of the work being brought into a condition of close cohesion by means of a liquid phase. It is also generally known that solid-phase joining including good adherence which accomplished through heating, pressurization, also distortion that occurs through the surfaces at work and also the interdependence of natural temperature where it seemed direct contact for a period ranging between work surfaces through the settlement and activation. It involves normal temperature tension threads that are being made to produce a full-contact interface between these materials by an energy supply from a source other than the thermal one and to create a joint in the interface in proximity to its normal temperature.

The driving force for a chemical reaction is the chemical potential of the atomic species involved. In many systems, chemical reaction is not expected if only the interaction of the metal with non-metallic elements of the ceramic is considered. However, when all the possible reaction potentials are considered, a net negative free energy can result, which indicates that a chemical reaction is thermodynamically favourable. Equilibrium thermodynamics can use to predict possible reactions at the interface. However, when there are more than three elements in a ceramic-metal system, the prediction of all the possible reactions based on the

http://dx.doi.org/10.5772/64524

phase diagram is nearly impossible. In addition, the extent and possibility of the reaction are limited by kinetics for which data is not readily available for ceramic-metal interfaces [30].

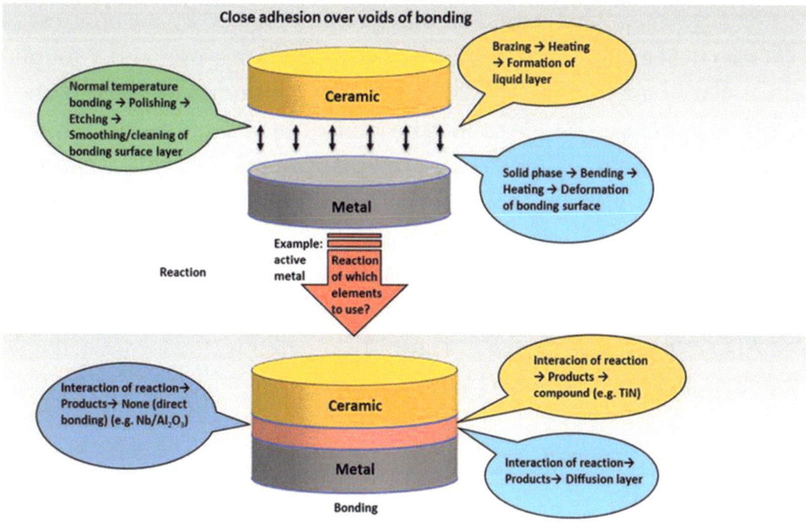

Figure 6. Chemical bonding of ceramics and metals [26].

Reaction phases such as brittle intermetallics and solid solutions often cause interfacial failure at very low stress [5]. An increase in bonding temperature and excessive time generally enhance chemical reactions and lead to thick reaction layer formation, which may decrease the joint strength. At interfaces where planar reaction layers form the thickness of the layer is often optimized by controlling bonding conditions to prevent interfacial debonding or brittle interfacial fracture along the reaction layer. In many ceramic-metal systems, it is observed that the growth of the reaction layer follows a parabolic rate law. It is found that the reaction product tends to be bonded to the ceramic with a coherent interface.

3.3.3. *Pores and unbonded areas on interface*

Unjoined area is frequently formed at the edge of a joint. This edge defect weakens the joint extremely as it works as a notch induced on the interface and one must control the formation of an edge unjoined band. The inhomogeneity in deformation of the metal layer will also reflect the strength. In the case of the reaction gas releasing system, the reaction in the outer region may be promoted by continuous evacuation. This will cause excess thinning of the ceramic at the edge region.

If the interface reaction emission gas as the reaction product, the pores loaded with the gas might be left on the interface bringing about the hindrance of contact. The Si_3N_4/Ni interface is one of the cases. This interface is feeble because of the nearness of pores along the interface.

At the point when nickel contains nitride forming elements, for example, chromium, no pore is formed at an interface and the strength is improved.

In the actual joining sequences, a perfect interface connection over the whole interface is hardly achieved within a certain joining period and temperature limited by the progress of interface reaction. Whereas the base surface roughness and applied pressure with a couple of critical features, which significantly affect the accomplishment of interfacial contact in the solid state bonding as well as in welding [31]. In solid-state joining, and advanced interfacial contact that plastic deformation in the next early stage that creep deformation and diffusion at a later stage. Basic effects pressure to achieve contact by plastic deformation at the elementary stage. Unjoined islands are formed inevitably on the interface joints under limited pressure. It will depend on the breadth of pressure, time, temperature and different material factors such as stress flow [32]. The relationship between the fracture stress and the unjoined area of the solid-state bonded Al_2O_3/Nb joint are shown in **Figure 7**. Apparently, the increase in unjoined area decreases the strength of the specimen [33].

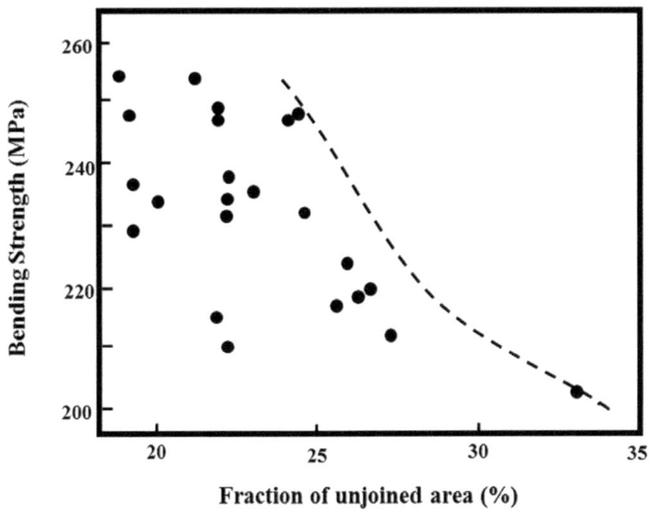

Figure 7. Bending strength of individual Al_2O_3/Nb joints as a function of unjoined area formed on interface [32].

3.4. Mechanical reliability of the joints

The important factor in mechanical reliability is a deviation in the mechanical strength. Knowing the distribution mode of the strength of ceramic/metal joints is significant for the evaluation of reliability. Ceramic materials have a strength distribution obeying the Weibull theory in general. Several researchers suggested the possibility of adopting the Weibull theory to the distribution of the strength of ceramic/metal joints [19].

In the Weibull theory, the cumulative distribution function of fracture, $F(\sigma)$, is written as [19, 34]

http://dx.doi.org/10.5772/64524

$$F(\sigma) = 1 - \exp(-\int_{\gamma}^{0}((\sigma - \sigma_u)/\sigma_O)^m \mathrm{d}V)) \tag{6}$$

where σ_u , σ_0 and m are the zero probability strength (location parameter), the scale parameter and the flaw density exponent (shape parameter). Below σ_u, the stress becomes zero; conventionally, σ_u is set to be zero.

Figure 8 illustrates the sample geometry and test configuration used in the mechanical characterization of ceramic/metal joints. This characterization of the interfacial strength by pull-off or shear-off tests has several limitations. The first one relates to the variety of techniques used by different research groups, making it difficult to establish a mutual comparison of results. The shear test provides an alternative way to assess the mechanical strength of interfaces. Samples are easily produced, but the results are generally lower than those obtained for bend and tensile tests. The selection of an appropriate method for measuring the bond strength is dictated by the purpose of testing, but the bonding process and parameters affecting the mechanical quality of the bond can be monitored by both fracture mechanics and conventional testing methods. The bond strength values obtained also depend on the testing technique chosen. Bend test values are generally higher than tensile test values for joints and for brittle ceramic materials. The shear stress test is one of the simplest techniques. However, the shear stress at the interface is not simple shear and it always contains a component of tensile stress that originates from a bending moment, which cannot be neglected. The influence of a slight change of the push position and the fixing condition on the stress distribution is very important. Therefore, the shear test is not recommended for the common evaluation method. Bending and tensile test has almost the same stress distributions as those derived from

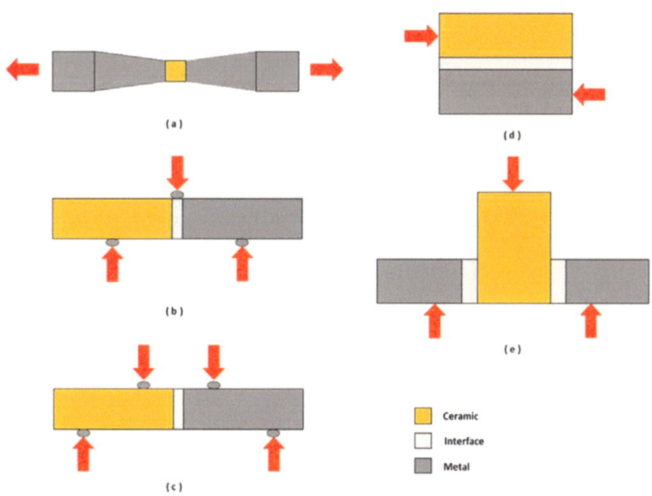

Figure 8. Sample geometry for mechanical tests of joining specimens: (a) tensile; (b) three-point bending, (c) four-point bending, (d) plain shear and (e) shear on ring/cylinder [1, 30].

analytical equations. However, the elastic constant mismatch between ceramics and metal induces inhomogeneity in stress distribution [23].

In the case of three-point bending, the peak stress occurs only along a single line on the surface of the test bar opposite the point of loading. The tensile stress decreases linearly along the length of the bar into the thickness of the bar, reaching zero at each bottom support and at the neutral axis, respectively. The probability of the largest flaw in the specimen being at the surface along the line of peak tensile stress is very low. Therefore, the specimen will fracture at either a flaw smaller than the largest flaw or a region of lower stress. Four-point bend testing results in lower strength values for a given ceramic material than does three-point bending. The peak of the stress distribution in a four-point bend specimen is present over the area of the tensile face between the load points. The tensile stress decreases linearly from the surface to zero at the neutral axis and from the load point to zero at the bottom supports. The area and volume under peak tensile stress or near peak tensile stress is much greater for four-point bending than for three-point bending, and thus the probability of a larger flaw being exposed to high stress is increased. As a result, the modulus of rupture (MOR) or bend strength measured in four-point is lower than that measured in three-point. Uniaxial tensile strength results in lower strength values for a given ceramic than does bend testing. **Figure 9** illustrates that in the case of uniaxial tension the complete volume of the gauge section of a tensile test specimen is exposed to the peak tensile stress. Therefore, the largest flaw in this volume will be the critical flaw and will result in fracture.

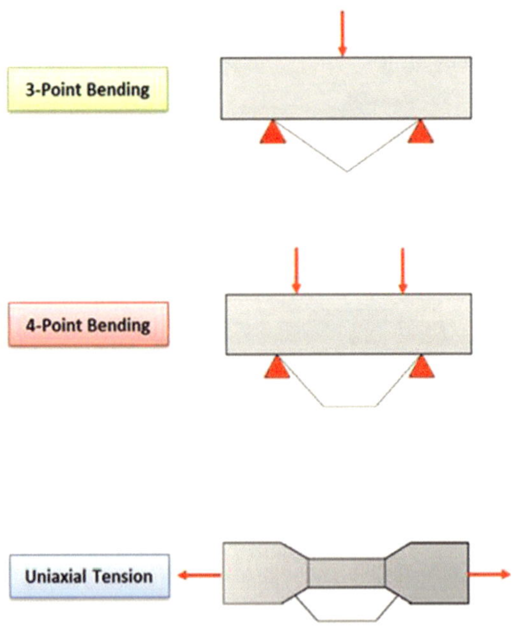

Figure 9. Comparison of the tensile stress distributions for three-point, four-point and uniaxial tensile test specimens [23].

The strength of metal/ceramic joint materials is generally characterized by bending tests, also referred to as flexure testing. The test specimen can have a circular, square or rectangular cross section and is uniform along the complete length. As shown in **Figure 10**, the test specimen is supported near the ends and the load is applied either at the centre, for three-point loading, or at two positions for four-point loading.

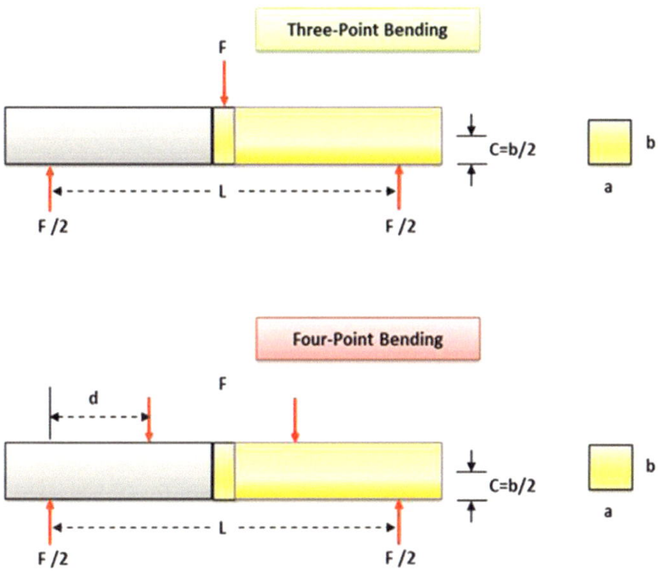

Figure 10. Derivation of the modulus of rupture equation for three-point and four-point bending [23].

The bend strength is defined as the maximum tensile stress at failure and is often referred to as the MOR. The bending strength of a rectangular test specimen can be calculated using the general flexure stress formula:

$$S = M.C / I \tag{7}$$

where M is the moment, C is the distance from the neutral axis to the tensile surface and I is the moment of inertia.

For a rectangular test specimen [23]:

$$I = a.b^3 / 12 \tag{8}$$

and

$$C = b / 2 \tag{9}$$

where b is the thickness and a is the width of the specimen.

From **Figure 10**, it is possible to illustrate the derivation of the three-point and four-point flexure formulas for rectangular bars. We can observe that: $M = (L/2)\cdot(F/2)$ in the case of three-point and $M = (F/2)\cdot d$ for four-point test.

Therefore, for three-point bending:

$$S = \sigma_{3P} = \frac{3.F.L}{2.a.b^2} \tag{10}$$

And for four-point bending test:

$$S = \sigma_{4Pt} = \frac{3.F.d}{a.b^2} \tag{11}$$

For most ceramic materials, the apparent strength will decrease when going from three-point to four-point to tensile testing and as specimen size increases.

Whatever joining processes are used, the successful formation of the joint depends on achievement of intimate contact between the base materials, conversion of the intimate contact into an atomic bonding/reaction, accommodation of residual stresses induced by different thermal and mechanical properties between the base materials undergoing temperature change. Each joining process is characterized by the methods and conditions employed to achieve intimate contact and to promote bond formation between the work pieces.

4. Ceramic-metal brazing problems

Brazing is a process for joining similar or dissimilar materials using filler metal [35, 36]. The filler metal is heated slightly above its melting point so it flows, but the temperature remains lower than the melting point of the ceramic metal joints (**Figure 11**). Flux or an inert atmosphere is utilized to keep two surfaces that have joined and brazing material from oxidation during the heating process. The filler material flows over the base metal and ceramic, and the entire assembly is then cooled to join the pieces together. Brazing forms very strong, permanent joints. Brazing is considered to be well-established commercial processes for ceramic metal joints also [37], where it is widely used in industry, in different parts, because almost every metallic and ceramic material can be joined by this process. Generally, brazing can easily be performed by manual techniques, but, in many cases, it can just as easily be automated if necessary.

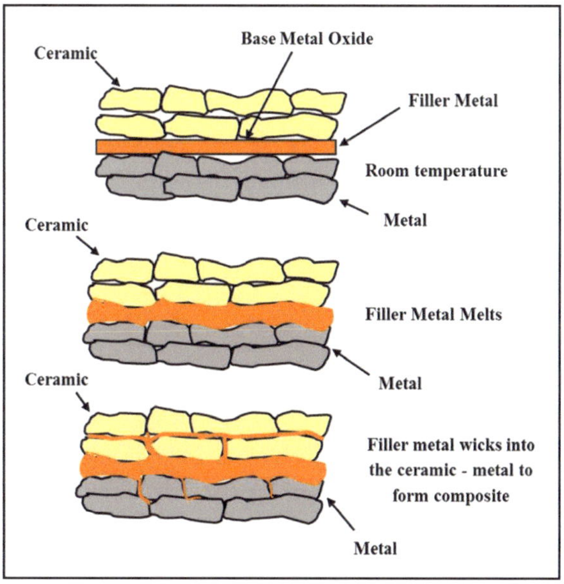

Figure 11. Brazing schematic [38].

Brazing has numerous focal points over other metal-joining methods, for example, welding [39]. Since brazing does not fuse the base metal of the joint, it permits much more tightly control over resilience and produces a perfect join without the requirement for optional wrapping up. Furthermore, dissimilar metals and ceramic can be brazed. When all is said in done, brazing likewise creates less thermal deformation than welding because of the uniform heating of a brazed piece [39]. Complex and multi-part assemblies can be brazed cost-effective. Another feature is that the brazing can be covered or clad for defensive purposes. Finally, brazing is effectively adjusted for large scale manufacturing and it is anything but difficult to mechanize on the grounds that the individual procedure parameters are less delicate to variety [40].

One of the major disadvantages is the absence of joint strength when contrasted with a welded joint because of the softer filler metals utilized. The strength of the brazed joint is liable to be not as much as base metals but more than the filler metal [41]. Another disadvantage is that brazed joints can be damaged under high temperatures. The brazed material joints require a high purity when done in an industrial environment. Also some applications for brazing require the utilization of satisfactory fluxing agents to control cleanliness. The colour of joint is frequently not quite the same as that of the base metal, making a stylish disadvantage.

The two major problems when the joining these materials by brazing process are firstly the differences in physical properties between ceramics and metals, and secondly the poor wettability of ceramics by most metals and alloys [42, 43]. The first problem in joining ceramics to metals for high-temperature results from the huge contrasts in thermal expansion behaviour. At the point when the thermal expansion of these materials get together is modified, these differences in the thermal expansion behaviour can prompt high stresses [42]. This condition is regularly subsequently heightened by thermal inclinations that rise as a result of thermal diffusivity contrasts between the metal and ceramic. Ceramic for the most part have high elastic moduli and low-relaxation characteristics which prevent relief or redistribution of the stresses. The low tensile strengths of most ceramics may then make them unable to resist fracture under such stresses [44, 45]. Another problem of wettability is overcome with the use of an activated braze alloy, where an active element, e.g. Al, Zr and Ti, alters the surface chemistry of the ceramic by the formation of intermediate reaction layer and lowers the wetting angle of the molten braze on the ceramic [46, 47]. The compounds that form are commonly spinels for the oxide ceramics and complex nitrides for the ceramic nitrides [39]. For the purpose of addressing this problem should be used high vacuum or high-purity reducing or inert-gas atmospheres are necessary for the successful brazing process.

Cazajus et al. [48] have studied the thermal stresses in the ceramic-metal joining after brazing process. The framework of this study is the thermomechanical analysis and the simulation of the brazing process of ceramic and metal joining. **Figure 12** gives the physical phenomena involved during brazing and their coupling relations. The brazing is a joining process which produces the coalescence of materials by heating them to a suitable temperature or by using a filler metal, having liquids under the solids of the base materials. The difference between ceramic and metal thermal expansion coefficient (CTE) leads to the development of residual thermal stresses during cooling from brazing process to room temperature which reduce the join strength. The design of joints in material engineering and the optimization of the industrial brazing process require to control and to examine such a phenomenon. The conclusions from that paper can be drawn for different parameter effects on residual stresses during the brazing process simulation. The mechanical behaviour and geometrical parameters have a significant influence on the residual stress distribution and their maximum values. The difference between ceramic and metallic material's CTE and the metallic materials elastoplastic properties are the most important parameters of the assembly mechanical behaviour. The ratio between the alumina height and stainless steel (H_A/H_{SS}) represents the most important geometrical factor (**Figure 13**). The cooling conditions and the filler metal yield stress evolution depending on temperature have only a significant impact on the residual stresses evolution during the brazing process and not on the final value [49, 50].

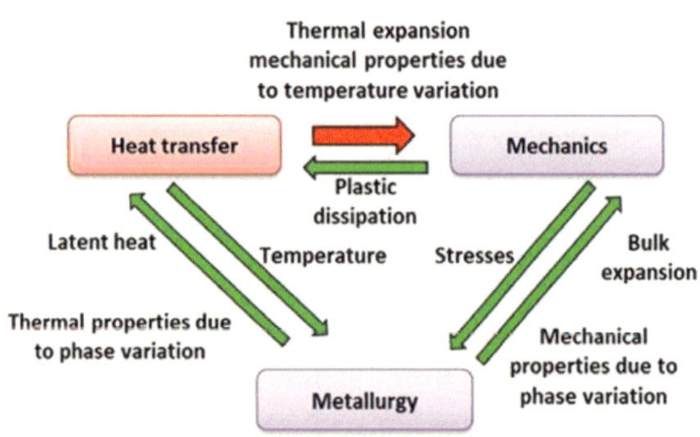

Figure 12. Physical phenomena during brazing process and their couplings [48].

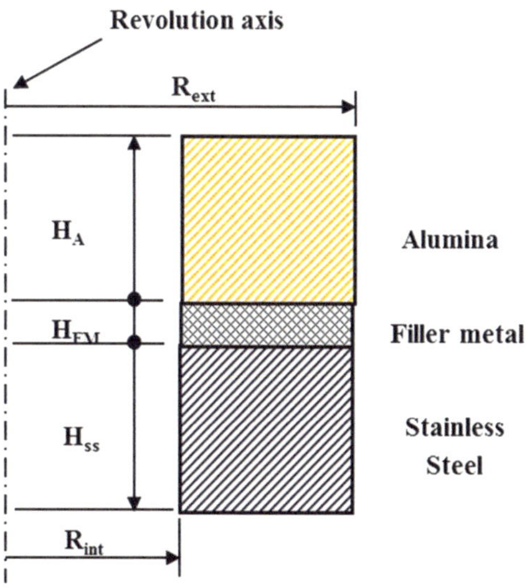

Figure 13. Geometrical parameters of the ceramic metal assembly [50].

Shirzadi et al. [51] developed the general method for brazing ceramics to metals using compliant metallic foam as a buffer layer. Using stainless steel foams, bonds between alumina and 316 Stainless Steel with shear strengths up to 33 MPa have been achieved. From this study, it is found that the utilization of metallic foam as a buffer layer between ceramic and metal could be an efficient method to avoid the mismatch that occurs in thermal expansion between the two materials when bonded together by brazing. They have been exhibited that the joints were tolerant to serious thermal cycling tests. The number of thermal cycles (200–800 °C in normal condition) to disappointment of 67 ± 3 through the thermal cycling test. According to shear test results, the fracture mode was ductile because of the flexibility in the region based on the layer of foam. The fracture surfaces of the samples brazed without and with foam after

Figure 14. Fracture surfaces of joints without and with metallic foam following thermal cycling between 200 and 800 °C in air. Number of cycles to failure were <1 and 60 ± 4 for samples without and with a foam interlayer, respectively [51].

thermal cycling (200–800 °C in air) as shown in **Figure 14**. It showed the former 'cup and cone' break inside the ceramic after the first cycle, while the latter failed from the interface of the ceramic foam after more than 60 cycles.

The article of Walker and Hodges [52] is intended to familiarize the designer with brazing methods commonly used to join metals to ceramics, discussed the advantages and disadvantages of two methods, and show the relative tensile strengths obtained from samples fabricated using these methods. In most article cases discussed, 94% alumina ceramic (6% glassy phase) ASTM-F19 tensile button samples were joined to Fe-29Ni-17Co alloy using a gold-or silver-based braze filler metal (**Figure 15**). The article is limited to the two metallization methods most commonly used for joining metals to ceramics: the molybdenum-manganese/nickel plating method and physical-vapour deposition or thin-film method.

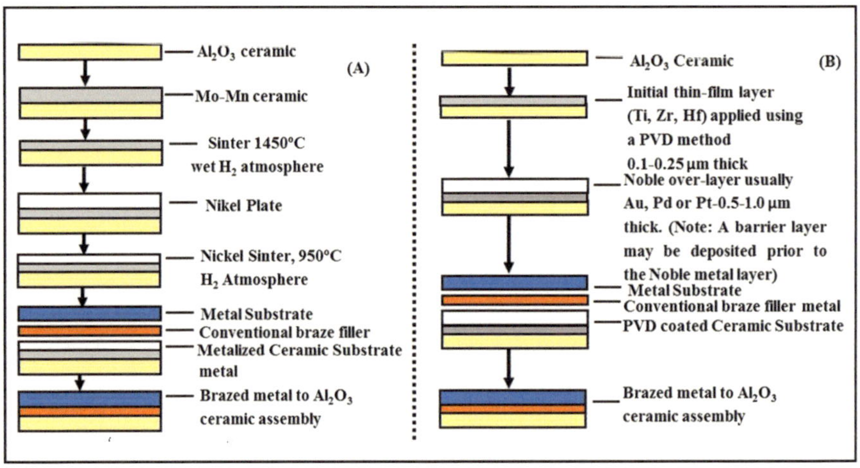

Figure 15. Commonly used ceramic metallization methods: (A) moly-manganese metallization process; (B) thin-film metallization process [52].

At the point when ceramics production are to be brazed, especially in those conditions where the coupling part is a metal, the individual differential thermal coefficient of thermal expansion between the coupling parts is of foremost significance. When in doubt engineers have a tendency to expect that the coefficient of thermal extension of a metal will be a few times that of a ceramic. This is not generally genuine, be that as it may. For instance ceramic, for example, silicon nitride and silicon carbide do, in fact, have low coefficient of thermal extension, and issues emerging from stresses produced amid the cooling stage can be normal in conditions where such materials are brazed to stainless steels or copper both of which being metals that have a high coefficient of thermal expansion. However, titanium, titanium alloys and some exceptional materials, for example, invar and kovar each have a coefficient of thermal expansion which is near that of alumina, while expansion coefficients of molybdenum and tungsten are near those of both silicon carbide and silicon nitride. This implies while picking the 'active' filler material that will be utilized for a specific employment it is fundamental to decide the differential coefficient of expansion that exists between the materials that are to be joined.

http://dx.doi.org/10.5772/64524

5. Ceramic–metal solid-state joining

Solid state joining is a gathering of joining procedures which produces cohesion at temperatures basically underneath the melting point of the base materials being joined, without the expansion of brazing filler metal. Pressure might possibly be utilized. These procedures are infrequently mistakenly called solid state bonding forms: this gathering of joining procedures incorporates friction welding, diffusion and laser welding. In all of these welding processes, the parameters such as temperature, time and pressure separately or together to produce ceramic metal joint without melting of the base metal. Solid-state bonding process contains some of very oldest processes and some of them new. The bonding processes provide certain advantages as the base metal do not melt and compose an interface. The materials that joined keep their original properties without the heat-affected zone problems included when there is no melting for base materials [53]. At the point when dissimilar metals are joined their thermal expansion and conductivity is of substantially less significance with solid state welding than with the arc welding processes. Time, temperature and pressure are included; nonetheless, in some processes the time component is to a great degree short, in the microsecond run or up to a few moments. In different cases, the time is reached out to a few hours. As temperature expands time is generally decreased. Since each of these processes is different each will be qualified.

5.1. Ceramic-metal friction welding

Friction welding is a solid state joining that produces a bond under the compressive force of one rotating workpiece to another stationary workpiece [54]. Heat is generated at the weld interface during the friction between two materials because of the non-stop rubbing for different contact surfaces, which is produced later in the softening of metal (**Figure 16**). Finally, the metal side at the weld interface begins to flow elastically and forms an axial shortening [55]. When a certain amount of forging had occurred, the rotation stops and the compressive force are maintained or slightly increased to consolidate the weld. Some of the important operational parameters in friction welding are friction time, friction pressure and rotation speed [56].

Friction welding, like any welding process, has its specific advantages and disadvantages. The following are some advantages of friction welding such as no filler metal is needed. Flux and shielding gas are not required. The process is environmentally clean, no arcs, sparks, smoke or fumes are generated by clean parts [58]. Surface cleanliness is not as significant, compared with other welding processes, since friction welding tends to disrupt and displace surface films. There are narrow heat affected zones. The process is suitable for welding most engineering materials and is well suited for joining many dissimilar material combinations. In some cases of weld, the strength of the joint is equal to or greater than the strength of the weaker of the two materials joined. The bond that is created by the mechanical intermixing and solidification of the two materials is strong and free from voids and porosity. It can be cost effective and offers design engineers many more options than other methods.

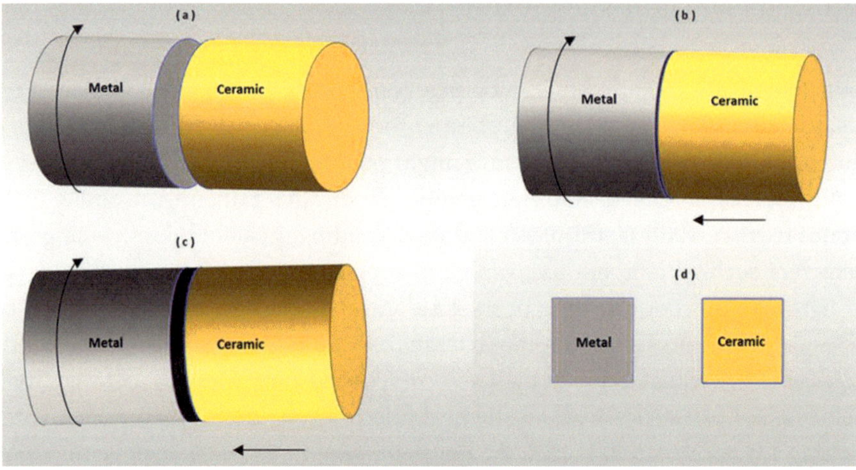

Figure 16. Basic steps of friction welding of ceramic and metal [57].

There are also some limitations of the process like; one workpiece must have an axis of symmetry and be capable of being rotated about that axis. Preparation and alignment of the workpieces may be critical for developing uniform rubbing and heating, particularly with diameters greater than 50 mm. Capital equipment and tooling costs are high. Dry bearing and non-forgeable materials cannot be welded. If both parts are longer than 1 m, special machines are required. Free-machining alloys are difficult to weld [59].

For a particular application, heating time is determined during the setup or from previous experience [57, 60]. Excessive heating limits productivity and increase wastes material. Similarly, uneven heating as well as entrapped oxides causing unbonded areas at the interface may be due to insufficient welding time. The ranges of effective pressure are not essentially slight for forging and heating, although the selected pressures should be reproducible for any specific process. Friction pressure has influence on the axial shortening distance and the temperature gradient in the weld zone [61]. The friction pressure depends on the materials being joined and the surface joint geometry [59]. Selected pressure must be high enough to hold the faying surfaces in intimate contact to avoid oxidation. Joint quality is improved in many metals, including steels, by applying a forging force at the end of the heating period.

On the other hand, the rotational speeds are related to the welding material and welding diameter in the interface. They may have different effects on the mechanical properties of the friction joint. Increase the rotational speed may lead to more frictional heat at the interface, thus leading to a greater amount of softening materials, recrystallization, or even increased intermetallic formation [62]. In addition, depending upon the type of materials joined or more accurately, the physical and mechanical properties involved the rotational speed of the production of the various effects on the quality of the joint Therefore, an appropriate rotation speed must be used to minimize any harmful effects and produce good quality of joints is an effective pressure range pressure.

Weiss [63] studied the residual stresses and strength of friction welded ceramic/metal joints. From this article, based on friction welding results of different ceramic materials to an aluminium alloy, the effect of residual stresses on the strength of ceramic-metal joints was calculated numerically. Heat conduction process calculations to evaluate the temperature distribution have been conducted by the method of finite element (FEM), using hardware experimental data for input. From this chapter, the theoretical analyses clearly show that edge geometry of the joint in the area of the interface (flash) has a strong effect on the weld joint strength. Improvement of weld joint strength seems to be possible by optimization of the geometry in the area of the weld interface. The effect of joining parameters on ceramic metal joint strength through residual stresses is comparatively low. However, the welding parameters may have more effect on the joint strength by means of the bonding process, resulting in higher or lower bonding strength [64].

Rombaut et al. [65] summarized of the literature review performed during the master thesis on friction welding on dissimilar materials. Of main interest in this work is the welding of steel to a ceramic material such as alumina (Al_2O_3). Because of the difficulties involved in the production of welding sound for this material combination, and not a lot of literature is available on this topic. This work begins with a discussion about the basics of friction welding and typical problems encountered in steel welding with ceramic. There are three major reasons related to joining problems noted for these materials combination. First, there is an important variance in the type of atomic bonds between metal like steel and ceramic. The joining in the ceramic is mostly from the nature of the ionic or covalent (usually a hybrid of these), while metals have a metallic bonding character. Second, there is often a very large difference in the thermal expansion between these materials. Ceramic is usually lower coefficient of thermal expansion. When two parts of materials cool down after joining, the thermal stresses will be push in the weld interface, this may lead to cracking after that. Third, the brittle mode and porous of ceramics makes it very hard to absorb fabrication defects. The strength of a ceramic is highly dependent of its grain size and surface roughness [66, 67].

Seli et al. [68–70] presented the evaluation of mechanical and interfacial properties of friction welded alumina-mild steel rods with use of A6061 sheet as an interlayer. A preliminary simulation was made to predict the deformation, stress, strain and temperature distribution during the joining operation using a fully coupled thermo-mechanical FE model. This paper also starts with a discussion on the basics of friction welding and typical problems encountered in welding the dissimilar materials. Problems related to friction welding of dissimilar materials are not only related with specific characteristics such as melting point and hardness, but also with the reactions that occur at the joint interface. Metals generally have a coefficient thermal expansion higher than ceramics. Therefore, when joining ceramics to metals using friction welding, it will be induced very large thermal stress and in many cases these large stresses cause joint failure In order to overcome this problem, the development of solid-phase bonding processes, which a metal or composite metal-ceramic layers are placed between the ceramic and metal surfaces to be joined [62, 71, 72].

Uday et al. [20] investigated the effect of welding speeds (630–2500 rpm) on the mechanical strength of friction welded joint of alumina-YSZ composite and 6061 aluminium alloy. From this study, alumina-0, 25 and 50 wt.% YSZ composite with 6061 aluminium alloy joints were welded successfully by friction welding. The bending strength values of alumina-25 wt% YSZ composite joint obtained were greater at a rotational speed of 630 rpm than at 2500 rpm. The bending strength values at the joints were smaller in the pure alumina joint at a rotational speed of 1250 rpm than at 2500 rpm. The joint with large thermal expansion mismatch decreased the strength. However, it occasionally happens that some specimen is strong but the other is weak even if they are of the same kind. This depends on the presence and distribution of internal flaws induced by thermal stress during the joining process. The ceramic composite (Al_2O_3-25 wt.% YSZ) joints were welded productively at the low rotational speed (630 rpm) compared with pure alumina when joining with aluminium alloy by friction welding. The frictional heat at low rotational speeds (630 rpm) [73] produced lower temperature gradients in the surface of friction, with temperature falling in the radial direction. Friction at high speed 2500 rpm produced more heating along the whole of the interface. The lower heating of the rod end-faces reduced stresses within the rod material [59, 74].

5.2. Ceramic-metal diffusion bonding

Diffusion bonding is a joining method where the principal mechanism for joint formation is diffusion solid-state. Coalescence of the faying diffusion surfaces is accomplished through the application of pressure at raised temperature. No melting and limited macroscopic deformation or relative motion of different parts occurs during bonding. A filler metal (diffusion aid) can or cannot be used between the faying surfaces [75]. It has involved interest as a means of joining ceramic and successes have been realized by controlling the microstructure of the interfaces formed. The first condition for diffusion bonding is to create an intimate linking between two surfaces to be joined to the atomic species comes into intimate contact. Furthermore, to a good connection, there must be enough diffused between the materials in a reasonable period of time. Pressure can be applied by hot press or hot isostatic press on a diffusion couple. **Figure 17** shows illustrations of events during metal/ceramic diffusion bonding in solid-state [76].

Diffusion bonding is primarily employed in the joining of dissimilar materials, i.e. dissimilar metals, metal-glass, metal-ceramic and ceramic-glass, either directly or through the use of interlayers [1, 30, 77, 78]. It offers numerous points of interest, mostly the strength of the bonding line, which is equivalent to the base metals. The microstructure at the bonded area is precisely the same as the origin materials. Otherwise, this point of interest joining process requires a few entirely controlled conditions: spotless and smooth contacting surfaces which are free from oxides, and so forth, high temperature condition to advance diffusion process [79–81]. Then again, diffusion holding requires a considerably more joining time. Also, the equipment expenses are high because of the mix of high temperature and pressure in vacuum situations. This frequently constrains the part measurements, which might be unfavourable from a financial viewpoint [79].

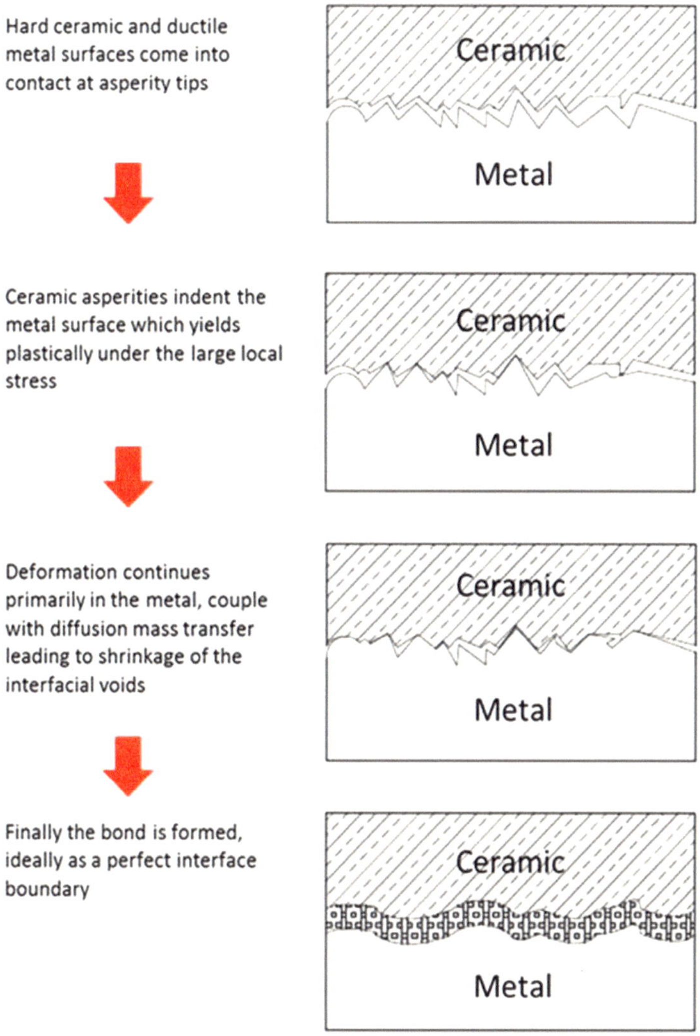

Figure 17. Sequence of events during metal-ceramic diffusion bonding [1].

The main process parameters which control the diffusion bonding process are temperature, time and pressure [82]. Process temperature parameter is the most important because of the way in a thermally activated process, a slight change in temperature will lead to a significant change in the kinetics of the process compared with other parameters, and almost all the mechanisms, including plastic deformation and disseminate sensitive to temperature [83]. The temperature chosen is typically in the region of 0.5–0.8 of the absolute melting temperature of the component having the lower melting point [84]. Hence, melting and melting related defects are avoided in diffusion bonding process [76, 79, 85].

Zhang et al. [86] have introduced the research and development of joining methods of ceramics to metals, especially brazing, diffusion bonding and partial transition liquid phase bonding.

From this article, the diffusion bonding is a technology to achieve compact joint by diffusion of atoms, even chemical reactions between materials or interlayer and materials. The diffusion of atoms in interface is carried out by several mechanisms, such as the replacement of near atoms, movement of clearance atoms and movement of vacancies, etc. The surface of the materials to be joined must be clean and flat (the roughness less than 0. 4 μm). Joining time can be a few hours at a mild temperature (0. $6T_m$, T_m is melting point of metal to be joined), also can be several minutes at high temperature (0. $8T_{m}$,). Diffusion bonding can be achieved with inserted interlayer [87–92] or without interlayer [93, 94]. The diffusion interlayer can reduce the cracking, relax the thermal residual stress and improve the joining strength. The diffusion interlayer is produce of different element active to ceramics, such as titanium, niobium and zirconium etc.

Burger and Ruhle [95] studied the material transport mechanisms during the diffusion bonding for niobium (Nb) to alumina (Al_2O_3). According to this chapter, the many different material transport phenomena may occur during the diffusion bonding process of a metal to a ceramic at high temperatures. The operating transport mechanisms depend on the selected combination of materials as well as on the bonding conditions. So from this work, the results were completed in which different faces in the niobium surface were bonded to a polished alumina surface. The niobium metal had either a very flat polished surface, or well-defined flaws of different shapes and dimensions and that were presented into the surface. The authors were found that the chemical reactions control the transport of materials and according to the conditions chosen for these experiments. As well as the interdependence of the diffusion joining of ceramics and metals requires that two couples have a near contact over the entire area of the joint interface. Even if all defects are detached, there may still be residual thermal stresses due to reaction layers, dislocations, facets, chemical gradients, dislocation arrangements, and precipitates formed during bonding. On Nb/Al_2O_3 interfaces, thermal stresses are expected to be rather small since the thermal expansion coefficients of both materials are very near. No reaction layer was perceived.

5.3. Ceramic-metal laser welding

Laser welds bonding technique is a new kind of welding technology [96]. It has been developed as an alternative to adhesive bonding and laser welding. Laser welding has a small heat effect zone, which has little effect on the adhesive bonding area [97, 98]. The adhesive in the fusion zone decomposes during the laser welding process, which produces little effect on the properties of the joint. Thus, it can be assumed that laser welding and adhesive bonding hardly affect each other (**Figure 18**). The advantages of laser welding and adhesive bonding are both included in the laser weld bonding technique. The adhesive provides excellent stress distribution over large bonding areas and laser welding improves the peel resistance of adhesives. Thus, a laser weld bonding joint has better mechanical properties than either a laser welded or adhesive bonded joint alone. Laser welds bonding is a new hybrid technique that combines metallurgical joining, mechanical joining and chemical bonding [96].

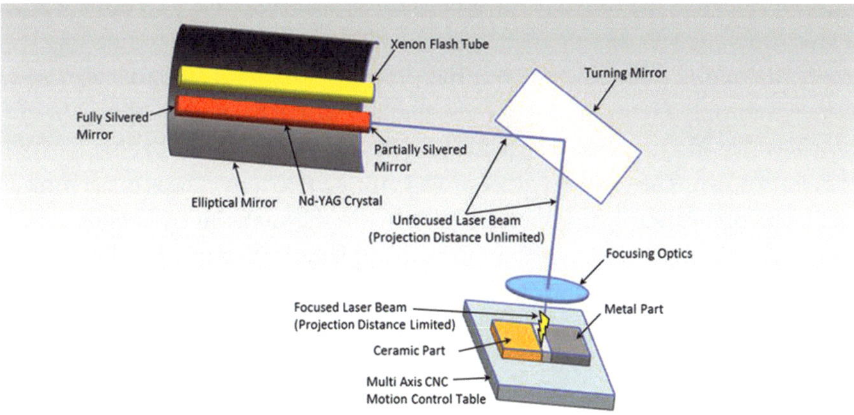

Figure 18. Schematic of the principle of laser welding process [99].

The development of more effective joining techniques for structural ceramics could also have a great impact on their use in mass-produced components. However, there are several challenges on component manufacturing by ceramic processing techniques and by the material themselves. Deformation densified ceramics to form complex shapes is practically impossible because most ceramic materials are brittle even at elevated temperatures. Moreover, ceramics are undesirable for mass production because of their high cost and machining difficulties. Effective ceramic joining techniques can play an important role in improving the reliability of ceramic structures as well [100]. Ceramics are very sensitive to flaws, due to the quality of raw materials used in their production and to the characteristics of various processing techniques, such as machining. Several techniques have been developed to join ceramics for structural application: brazing with filler metals; diffusion bonding; microwave joining; and the use of interface layers designed to form a thin transient liquid phase at a relatively low bonding temperature [97].

Many studies have been previously conducted on laser interactions with various metals and semiconductors, but few have been done in the processing of ceramics with lasers [101]. The advanced ceramic composite technology has offered more opportunity to fabricate complex structures of composite ceramic lasers, due to the availability of perfect inherent interface characteristics. One of the main problems in fusion welding of ceramics is to control cracking because of the residual thermal stresses. The result has been to give extra heating in a more extensive region around the zone of weld so that the net thermal slope of the extra heating and the joining source is presently adequately low so that no residual thermal stresses sufficiently high to cause cracking when reached [102]. This extra heating also allows the part to be heated and cooled very slowly enough to avoid thermal shock. In order to avoid weld cracking, and heated ceramic samples with radiant energy formed by halogen lamps, which have been collected by the indicators [103].

Exner and Nagel [104] have investigated about the laser welding of functional and constructional ceramics for microelectronics. They presented successful method of a laser welding technology developed in the Laser Institute of Mittelsachsen (Germany). The investigations of alumina laser welding with a purity of 97% showed that in general the technology is suitable. Furthermore, it enables them to carry out the procedure without furnaces and in a natural atmosphere within only a few minutes. It was established, that the high quality of laser welding joints are achievable. Homogeneous structure and lead to no loss of power also, loss of tangible property is not known. The technology permits joining up to a thickness of 3.5 mm. Through using particular preheating it is conceivable to settle the material by metals. The shortest distance from the joining area is more than 25 mm. Implementation of the technology develops the application of ceramic dramatically. All the outstanding advantages of the laser material processing are useful: touchlessness, flexibility , precision and high velocity [104].

6. Conclusion

Advanced ceramics are key materials and are widely used in the electronics, fuel cell, sensor, insulator and Bioengineering fields. The joining of ceramics to metal is necessary and un-avoidable in the miniature manufacturing field. Ceramic-metal joining processes and their resulting interfaces have been extensively studied over the years. New developments in the field have granted structural ceramics new horizons in applications involving adverse conditions and reliable materials. However, there still remain several unknown problems. Further experimental evidences could allow a more detailed understanding of the joining mechanism. A small size component (up to approximately 15 mm in diameter) can be joined by using a soft metal or a laminated interlayer for limited kinds of ceramics. How to join large size one with a metal is, however, still one of serious problems because the size dependence of residual stress is so severe. Since most of structural components will be used at elevated temperatures, the examinations on high temperature properties such as strength, oxidation, thermal expansion and thermal stress are required.

Author details

Uday M.B.[1*], Ahmad-Fauzi M.N.[2], Alias Mohd Noor[1] and Srithar Rajoo[1]

*Address all correspondence to: ummb2008@gmail.com

1 Faculty of Mechanical Engineering, UTM Centre For Low Carbon Transport In Cooperation With Imperial College London, Institute For Vehicle Systems and Engineering, University of Technology Malaysia, Johor, Malaysia

2 School of Materials and Mineral Resources Engineering, Engineering Campus, Universiti Sains Malaysia, Penang, Malaysia

References

[1] Lemus-Ruiz J, Ceja-Cárdenas L, Bedolla-Becerril E, López-Morelos VH. Production, characterization, and mechanical evaluation of dissimilar metal/ceramic joints. Nanocomposites with Unique Properties and Applications in Medicine and Industry. InTech, Croatia. 2011:p. 205–24.

[2] Loehman, R., Characterization of Ceramics, Momentum Press, 2009, p. 212–216.

[3] Jadoon, A.K., *Employing reactive synthesis for metal to ceramic joining for high temperature applications*. Journal of Materials Science, 2004. 39(2): p. 593–604.

[4] Palit, D., Meier, A., *Reaction kinetics and mechanical properties in the reactive brazing of copper to aluminum nitride*. Journal of Materials Science, 2006. 41(21): p. 7197–7209.

[5] Park, J.-W., A Framework for Designing Interlayers for Ceramic-to-Metal Joints, Massachusetts Institute of Technology, 2002, p.18–24.

[6] Wiese, J.L., Strength of Metal-to-Ceramic Brazed Joints, Massachusetts Institute of Technology, Department of Materials Science and Engineering, 2001, p.1–26.

[7] Zhou, Y., Microjoining and Nanojoining, CRC Press, Inc, 2008, p.758–785.

[8] Raić, K.T., *An estimate of* Si_3N_4 *diffusion into Cu based filler metal*. Ceramics International, 2000. 26(1): p. 19–24.

[9] Elssner, G., Petzow, G., *Metal/ceramic joining*. ISIJ International, 1990. 30(12): p. 1011–1032.

[10] Somiya, S., Handbook of Advanced Ceramics: Materials, Applications, Processing, and Properties, Academic Press, 2013, p.23–65.

[11] Lemus-Ruiz, J., Ceja-Cárdenas, L., Verduzco, J., Flores, O., *Joining of tungsten carbide to nickel by direct diffusion bonding and using a Cu–Zn alloy*. Journal of Materials Science, 2008. 43(18): p. 6296–6300.

[12] Suganuma, K., *Recent advances in joining technology of ceramics to metals*. ISIJ International, 1990. 30(12): p. 1046–1058.

[13] Khor KA, Wang M, Zhou W, Boey F, Srivatsan TS, editors., Processing and Fabrication of Advanced Materials VIII. World 26 Scientific, 2001, p. 438–442.

[14] Anderson, R., *Testing advanced ceramics*. Advanced Materials and Processes, 1989. 135: p. 31–36.

[15] Mizuhara, H., Huebel, E., Oyama, T., High-reliability joining of ceramic to metal. American Ceramic Society Bulletin (USA), 1989. 68(9), p.1591–1599.

[16] Olofsson, J., Friction and Wear Mechanisms of Ceramic Surfaces: With Applications to Micro Motors and Hip Joint Replacements, Uppsala University, 2011, p.300–337.

[17] Çelik, B., Distortion, Thermally Induced Stresses, Residual Joints, in Welded, in Institute of Natural and Applied Sciences, Cukurova University, 2009, p.1–29.

[18] Lancaster, J.F., Metallurgy of Welding, William Andrew Pub., 1999, p.37–51.

[19] Suganuma, K., Miyamoto, Y., Koizumi, M., *Joining of ceramics and metals*. Annual Review of Materials Science, 1988. 18(1): p. 47–73.

[20] Uday, M.B., Ahmad-Fauzi, M.N., Zuhailawati, H., Ismail, A.B., *Effect of welding speed on mechanical strength of friction welded joint of YSZ–alumina composite and 6061 aluminum alloy*. Materials Science and Engineering: A, 2011. 528(13): p. 4753–4760.

[21] Bellosi, A., Kosmac, T., Tomsia, A.P., *Interfacial Science in Ceramic Joining*. Vol. 58. 1998, Springer.

[22] Suganuma, K., Okamoto, T., Koizumi, M., Kamachi, K., *Influence of shape and size on residual stress in ceramic/metal joining*. Journal of Materials Science, 1987. 22(10): p. 3561–3565.

[23] Lemus-Ruiz, J., *Diffusion bonding of silicon nitride to titanium*, in *Department of Mining and Metallurgical Engineering*. 2000, McGili University: Montréal, Canada.

[24] Hadian, A.M., *Joining of silcon nitride to silicon nitride and to molybdenum for high temperature applications*, in *Department of Mining and Metallurgical Engineering*. 1993, McGill University: Montreal, Canada.

[25] Wunderlich, W., *The atomistic structure of metal/ceramic interfaces is the key issue for developing better properties*. Metals, 2014. 4(3): p. 410–427.

[26] Nakahashi, M., *Joining of ceramics to metals (1): Interfacial reactions between ceramics and metals*. Welding international, 1996. 10(10): p. 765–770.

[27] Saiz, E., Cannon, R.M., Tomsia, A.P., *High-temperature wetting and the work of adhesion in metal/oxide systems*. Annual Review of Materials Research., 2008. 38: p. 197–226.

[28] Finnis, M., *The theory of metal-ceramic interfaces*. Journal of Physics: Condensed Matter, 1996. 8(32): p. 5811.

[29] Passerone, A., Muoio, M.L., *Joining technology in metal-ceramic systems*. Materials and Manufacturing Processes, 2000. 15(5): p. 631–648.

[30] Nascimento, R.M., Martinelli, A.E., Buschinelli, A.J.A., *Review article: recent advances in metal-ceramic brazing*. Cerâmica, 2003. 49: p. 178–198.

[31] Abbas M. R., Uday M. B., Alias Mohd Noor, Norhayati Ahmad, and Srithar Rajoo. Microstructural evaluation of a slurry based Ni/YSZ thermal barrier coating for automotive turbocharger turbine application. Materials & Design. 2016, 109, p. 47–56.

[32] Suganuma, K., *Reliability Factors in Ceramic/Metal Joining*. 1993, National Defense Academy Yokosuka, Japan.

[33] Vila, M., Prieto, C., Zahr, J., Pérez-Castellanos, J.L., Bruno, G., Jiménez-Ruiz, M. Miranzo, P., Osendi, M.I., *Residual stresses in ceramic-to-metal joints: diffraction measurements and finite element method analysis*. Philosophical Magazine, 2007. 87(35): p. 5551–5563.

[34] Ekwaro-Osire, S., Khandaker, M., Gautam, K., Accounting for high stress gradient by a modified Weibull failure theory. Journal of Engineering Materials and Technology, 2008. 130(1), p.1–8.

[35] Joints, E.B., ed. Brazing Handbook, American Welding Society (AWS) C3 Committee on Brazing and Soldering, 2007, p.1–100.

[36] Angelescu, D.E., Highly Integrated Microfluidics Design, Artech House Publishers, 2011, p.1–41.

[37] Lugscheider, E., Tillmann, W., *Methods for brazing ceramic and metal-ceramic joints*. Material and Manufacturing Process, 1993. 8(2): p. 219–238.

[38] Abbott Furnace Company, Brazing Process Fundamentals; Available from: http://www.abbottfurnace.com/technology/brazing-fundamentals

[39] Schwartz, M.M., Brazing. ASM International, 2003, p.7–13.

[40] Hunt, R.M., *Diffusion Bonding Beryllium to Reduced Activation Ferritic Martensitic Steel: Development of Processes and Techniques*. 2012, University of California, Los Angeles.

[41] Belohlav, A., *Understanding brazing fundamentals*. Welding Journal (USA), 2000. 79: p. 11–14.

[42] Blugan, G., Kuebler, J., Bissig, V., Janczak-Rusch, J., *Brazing of silicon nitride ceramic composite to steel using SiC-particle-reinforced active brazing alloy*. Ceramics International, 2007. 33(6): p. 1033–1039.

[43] Churchill RJ, Varshney U, Groger HP, Glass JM, inventors; American Research Corporation Of Virginia, assignee. Laser brazing for ceramic-to-metal joining. United States patent US 5, 407, 119. 1995 Apr 18.

[44] Bissig, V., Galli, M., Janczak-Rusch, J., *Comparison of three different active filler metals used for brazing ceramic-to-ceramic and ceramic-to-metal*. Advanced Engineering Materials, 2006. 8(3): p. 191–196.

[45] Galli, M., Botsis, J., Janczak-Rusch, J., *Relief of the residual stresses in ceramic-metal joints by a layered braze structure*. Advanced Engineering Materials, 2006. 8(3): p. 197–201.

[46] Nono, M., Barroso, J.J., Castro, P.J., *Mechanical behavior and microstructural analysis of alumina-titanium brazed interfaces*. Materials Science and Engineering: A, 2006. 435: p. 602–605.

[47] M. Singh, T. Ohji, R. Asthana, S. Mathur, Ceramic Integration and Joining Technologies: From Macro to Nanoscale, Wiley.com, 2011, p.91–107.

[48] Cazajus, V., Seguy, S., Lorrain, B., Welemane, H., Prat, C., Karama, M., Thermal stresses in a ceramic-metal composite after brazing process, ICSAM 2005, University "POLITEHNICA" of Bucharest, Romania, 15–17 Sept. 2005.

[49] Cazajus, V., Lorrain, B., Welemane, H., Karama, M., *Residual stresses in ceramic metal assembly after brazing process*. Advances in Science and Technology, 2006. 45: p. 1543–1550.

[50] Cazajus, V., Seguy, S., Welemane, H., Karama, M., *Residual stresses in a ceramic-metal composite*. Applied Mechanics and Materials, 2012. 146: p. 185–196.

[51] Shirzadi, A., Zhu, Y, Bhadeshia, H.K.D.H., *Joining ceramics to metals using metallic foam*. Materials Science and Engineering: A, 2008. 496(1): p. 501–506.

[52] Walker, C., Hodges, V., *Comparing metal-ceramic brazing methods*. Welding Journal, 2008. 87(10): p. 43–50.

[53] Uday, M.B., et al., *An insight into microstructural evolution during plastic deformation in AA6061 alloy after friction welding with alumina-YSZ composite*. Mechanics of Materials, 2015. 91: p. 50–63.

[54] Akbarimousavi, S., Goharikia, M., *Investigations on the mechanical properties and microstructure of dissimilar cp-titanium and AISI 316L austenitic stainless steel continuous friction welds*. Materials & Design, 2011. 32(5): p. 3066–3075.

[55] Mousavi, S., Kelishami, A.R., *Experimental and numerical analysis of the friction welding process for the 4340 steel and mild steel combinations*. Welding Journal, 2008. 87(7): p. 178.

[56] Özdemir, N., SarsIlmaz, F., HasçalIk, A., *Effect of rotational speed on the interface properties of friction-welded AISI 304L to 4340 steel*. Materials & Design, 2007. 28(1): p. 301–307.

[57] Uday, M.B., Ahmad-Fauzi, M.N., Zuhailawati, H., Ismail, A.B., *Advances in friction welding process: a review*. Science and Technology of Welding & Joining, 2010. 15(7): p. 534–558.

[58] Kumar, G.K., Kishore, K., GopalKrishna, P.V., Author, C., *Investigating the capabilities of medium duty lathe for friction welding*. Journal of Emerging Trends in Engineering and Applied Sciences (JETEAS), 2010. 1(1): p. 36–39.

[59] Uday, M.B., Friction welding of 6061 aluminum alloy with YSZ-alumina composite for improved mechanical and thermal properties, in School of Materials and Mineral Resources Engineering, Engineering Campus, USM, 2013, p. 11–24.

[60] Uday M.B., Ahmad-Fauzi, M.N., Ismail, A.B., Zuhailawati, H., *Effect of friction time on the properties of friction welded YSZ-alumina composite and 6061 aluminium alloy*. QScience Connect, 2013. (2013): p. 8.

[61] Khan, I.A., Experimental and Numerical Investigation on the Friction Welding Process, Jawaharlal Nehru Technological University, 2012, p.208–155.

[62] Yeoh, C., Shamsul, B.J., Ahmad, Z.A., *The effect of varying process parameters on the microhardness and microstructure of Cu-steel and Al-Al_2O_3 friction joints*. Journal Teknologi, 2004. 41(A): p. 85–95.

[63] Weiss, R., *Residual stresses and strength of friction welded ceramic/metal joints*. Welding Journal, 1998. 77(3): p. 115.s–122.s.

[64] Weiss, R. Sassani, F., *Strength of friction welded ceramic–metal joints*. Materials Science and Technology, 1998. 14(6): p. 554–560.

[65] Rombaut P, De Waele W, Faes K. Friction welding of steel to ceramic. InSustainable Construction and Design 2011 (SCAD), Ghent University, Laboratory Soete, 2011, 2(3), p. 448–457.

[66] Ikeuchi, K., Yamamoto, N., Tkahashi, M., Aritoshi, M., *Effect of interfacial reaction layer on bond strength of friction-bonded joint of Al alloys to steel*. Transactions of JWRI, 2005. 34(1): p. 1–10.

[67] Davidge, R.W., Mechanical Behaviour of Ceramics, Cambridge University Press, 1979, p. 102–134.

[68] Seli H, Awang M, Ismail AI, Rachman E, Ahmad ZA. Evaluation of properties and FEM model of the friction welded mild steel-Al6061-alumina. Materials Research. 2013 Apr; 16(2), p. 453–67.

[69] Seli, H., Ismail, A.I.Md., Rachman, E., Ahmad, Z.A., *Mechanical evaluation and thermal modelling of friction welding of mild steel and aluminium*. Journal of Materials Processing Technology, 2010. 210(9): p. 1209–1216.

[70] Seli, H., Rachman, E., Ahmad, Z.A., *Temperature modeling for friction welding process between ceramic and metal*. Advances in Science and Technology, 2011. 64: p. 115–124.

[71] Noh, M.Z., Studies On The Performance Of Mild Steel-Alumina Joining Fabricated Via Friction Welding [TA470. Z21 2008 f rb], Universiti Sains Malaysia, 2008, p.10–57.

[72] Noh, M.Z., Hussain, L.B., Ahmad, Z.A., *Alumina–mild steel friction welded at lower rotational speed*. Journal of Materials Processing Technology, 2008. 204(1): p. 279–283.

[73] Uday, M., Ahmad-Fauzi, M., *Joint properties of friction welded 6061 aluminum alloy/YSZ–alumina composite at low rotational speed*. Materials & Design, 2014. 59: p. 76–83.

[74] M. B. Uday, M. N. Ahmad-Fauzi, Z. Hussain, A. B. Ismail, Effect of deformation behavior on the grain size of the 6061 aluminum alloy joint with alumina by friction welding. in Applied Mechanics and Materials, Trans Tech Publ., 2011, p.97–102.

[75] Kearns, W.H., Welding Processes-Resistance and Solid-State Welding and Other Joining Processes. Vol. 3. 1980, American Welding Society, p. 20–145.

[76] Nicholas, M., Mortimer, D., *Ceramic/metal joining for structural applications*. Materials Science and Technology, 1985. 1(9): p. 657–665.

[77] Quillen, B.G., Silicon Nitride Joining Using a Sr-Celsian-Based Glass Interlayer, DTIC Document, 1996, p.1–88.

[78] Kumar, V., Studies on some glass sealants for solid oxide fuel cells, 2006, School of Physics and Materials Science, Thapar Institute Of Engineering And Technology, Deemed University.

[79] Akselsen, O., *Diffusion bonding of ceramics*. Journal of Materials Science, 1992. 27(3): p. 569–579.

[80] Rusnaldy, R., *Diffusion bonding: an advanced of material process*. Rotasi "Jurnal Teknik Mesin", 2001. 3(1): p. 23–27.

[81] Garrett, B.R., Blank, G., Ranadive, A., Broad Applications of Diffusion Bonding, DTIC Document, 1966, p.1–176.

[82] Balasubramanian, M., *Development of processing windows for diffusion bonding of Ti–6Al–4V titanium alloy and 304 stainless steel with silver as intermediate layer*. Transactions of Nonferrous Metals Society of China, 2015. 25(9): p. 2932–2938.

[83] Sanfandila, P.T.Q., *Joining and characterizations of silicon nitride (Si3N4) to titanium (Ti) using a Cu-foil and Cu-Zn braze alloy*. Revista Mexicana de Física, 2009. 55(1): p. 25–29.

[84] Kumar, S.S., et al., *Methodology to evaluate the quality of diffusion bonded joints by ultrasonic method*. Indian Journal of Engineering and Materials Sciences, 2009. 16(5): p. 331.

[85] Jacobson, D.M., Humpston, G., Principles of Brazing, ASM International, 2005, Ch.1, p. 1–47.

[86] Zhang, Y., Feng, D., He, Z.-y., Chen, X.-c., *Progress in joining ceramics to metals*. Journal of Iron and Steel Research, International, 2006. 13(2): p. 1–5.

[87] Larker, R., Wei, L.Y., Loberg, B., Olsson, M., Johansson, S., *AEM investigation of ceramic/lncoloy 909 diffusional reactions after joining by HIP*. Journal of Materials Science, 1994. 29(16): p. 4404–4414.

[88] Locatelli, M.R., Dalgleish, B.J., Nakashima, K., Tomsia, A.P., Glaeser, A.M., *New approaches to joining ceramics for high-temperature applications*. Ceramics International, 1997. 23(4): p. 313–322.

[89] Takaki, K., Fujimaki, Y., Takada, Y., Itagaki, M., Fujiwara, T., Ohshima, S., Oyama, K., Takahashi, I., Kuwashima, T., *A capacitor discharge technique with optimized energy for joining ceramics*. Vacuum, 2002. 65(3–4): p. 457–462.

[90] Pan, H.H., Iijima, K., Liu, Y.J., Itoh, I., *Diffusion bonding of ZrB2 ceramic and Al-Sn-Mg alloys at atmosphere*. Advanced Materials Research, 2011. 189: p. 3249–3252.

[91] Janghorban, K., *Multilayer diffusion bonding of polycrystalline Al_2O_3 to Al alloys containing Sr, Mg, and Si*. Iranian Journal of Science and Technology Transaction B-Engineering, 2009. 33(B3): p. 241–251.

[92] Juan, W., Yajiang, L., Yansheng, Y., *Microstructural characterization in diffusion bonded TiC-Al_2O_3/Cr18-Ni8 joint with Ti interlayer*. Bulletin of Materials Science, 2006. 29(2): p. 155–158.

[93] Martinelli, A.E., Drew, R.A.L., *Microstructural development during diffusion bonding of α-silicon carbide to molybdenum*. Materials Science and Engineering: A, 1995. 191(1–2): p. 239–247.

[94] Firmanto, H., Hussain, P., Mamat, O., *Reaction layers of the diffusion-bonded sialon and high-chromium steel*. Journal of Applied Sciences, 2011. 11(12): p. 1809–1814.

[95] Burger, K., Rühle, M., *Material transport mechanisms during the diffusion bonding of niobium to Al_2O_3*. Ultramicroscopy, 1989. 29(1–4): p. 88–97.

[96] Sun, X., Failure Mechanisms of Advanced Welding Processes, Woodhead Pub, 2010, p. 190–212.

[97] Ion, J., Laser Processing of Engineering Materials: Principles, Procedure and Industrial Application, Butterworth-Heinemann, 2005, p.1–259.

[98] Mikhailova, G., Mikhailov, B., Troitskii, A., *Laser welding of HTSC ceramics*. Laser Physics Letters, 2004. 1(10): p. 525–527.

[99] Welding Miniature Devices, Decisions to make when choosing a joining process, Available from: http://www.thefabricator.com/article/arcwelding/welding-miniature-devicesDecisions to make when choosing a joining process.

[100] Lee, C.S., Zhang, X.F. Thomas, G., *Novel joining of dissimilar ceramics in the Si_3N_4–Al_2O_3 system using polytypoid functional gradients*. Acta Materialia, 2001. 49(18): p. 3775–3780.

[101] Kuramoto NO, Taniguchi HI., Transparent Ceramics, Springer, 2015, p.1–25.

[102] De Paris, A., Robin, M., Fantozzi, G., *Welding of ceramics SiO_2-Al_2O_3 by laser beam*. Le Journal de Physique IV, 1991. 1(C7): p. C7-127–C7-129.

[103] Riviere, C., Robin, M., Fantozzi, G., *Comparison between two techniques in laser welding of ceramics*. Le Journal de Physique IV, 1994. 4(C4): p. 4–4.

[104] Exner, H., Nagel, A.-M., *Laser welding of functional and constructional ceramics for Microelectronics*. in *Optoelectronics' 99-Integrated Optoelectronic Devices*. 1999, International Society for Optics and Photonics.

Chapter 9

Diffusion Bonding: Influence of Process Parameters and Material Microstructure

Thomas Gietzelt, Volker Toth and Andreas Huell

Additional information is available at the end of the chapter

http://dx.doi.org/10.5772/64312

Abstract

Diffusion welding is a solid joining technique allowing for full cross-section welding. There is no heat-affected zone, but the whole part is subjected to a heat treatment. By diffusion of atoms across the bonding planes, a monolithic compound is generated.

The process takes place in a vacuum or inert gas atmosphere at about 80% of the melting temperature and is run batch-wisely. Hence, it is rarely used despite its advantages to achieve holohedral joints and is widespread in the aerospace sector only.

The quality of a diffusion-welded joint is determined by the three main parameters bonding temperature, time, and bearing pressure. The difficulty tailoring the process is that they are interconnected in a strong nonlinear way.

Several additional factors may influence the result or may change the material, e.g. surface roughness and passivation layers, all kinds of lattice defects, polymorphic behaviour, and formation of precipitations at grain boundaries, design of the parts to be welded and its aspect ratio as well as mechanical issues of the welding equipment. Hence, experiments are necessary for almost each special part.

In this chapter, an overview about the experience of diffusion welding is given. Influences are discussed in detail and conclusions are derived.

Keywords: diffusion welding, diffusion bonding, lattice defects, grain growth, precipitation, sensitization, passivation layer

1. Introduction

Diffusion welding is the only welding technique by means of which full cross-sectional welds, also of internal structures, can be obtained. Normally, there is no liquid phase and the monolithic compound is formed completely under solid-state conditions.

For the conditions to be appropriate, mechanical properties across the joined part are comparable to the bulk material. Due to heating of the whole parts, no distinct heat-affected zone (HAZ) is formed. However, properties are changed compared to the as-delivered conditions of the material. This may cause problems in some cases.

For diffusion welding, special and expensive equipment is required: the parts have to be mated at high temperatures by applying high forces depending on the size and cross-section to be welded under a vacuum or inert gas environment. Equipment and parts are heated mostly indirectly by radiation. To limit thermal stress, the heating rates are restricted to some 10 K/min.

The welding process takes place in vacuum and cannot be performed on site. Mating surfaces must be free of any impurities and have a low surface roughness without deep scratches. Joining of multiple layers is possible in one step.

Diffusion welding is always accompanied by a certain deformation of the parts. This deformation depends mainly on bonding temperature, bonding time and bearing pressure. Unfortunately, influences of temperature and bearing pressure are non-linear, making it difficult to predict the deformation of a new design. Additionally, secondary impacts on deformation and the quality of joining may be due to specific geometric parameters, e.g., the aspect ratio, the number of layers, the micro-structure of the material itself and surface layers.

Recently, thin coatings of other metals, forming a temporary liquid phase (TLP) by passing a eutectic composition, or multiple layers of different metals of nanometre thickness exploiting the enormous interfacial energy of such compounds were investigated.

In contrast to conventional welding techniques, such processes are highly complex. The process has to be optimised for each material and even for different compositions of alloys depending on the geometry. For this reason, application of diffusion welding is limited to the aerospace industry or special applications where other welding techniques fail. For example, large- and thin-walled titanium sheets are joined to reinforcing structures and internal cooling channels for injection moulding tools and nozzles of rocket engines.

Unfortunately, not all the information necessary for reproducing the results, e.g., material, procedure of sample preparation and process parameters, is given in the literature.

For joining micro-structured components, additional aspects must be taken into account.

The aim of this chapter is to summarise knowledge on diffusion welding in conjunction with the fundamental processes taking place inside of the micro-structure of a material. For this, lattice defects are discussed according to their dimensionality.

http://dx.doi.org/10.5772/64312

2. Micro-structure of metals and the impact of lattice defects on diffusion welding

2.1. Atoms in the lattice of metals

To minimise the energy of a system, isolated metal atoms tend to arrange in a regular lattice at positions according to the annihilation of attractive and repulsive forces (**Figure 1**). The positions are well-defined and specific of each metal. Hence, they can be used, e.g., for determining the composition of an alloy by means of WDX (wavelength dispersive X-ray). When forming a compound, atoms split up into positively charged atomic nuclei, while valence electrons are transferred to the so-called electron gas and can move freely within the lattice. Consequently, metals are good conductors of heat and electricity.

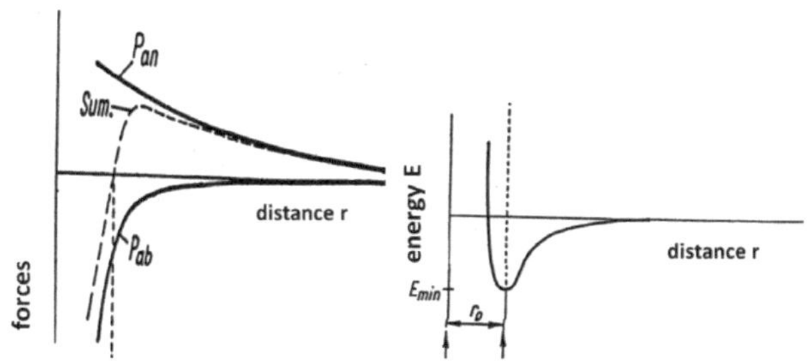

Figure 1. Equilibrium of attractive and repulsive forces in the metallic lattice [1].

2.1.1. Thermal expansion

Depending on the thermal energy of the whole system, the positively charged atomic nuclei oscillate around their position, leading to a thermal expansion (**Figure 2**). According to *Grüneisen's rule*, linear expansion is in the range of 2% and volumetric expansion is 6–7% up to the melting point of a metal [1]. Hence, the melting point can be used to estimate the thermal coefficient of expansion. Below the melting temperature, the oscillation amplitude is about 12% of the lattice constant [2].

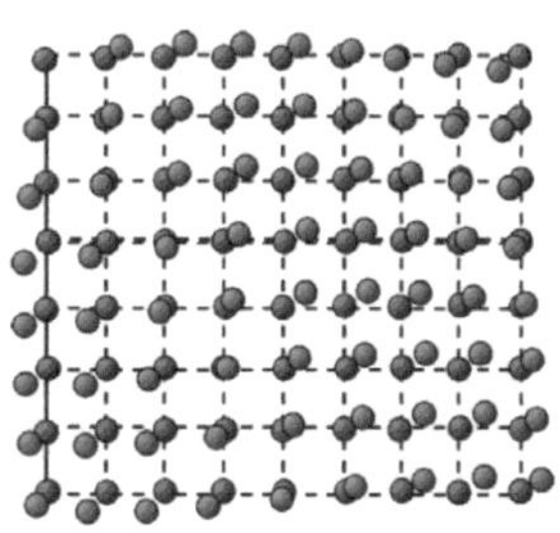

Figure 2. Thermal oscillation of atoms.

2.1.2. Thermal activation, diffusion, polymorphism and zero-dimensional lattice defects

With increasing thermal oscillation, not only the amplitude increases but also the energy of collisions between atoms. Gradually, some atoms are facilitated to leave its lattice sites and a vacancy is left leading to a punctual stress state (**Figure 3**). With increasing temperature, an exponentially increasing number of atoms is displaced from the lattice sites and the density of vacancies is considerably enhanced (Eq. (1)):

$$c_V = \frac{n}{N} * \exp.\left(\frac{-\Delta U}{RT}\right) \tag{1}$$

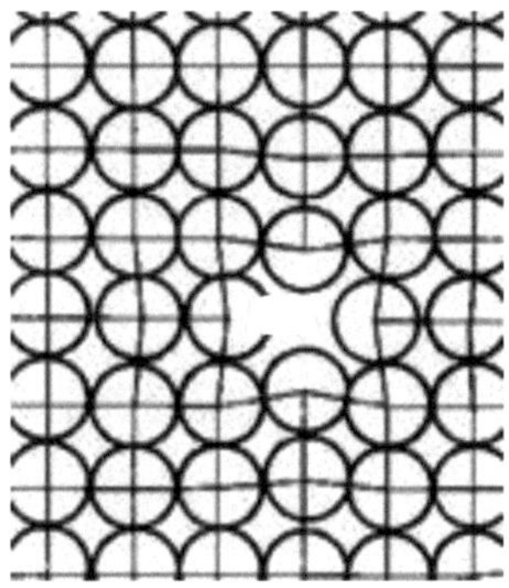

Figure 3. Vacancy in the lattice causing punctual stress.

where cV is the concentration of vacancies (cm^{-3}), n is the number of vacancies, N is the number of sites in the metallic lattice, U is the energy of formation of vacancies (for metals 80–200 J/mol), R is the gas constant (J/mol*K) and T is the temperature (K).

Vacancies are regular lattice sites not occupied by an atom. Due to a missing atom, the surrounding atoms tend to fill the gap and the lattice is distorted at this point, representing a zero-dimensional defect.

According to [3], the density of vacancies is 10^{-12} at room temperature and increases to 10^{-4} below the melting temperature.

Vacancies strongly facilitate the diffusion of atoms between different sites of the lattice and, hence, concentration facilitates the formation of a monolithic compound during diffusion welding. As a consequence, the coefficient of diffusion increases exponentially with temperature (Eq. (2)). An increase in bonding temperature by 20 K may result in a doubling of the diffusion coefficient, thus illustrating the strong non-linear influence of temperature on diffusion welding:

$$D = D_0 * \exp.\left(\frac{-\Delta U}{RT}\right) \tag{2}$$

where D is the diffusion coefficient (m^2/s), D_0 is the frequency factor (material constant) (m^2/s) and U is the energy of formation of vacancies (J/mol).

The number of vacancies versus temperature can be plotted as a logarithmic function, the so-called Arrhenius plot (**Figure 4**).

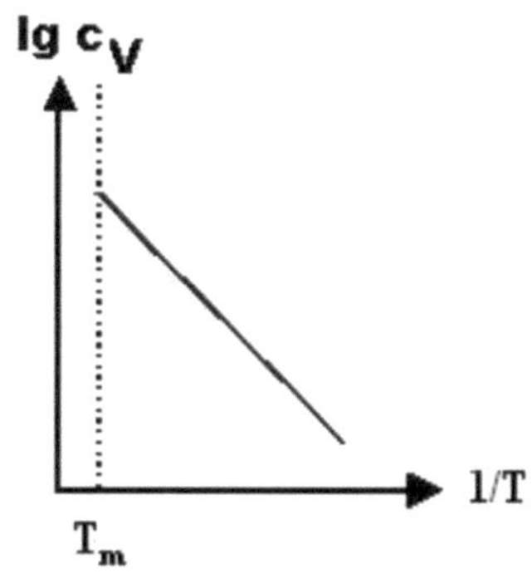

Figure 4. Arrhenius plot. The density of vacancies increases with a logarithmic dependency with temperature.

Depending on the real micro-structure of technical materials, different types of diffusion can be distinguished corresponding to different activation energies for different lattice defects. Straight lines for different diffusion paths can be plotted for surface, grain boundary and volume diffusion, respectively (**Figure 5**). For diffusion welding, grain boundary diffusion predominates at low and medium temperature. As the cross-section of grain boundaries is related to the volume and the density of vacancies increases exponentially, volume diffusion becomes predominant at high temperature.

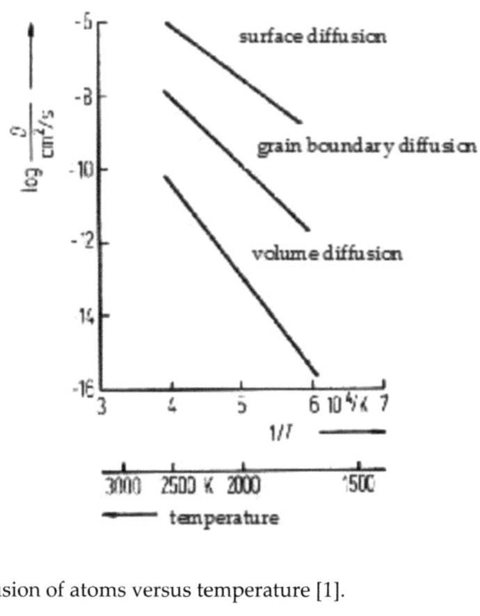

Figure 5. Different modes of diffusion of atoms versus temperature [1].

At the same time, grain growth takes place at high temperatures, which minimises the interfacial energy of the system. If the material shows no polymorphic transformation or the grain boundaries are not pinned by insoluble intra-granular precipitations (e.g., for ODS alloys), diffusion welding will be accompanied by grain growth.

Technical materials are no pure metals, but also contain other sorts of atoms, e.g., alloying elements like manganese, chromium or carbon for steel. Similar to vacancies, these atoms are integrated into the basic lattice as zero-dimensional defects. If they form the same type of lattice (e.g., cubic face-or cubic space-centred), and if the difference in atomic radii is less than 15%, they can occupy regular sites of the host lattice [4]. Small non-metallic atoms with an atomic radius smaller than 59% of the host atoms can be dissolved interstitially like carbon in iron [1].

Although solubility of interstitial atoms is low, they have diffusion coefficients higher by some orders of magnitude in the lattice, since more suitable gaps are available.

In case of low temperature rates during cooling down from diffusion welding temperature, this may be of relevance to the formation of undesired precipitations. The dwell time during diffusion welding should always be kept in the range of solution annealing for an alloy. This may conflict with a low temperature to limit grain growth.

If a metal is polymorphic, abrupt changes in solubility and in the diffusion coefficient may occur. For iron, e.g., these parameters change by two orders of magnitude (**Figure 6**). Reasons are different solubilities for foreign atoms and different sizes of gaps between the atoms in the lattice. For example, maximum solubility of carbon in α-ferrite (cubic space-centred) is 0.02% at 723°C, whereas the solubility of carbon in Y-ferrite (cubic face-centred) is 2.06% at 1143°C, i.e., higher by a factor of about 100 [1].

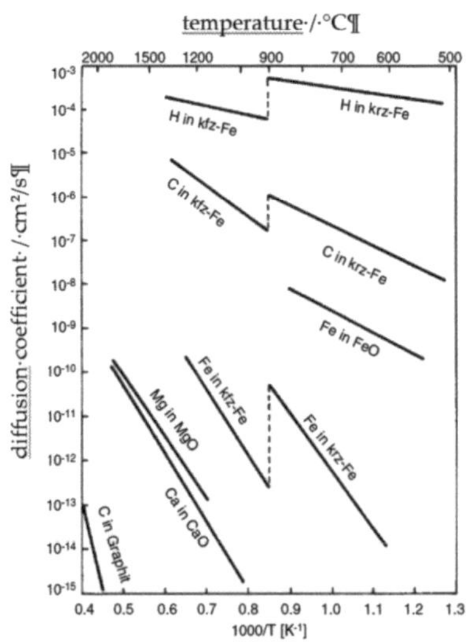

Figure 6. Diffusion coefficients of different sorts of atoms and depending on the type of lattice [5].

Additionally, polymorphism is accompanied by a complete new formation of the microstructure, and grain size is reduced. Although grain growth occurs at high temperature, the same transformation happens when cooling down. This is the reason why normal steels, showing an a⇔Y transformation, can be diffusion welded easily with a finely grained microstructure (**Figure 7**).

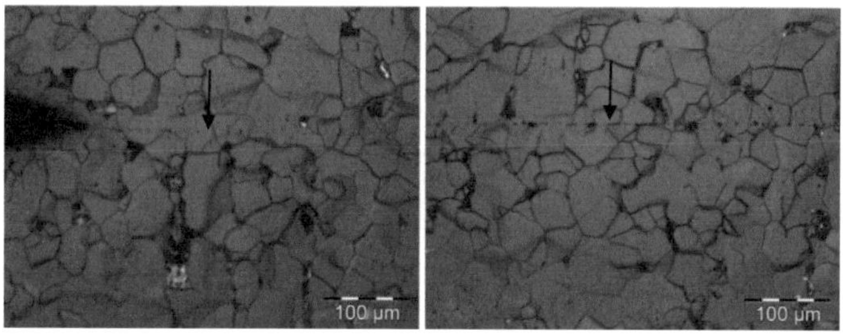

Figure 7. Diffusion weld of St 37 (1.0254), T = 1075°C, t = 1 h, p = 10 MPa, normal $\alpha \Leftrightarrow \gamma$-transformation, deformation: 3.13%.

On the other hand, a diffusion weld of austenitic steel is displayed in **Figure 8**. The impact of the four times longer dwell time on the grain size can be seen clearly.

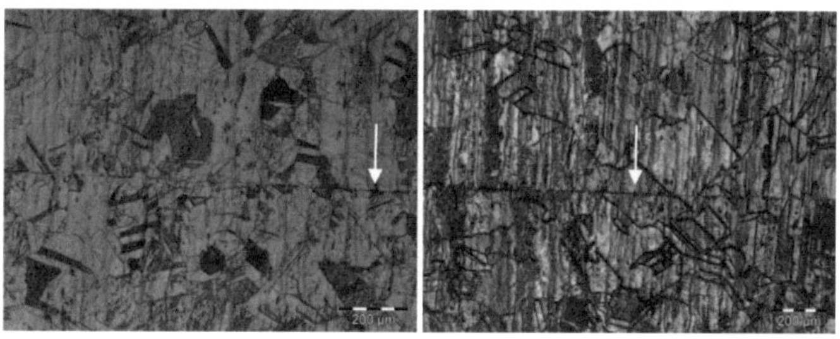

Figure 8. Diffusion weld of austenitic stainless steel AISI 304 (1.4301) at T = 1075°C, p ≈ 15 MPa. Left: t = 1 h, deformation: 2.75%. Right: t = 4 h, deformation: 7.04%.

Although the bearing pressure for 1-h dwell time is 50% higher than for 1.0254, deformation is comparable due to the lower diffusion coefficient in the cubic face-centred lattice. When dwell time is increased by a factor of 4, however, deformation increases by a factor of about 2.5, see [6].

In **Figure 9**, ten 1-mm layers with a diameter of 40 mm of a fully ferritic stabilised stainless steel were diffusion welded. Welding at T = 1075°C, t = 1 h and p = 10 MPa for comparison failed due to excessive deformation. Even at a reduced temperature of T = 1000°C and reduced bearing pressure of p = 6 MPa [7], the deformation was huge at 14.6%. For T = 950°C, t = 1 h, p = 6 MPa, the deformation was still 3.8%.

These high deformations under these mild conditions have to be attributed to the high diffusion coefficient in ferrite, see **Figure 6**. However, despite the high deformation and excessive grain growth due to lacking polymorphism, only very little grain growth across the bonding planes is visible in **Figure 9**, illustrating the role of surface passivation layers, see Section 2.2.

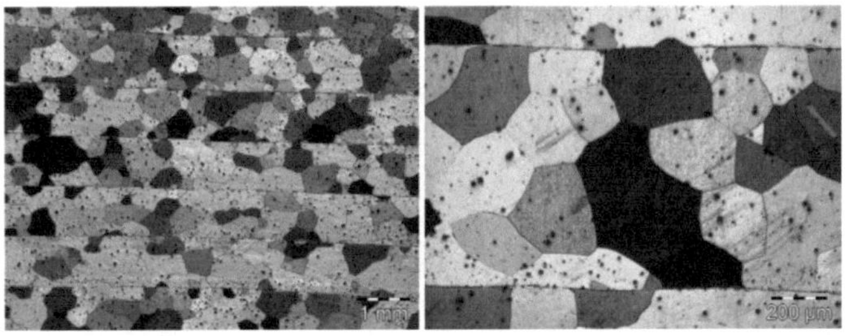

Figure 9. Diffusion weld of a fully ferritic stabilized stainless steel Crofer 22 APU (1.4760) at T = 1000°C, t = 1 h, p = 6 MPa. Deformation: 14.6%.

2.1.3. One-dimensional defects: impact of dislocation density on mechanical properties

Dislocations represent an inserted plane in a metallic lattice (**Figure 10**).

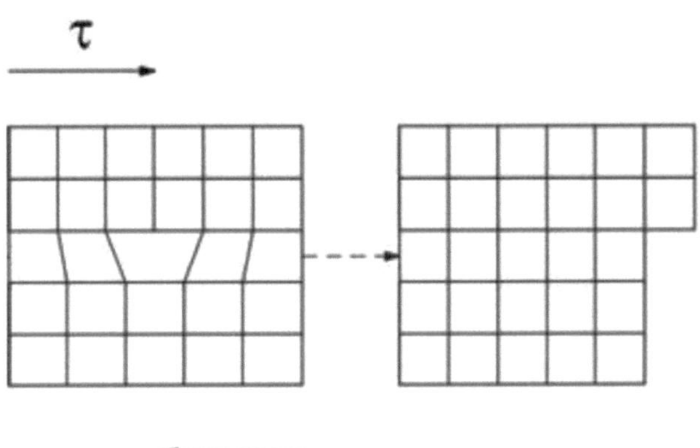

Figure 10. Plastic deformation by the movement of a dislocation across the lattice.

Of course, the inserted plane does not end at a constant level in different layers of the third dimension, but at an arbitrary depth, leading to complex stress conditions. If adequate shear stress appears, the dislocations are moved through the lattice in a step-wise manner, thus causing a plastic deformation. However, the dislocation density is not dropping, despite the dislocations leave the material at the surface. In opposite, it increases exponentially during cold working due to the so-called Frank-Read mechanism. Dislocation density can be given as a length of the dislocation line per unit of volume and can reach as much as 10^{12} cm^{-2} [2]. Only by cold work hardening alone can the mechanical strength of a material be multiplied (**Figure 11**).

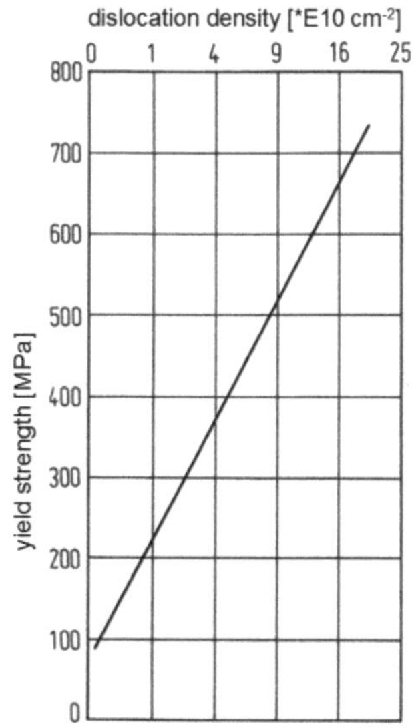

Figure 11. Increase of yield strength of pure iron by cold work hardening [8].

At room temperature, dislocation movement is the predominant deformation mechanism in metals. Dislocations represent a one-dimensional lattice defect.

Dislocations mean an energy excess compared to an undistorted lattice. Hence, at elevated temperatures of about 40% of the melting temperature of pure metals and 50% for alloys, recrystallization takes place [9]. For metals showing no polymorphism, cold work hardening and subsequent recrystallization is the only way to reduce the original grain size. However, it is applicable to half-finished products only.

Hence, when cold-worked material is diffusion welded, recrystallization will be included and affect the grain size.

2.1.4. *Two-dimensional defects: grain and phase boundaries; interfacial layers and their influence on diffusion welding*

Two-dimensional defects of a metallic lattice are reflected, e.g., by grain boundaries. They can be described as an interfacial area per unit of volume and can vary over a wide range, whereas the grain size of technical alloys is in the range of about 5–200 μm. Coating technologies such as galvanic deposition, physical vapor deposition (PVD) or chemical vapor deposition (CVD) processes lead to amorphous or nanocrystalline micro-structures possessing a high internal energy.

Two-dimensional defects affect diffusion welding in several ways: first, the dislocation movement is limited according to the grain size, since grain boundaries are obstacles for movement through the lattice. This means that for a constant strain, deformation by dislocation movement will be smaller for a material with a small grain size. At elevated temperature, however, grain growth occurs and the driving force is larger for a fine-grained material.

A similar effect is observed when using the so-called nanofoils, a thin stack of multiple nanometre layers of different materials possessing a very high interfacial energy in a metastable state [10, 11]. As a result, very high temperatures can be achieved temporarily.

Secondly, deformation at elevated temperatures is governed by grain boundary sliding (Coble creep) or the flow of vacancies through the volume (Nabarro-Herring creep) [12]. This means a coarsely grained material will tend to a larger deformation during diffusion welding because there are less obstacles for dislocation movement and grains tend to slide against each other.

In summary, it can be stated that the degree of deformation and the creep rate for a material during diffusion welding will depend on its grain size and will be very sensitive to the temperature used.

More complex deformation behaviour may result from multi-phase materials: phase boundaries can occur in a wide range of orders of magnitude, either between grains or within, e.g., as thin lamellas in grains of eutectic or eutectoid composition, such as perlite for steel.

Temporarily liquid phases (TLP) can be formed, e.g., by galvanic or PVD deposition of thin layers of two or more different metals, forming a low melting alloy during diffusion welding. Since the inter-layer diffuses into the bulk material, ideally a homogeneous material is left after finishing the process, which is insusceptible to inter-crystalline corrosion.

The opposite happens when different metals form inter-metallic compounds that are brittle and have a high melting temperature. In this case, bonding temperature and time should be limited, such that a thin layer only can be formed between both materials, which do not exhibit any excessive brittleness.

2.1.5. Three-dimensional defects: precipitation

As regards precipitations, it must be distinguished between soluble and insoluble species at diffusion welding temperature. Precipitation may be formed, e.g., due to a low cooling rate after diffusion welding in the range of solution annealing temperature. As a consequence, a two-phase micro-structure with coarse precipitations is formed at the grain boundaries. It is subjected to inter-crystalline corrosion (**Figure 12**). Examples are nickel-based alloys that lose their favourable corrosion resistance.

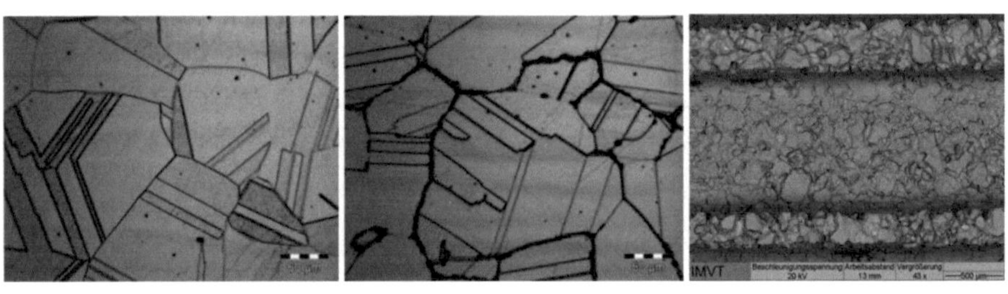

Figure 12. Micro-structure of Hastelloy C-22 (2.4602). Left: after quenching from 1100°C/70 min in water. Middle: after cooling from 1100°C with a rate of 3 K/min (1100°C ≥ 650°C = 2.5 h). Right: corrosion attack after diffusion welding in 95–97% sulphuric acid at 100°C and 1008 h.

Again, the size of precipitations determines the consequences. In case of nanoscaled, insoluble precipitations, e.g., for ODS materials, dislocation movement and grain growth is restricted very effectively [13].

For example, a pure OF-Cu showed good results at T = 850°C (**Figure 13**). The dimension of the material was 28 × 15 mm^2, consisting of six micro-structured layers with a thickness of 3.04 mm and an overall height of 13.04 mm, respectively.

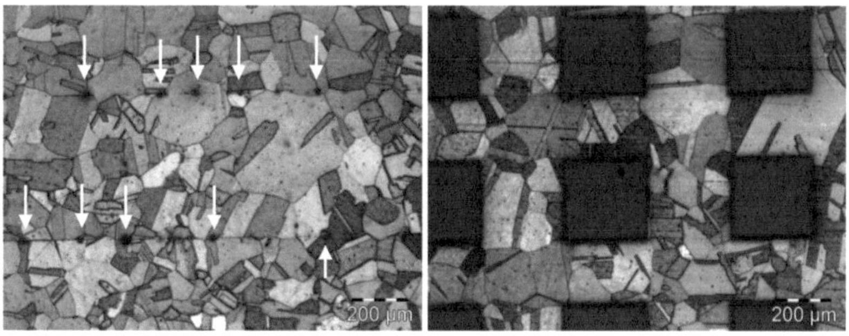

Figure 13. Diffusion welding OF copper. T = 850°C, t = 4 h, p = 2 MPa, micro-structured stack: 18.2%, overall: 4.2%.

Especially in thin-walled micro-structures, perfect grain growth across the bonding planes can be seen. However, in the massive border area, pores remain and grain growth is not as pronounced. The reason probably is a local excess of bearing pressure at the thin walls. However, in massive areas, the bearing pressure of 2 MPa is too low to deform asperities and fill pores sufficiently at this temperature.

Comparative diffusion welding experiments were made using two discs made of ODS copper Discup C3/80 with a diameter of 40 mm and an overall height of 6.88 mm (**Figure 14**). The surfaces were flycut using a polycrystalline diamond tool at a feed rate of 240 mm/min and 3000 rpm, giving a period of 80 µm feed per revolution at a very low roughness in the range of R_t = 1–1.5 µm, R_a = 0.2 µm. The roughness patterns of the discs were not aligned to each other.

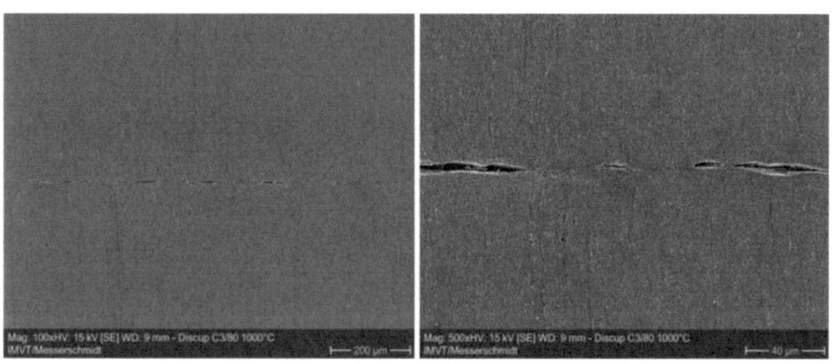

Figure 14. Diffusion welding experiment using Discup C3/80, an oxide-dispersion-strengthened copper alloy. T = 1000°C, t = 4 h, p = 6 MPa, deformation: 0.8%.

Discup alloys consist of pure copper containing a few tenths percent of sub-micron disperoids generated by reactive milling. Afterwards, the material is strongly deformed by extruding. The melting temperature is 1083°C like for pure copper. Despite a much higher temperature and bearing pressure compared to the OF copper sample shown in **Figure 13**, the deformation is as low as 0.8%. SEM images are taken, illustrating very poor joining of the mating surfaces. Grain boundaries are not visible. However, lamellar enrichment of dispersoids can be seen. Similar experiments were done using similar materials in [14].

2.2. Surface effects

For diffusion welding, a very good quality of surfaces is a pre-requisite. Surfaces must be free of single deep scratches preventing vacuum-tight joints and of impurities from machining. Careful cleaning using surfactants and subsequent rinsing with ethanol or acetone are required. Gloves free of powder should be used for handling.

The number of stacked layers will also influence deformation at the given diffusion welding parameters, since multiple surfaces are approached and levelled. Hence, it is not possible to give a certain percentage of deformation to achieve highly vacuum-tight joints. Deformation also depends on the composition of the material.

2.2.1. Influence of roughness

A pre-requisite for solid-state diffusion is a very good contact of the mating surfaces on the atomic level. Often, a "low surface roughness" that is not specified otherwise is required in the literature.

The diffusion welding process can be divided into several phases. In the beginning, surfaces are approached by the deformation of asperities. At local spots, diffusion starts on the atomic level. Between these centres, pores remain which must be closed subsequently by volume diffusion. For this, the density of vacancies and, hence, the temperature and bonding time are essential. Additionally, temperature affects the grain growth and deformation.

Several authors distinguish variable numbers of phases of the bonding process. An overview of the historical development of theoretical models can be found in [15].

Roughness influences the formation of a monolithic bond by the height and shape of asperities and the distance in between, forming temporary pores that must be filled.

Perfectly smooth surfaces made, e.g., by diamond fly cutting may prevent local deformation because asperities are lacking. Shape and height of asperities, in conjunction with bearing pressure, define the local deformation behaviour.

Asperities may also help penetrate surface passivation layers, thus producing local initial metallic contact.

2.2.2. Passivation layers

Some metals and alloys like aluminium, stainless steel, nickel-based alloys or titanium spontaneously form surface passivation layers. They consist mainly of oxides of the base metal, some alloying elements may be enriched. Often, oxygen is blocked to prevent further oxidation and the passivation layers are responsible for the good corrosion resistance in aqueous media or hot gases. Especially for aluminium, formation of passivation layers cannot be avoided completely. The thickness of these passivation layers is in the range of 2–20 nm depending on the type of metal and the content of alloying elements [16, 17]. Of course, composition, thickness and nature of passivation layers differ for normal austenitic stainless steel, heat-resistant steels or nickel-based alloys. Hence, the diffusion welding process must be optimised and the joint must be checked for grain growth across the bonding plane (**Figure 15**). High vacuum tightness is a necessary but not a sufficient criterion.

Figure 15. Diffusion-welded joint of Hastelloy C-22 (2.4602). T = 1100°C, t = 70 min, p = 12 MPa ≥ high vacuum tightness. Subsequently, solution annealed at T = 1125°C, t = 1 h, water-quenched ≥ leaky.

As mentioned above, a certain roughness may help penetrate this layer by local deformation. Hence, the passivation layer comes into contact with matrix material. Passivation layers may be removed by chemical pickling. Even if subsequent formation of a new passivation layer may not be prevented, at least a reproducible surface condition is created.

Long bonding durations and high temperatures above 80% of the starting melting temperature should be preferred in this case.

Another approach is to remove the surface passivation layer, e.g., by sputtering with argon ions. Subsequently, a layer of a different metal may be deposited, which is not that susceptible to oxidation, e.g., gold or silver, and may temporarily form a low-melting alloy helping to create a bond.

For titanium, the passivation layer is soluble in the matrix material and diffusion welding of titanium is widely used, e.g., in the aerospace industry [18, 19]. In **Figure 16**, a very good bond between thin micro-structured layers can be seen. For parts consisting of multiple thin sheets, however, it has to be kept in mind that grades 1–4 differ slightly only in terms of the contents of nitrogen, oxygen and iron, while the mechanical properties are changed dramatically [20]. Consequently, the properties of diffusion-welded parts may be changed for inappropriate ratios of surface layers to bulk material.

Figure 16. Diffusion weld of titanium, grade 2 at $T = 850°C$, $t = 4$ h, $p = 13$ MPa.

2.3. Influence of temperature, bearing pressure, bonding time and design

2.3.1. *Influence of bonding temperature*

For diffusion welding, the joining temperature is normally set in the range of about 80% of the melting temperature for a pure metal or of the starting melting temperature for alloys. Temperature is calculated in Kelvin. Obviously, similar to the appropriate temperature for recrystallization, the temperature for alloys should exceed this level. For materials with surface passivation layers, temperature should be even higher and the time longer, which changes the whole process in terms of creep rate and appropriate bearing pressure. When comparing diffusion welding of, e.g., pure aluminium ($T_s = 660°C$) and AlMg3 ($T_s = 610–640°C$), the whole process has to be optimised. Otherwise, welding will fail due to excessive deformation [21].

The influence of temperature is strongly non-linear. Keeping in mind the dependence of vacancy density on temperature, an increase of about 20 K can double the diffusion coefficient and, hence, drastically increase the creep rate for a given bearing pressure.

2.3.2. *Influence of bearing pressure*

Bearing pressure is responsible for joining the mating surfaces. Influence of the bearing pressure is contrary to that of temperature. When increasing the bonding time from 1 to 4 h, the deformation is not proportional but increased by a factor of about 2.5 [5].

Obviously, a certain minimum bearing pressure is necessary to facilitate the deformation of local contact areas of the sample depending on the temperature applied.

Additionally, deformation under the given conditions strongly depends on the aspect ratio of the part and on the frictional cross-section between the part and the die applying the load. Large format parts of low thickness are difficult to weld with a reproducible deformation.

http://dx.doi.org/10.5772/64312

If the parts contain internal thin-walled micro-structures, the deformation behaviour may be affected by grain boundary sliding. For comparison to the part displayed in **Figure 13**, a similar part with a format of 40 × 30 mm, containing 12 layers of unstructured foils with a thickness of 3.6 mm and an overall height of 13.84 mm was welded under the same conditions ($T = 850°C$, $t = 4$ h, $p = 2$ MPa). Deformation of the 12 layers was 1.3% only, while overall deformation was 0.35%, showing the influence of both micro-structures and aspect ratio.

The effective bonding area of a part should be distributed uniformly across the part. Otherwise irregular deformation or sink marks may occur (**Figure 17**). To prevent this, compensating areas may be helpful.

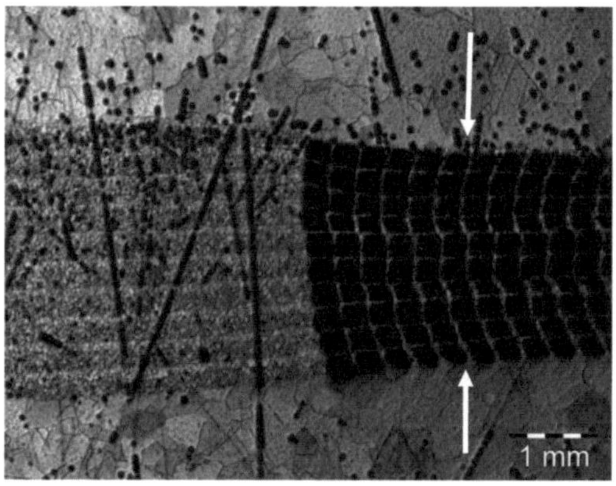

Figure 17. Irregular deformation on a part made of titanium with micro-structured sheets stacked in the same direction at $T = 850°C$, $t = 4$ h, $p = 10$ MPa.

2.3.3. *Influence of bonding time*

Bonding time is required for conducting the diffusion process. After the initial step of approaching mating surfaces, time is needed to fill the pores left in between the local contact areas. Hence, a sufficient long bonding time is required.

Bonding time, together with temperature, affects deformation. However, as mentioned above, its influence is non-linear. As soon as creep takes place during diffusion bonding, a long bonding time makes it difficult to control deformation and design changes may have a major impact.

Therefore, the diffusion welding process should be optimised for each serial application. It is difficult to weld prototypes of varying designs or materials without profound experience.

It is also hard to give a certain percentage of deformation to obtain a good diffusion bond. In fact, bonding quality depends on the number of layers to be bonded.

In any case, the deformation behaviour depends not only on the composition of a material but also on its micro-structure.

3. Special factors to be considered in diffusion welding of micro-devices

Some aspects relating to micro-structures have already been mentioned in the sections above. From this, it can be concluded that bearing pressure should be kept as low as possible, while, on the other hand, it must be sufficient to deform asperities and to increase the contact area during the bonding process.

The temperature should be sufficient for a high density of vacancies and for filling the pores by volume diffusion, which also depends on the bonding time.

Micro-devices mainly consist of micro-structured multiple sheets. Channels may run in the same direction or cross-wisely, and the load-bearing structures may not proceed over the whole thickness for technical reasons (**Figure 18**).

Figure 18. Displaced micro-structure with offsets made of 1.4301 diffusion-welded at locally varying bearing pressure at T = 1075°C, t = 4 h.

Bottom and top are often closed by discs of a few millimetres in thickness, having coarse grains. For thin sheet material, however, the grain size is about one order of magnitude smaller due to cold work hardening and recrystallization. The micro-structured stack and thick plates will deform completely differently and the deformation will be concentrated mostly on the micro-structured section. An intelligent design may help achieve reasonable results.

3.1. Shapes of thin walls in micro-structures

The cross-section of thin walls may be important to the deformation behaviour: if the bearing pressure forces the material to creep, cross-sections of rectangular wall may be bent or deformed to a barrel shape, as can be seen in the left section of **Figure 19**. In the SEM of material with etched micro-channels on the right, however, the part is stabilised, since the bonding cross-section increases when deformation occurs.

http://dx.doi.org/10.5772/64312

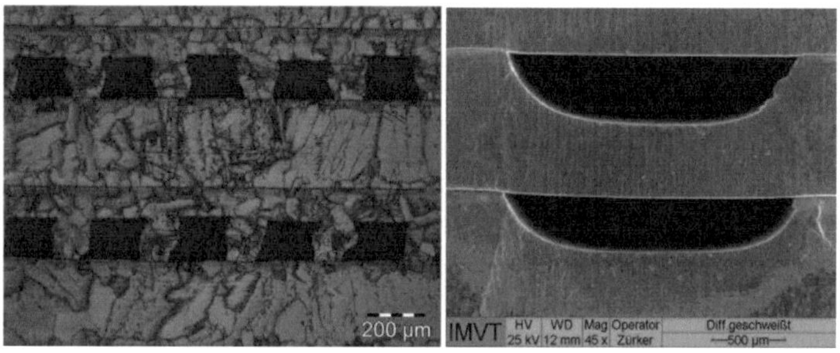

Figure 19. Impact of the cross-section of thin walls. Left: rectangular cross-section of thin walls. Right: self-reinforcing cross-section due to etched micro-channels.

The dimension of the walls should exceed the grain size of the material: in most cases, walls should be at least 100–200 µm in width. The aspect ratio should not exceed 1:1 for stability reasons, e.g., to avoid bending.

3.2. Impact of the design of mechanical micro-structures, the aspect ratio and the number of layers on the deformation

Moreover, the geometry of the micro-structured foils to be proper is important: the ratio between the thickness of the remaining bottom and the width of a trench should not exceed 1:1 to transmit sufficient bearing pressure to the next layer and to prevent lacking fusion (**Figure 20**, left).

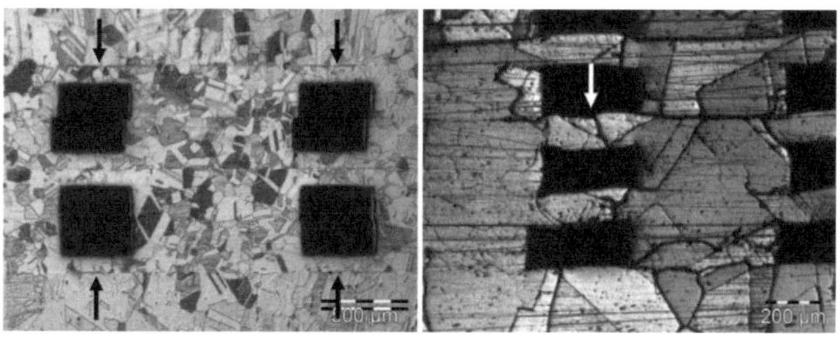

Figure 20. Left: 1.4301, T = 1075°C, t = 4 h, p = 8 MPa, incomplete fusion due to insufficient thickness of the bottom in relation to the width of trenches. Right: 1.4876, T = 1250°C, t = 1 h, p = 8 MPa, distortion due to grain boundary sliding within a thin bottom.

Depending on the application, a grain boundary crossing the remaining thickness of the bottom of a micro-channel should be avoided. In case of corrosion, this would be a favourable path for failure. During diffusion welding it causes local grain boundary sliding and distortion of the mechanical micro-structure (**Figure 20**, right).

Another topic is the aspect ratio of the parts to be welded, e.g., due to the different thermal expansion coefficients of the TZM-stamps (see Section 4), and the parts and deformation during the welding process, friction between both occurs. For a high aspect ratio in the range of one or more, a barrel-shaped profile results, accompanied by a high percentaged deformation. Flat parts, however, possess a low deformation at the same conditions. For example, for disks of 160 mm in diameter, a deformation of 10% was obtained for a height of 10 mm for $T = 1075°C$, $t = 4$ h, $p = 25$ MPa. For a height of 150 mm, however, the deformation was more than 33% [22].

The number of layers affects the deformation obtained at the same conditions since the roughness of more surfaces must be levelled. For example, a conical sample consisting of 51 layers had a deformation more than 30% higher than the same sample geometry consisting of five segments only (**Figure 21**).

Figure 21. Conical samples made of 1.4301, T = 1000°C, t = 4 h, F = 17.55 kN, corresponding to 15–25 MPa. Left: Before diffusion welding. Middle: Five segments, deformation: 5.41 and 5.11%, respectively. Right: Sample made of 51 layers; deformation: 8.34% [6].

4. Equipment for diffusion welding

Diffusion bonding can be carried out using hot isostatic pressing (HIP) at a high isostatic pressure applied by argon of up to 2.500 bar or using a heated press with uniaxial load. For HIP, the parts must be placed inside a steel shield container which is evacuated before sealing. This makes the handling of the parts and the process itself rather expensive.

Additionally, sticking of the parts to the container must be prevented, e.g., by rock wool layers in between or boron nitride spray, or the container has to be machined off afterwards. When using fibrous materials, desorption from a high specific surface area at high temperatures has to be considered. However, also parts with an irregular bonding plane can be welded by HIP, since homogeneous pressure is applied. HIP is widespread and offered by service providers, e.g., ABRA Fluid AG [23].

Diffusion bonding using uniaxial heated presses is performed under a protective inert gas atmosphere or high vacuum. Only a few companies supply equipment for diffusion welding, e.g., PVA TePla AG, FormTech GmbH, TAV VACUUM FURNACES SPA and Centorr Vacuum Industries. Other companies such as MAYTEC Mess- und Regeltechnik GmbH and SYSTEC Vacuum Systems GmbH & Co.KG modify equipment like tensile testing machines or produce equipment for special needs (**Figure 22**). For this, a water-cooled vessel with a vacuum-sealed feedthrough for the dies is installed. The oven is heated indirectly by metallic heaters, and a vacuum in the order of 1E-05 Pa must be maintained for the protection of the heaters. Temperatures of not more than 1400°C are sufficient for the most commonly used materials.

Figure 22. Diffusion bonding furnaces. Left: Maytec diffusion bonding furnace, maximum force 20 kN. Right: Systec diffusion bonding furnace, maximum force: 2 MN.

The stamps are often made of TZM, a molybdenum ODS-alloy, possessing still a high mechanical stiffness at high temperatures [24]. However, the stability also depends on the thickness-to-diameter ratio and must be adapted to the forces transferred to the sample to prevent irregular deformation of the parts to be welded.

Due to the thermal mass of the equipment and to limit thermal stress, the heating rate and especially the cooling rate are low. PVA TePla AG also offers a rapid cooling technology for decreasing the cycle time [25].

During diffusion welding of stainless steel and nickel-based alloys under vacuum, chromium depletion takes place at the surface due to high partial pressure of chromium oxide [26, 27]. Hence, corrosion properties differ from a heat treatment in inert gas or air. For these materials, also enrichment of carbon must be prevented. Hence, unshielded heaters made of graphite are unsuitable.

5. Discussion and outlook

Diffusion welding is the only welding process allowing for full cross-sectional welding, mostly without any liquid phase formation. Since the whole part is subjected to a heat treatment, attention must be paid to undesired material changes. Any cold work hardening effect disappears and the grain size will be larger than before.

With reasonable efforts, high-melting metals, e.g., tungsten or tantalum, cannot be welded.

The equipment is expensive. The process mostly runs batch-wise. Depending on the machinery and the geometry of the parts to be joined, the output is relatively low. Mostly, costs are high.

The process has to be optimised with respect to temperature, bearing pressure and time, taking into account the composition of an alloy and the mechanical history of the semi-finished product. This makes it an interesting field of research for materials scientists. High temperatures and long bonding times are favourable as long as grain growth is not important.

Overall, the result of diffusion welding is difficult to control and depends on many other geometrical factors as well. Therefore, it is used mainly for special applications or in the aerospace industry where cost pressure is lower.

The design of a part must be adapted to diffusion welding, e.g., in terms of a constant distribution of the bonding net cross-section across the part to prevent sink marks. High vacuum tightness is a necessary but not a sufficient criterion for diffusion welding of apparatuses.

To obtain good welding results, a certain deformation always must be accepted. It depends on, e.g., the aspect ratio of the parts and the number of layers to be joined. Obviously, microchannels inside a part will affect the amount of deformation additionally. For multi-layered parts, a higher deformation is required to achieve high vacuum tightness since more surfaces have to be levelled. In consequence, it needs a lot of experience to define appropriate parameters, especially for the bearing pressure, to ensure a sufficient deformation related to the number of layers. Hence, it is not possible to give an exact value of deformation necessary to obtain high vacuum tightness for a material itself.

Since a long bonding time makes it more difficult to control the deformation at a constant bearing pressure, a short increase of bearing pressure for approaching the surfaces may be helpful. Time should be given in between for closure of remaining pores at a reduced constant bearing pressure without a steady strain rate.

As shown for different types of steel, also the material properties and surface passivation layers may have high impact on the behaviour during diffusion welding. Not all materials of the same class can be welded at the same temperature since passivation layers may possess different thermal stability. Often an increased temperature is required to achieve grain growth across the bonding planes, depending on the alloying elements and its content.

Author details

Thomas Gietzelt*, Volker Toth and Andreas Huell

*Address all correspondence to: thomas.gietzelt@kit.edu

Karlsruhe Institute of Technology, Institute for Micro Process Engineering, Karlsruhe, Germany

References

[1] W. Schatt (ed.): "Einführung in die Werkstofftechnik", 6th ed., Leipzig, Verlag für Grundstoffindustrie, 1972, p. 446.

[2] E. Roos, K. Maile: "Werkstoffkunde für Ingenieure", 4th ed., Berlin, Heidelberg, Springer, 2011, p. 55.

[3] H.-J. Bargel, G. Schulze (eds.): "Werkstoffkunde", 11th ed., Berlin, Heidelberg, Springer-Vieweg, 2012, p. 6.

[4] W. W. Seidel, F. Hahn: "Werkstofftechnik", 8th ed., Munich, Carl Hanser Verlag, ISBN 978–3–446–42064–9, 2010, p. 70.

[5] D. R. Askeland: "The science and engineering of materials", Boston, Cengage Learning, 7th ed. ISBN 978-1-305-07676-1, 1989, 153 pages.

[6] T. Gietzelt, V. Toth, A. Hüll, F. Messerschmidt, R. Dittmeyer: "Systematic investigation of the diffusion welding behavior of the austenitic stainless steel 304 (1.4301)", *Advanced Engineering Materials,* 2014, vol. 16, issue 11, pp. 1381–1390.

[7] J. Wilden, S. Jahn, W. Beck: "Some examples of current diffusion bonding applications", Mat.-wiss. u. Werkstofftech., 2008, 39: pp. 349–352. doi: 10.1002/mawe.200800304.

[8] A. S. Key: "Direct observation of imperfections in crystals", J. B. Newkirk, J. H. Wernick (eds.), New York, London, 1962, *Metallurgical Society Conferences,* vol. 14, pp. 213–238.

[9] M. Merkel, K.-H. Thomas: "Taschenbuch der Werkstoffe", 7th ed., Carl-Hanser, Munich, ISBN 978-3-446-41194-4, 2008, p. 92.

[10] X. Sun (ed.): "Failure mechanisms of advanced welding processes", Woodhead Publishing Limited, Abington Hall, 2010, ISBN 978-1-84569-536-1, 307 pages.

[11] NanoFoil, reactive multi-layer foil, 2016, Available from: http://www.indium.com/nanofoil/ [Accessed: 2016–06–23].

[12] H. J. Frost, M. F. Ashby: "Deformation-Mechanism Maps, The Plasticity and Creep of Metals and Ceramics", Pergamon Press, Oxford, ISBN 978–0080293387, 1982, chapter 7, e.g. Fig. 7.2. Nickel-20 at.% chromium of grain size 100 μm, Available from: http://engineering.dartmouth.edu/defmech/ [Accessed: 2016–06–23] [13] CEP DISCUP high-performance copper alloys, 2016, Available from: http://www.cep-freiberg.de/media/pdf/CEP-Product-Information_english.pdf [Accessed: 2016-06-23]

[13] Data sheet of the ODS copper alloy Discup, see under http://www.cep-freiberg.de/media/pdf/CEP-Product-Information_english.pdf, last access: 02.02.2016.

[14] P.K. Samal: "Brazing and Diffusion Bonding of Glidcop Dispersion Strengthened Copper", The Metal, Science of Joining, edited by M.J. Cieslak et al., Minerals, Metals & Materials Society, 1992, Available from: http://www.aps.anl.gov/APS_Engineer-

ing_Support_Division/Mechanical_Operations_and_Maintenance/Miscellaneous/tech_info/Glidcop/glidcop_brazing_samal.pdf, [Accessed: 2016–06–23].

[15] S. Jahn: "Technologieentwicklung zur Herstellung variantenreicher innenstrukturierter Bauteile und Werkzeuge", PhD thesis, TU Ilmenau, 2007, ISBN 3–86130993–9, Verlagshaus Mainz GmbH Aachen.

[16] P. Gümpel (ed.): "Rostfreie Stähle", expert-Verlag, 4th ed., 2008, ISBN 978–3–8169–2669–4, pp. 51–52.

[17] Beizen von Edelstahl, Merkblatt 826, Informationsstelle Edelstahl Rostfrei (ed.), 2016, Available from: http://www.edelstahl-rostfrei.de/downloads/iser/MB_826.pdf [Accessed: 2016–06–23].

[18] H.-S. Lee, J.-H. Yoon, J.-T. Yoo: "Manufacturing of aerospace parts with diffusion bonding technology", *Applied Mechanics and Materials,* 2011, vol. 87, pp. 182–185.

[19] L. D. Hefti: "Advances in fabricating superplastically formed and diffusion bonded components for aerospace structures", *Journal of Materials Engineering and Performance,* 2004, vol. 13, issue 6, pp. 678–682.

[20] M. J. Donachie: "Titanium, a Technical Guide", 2nd ed., ASM International, Materials Park, Ohio, 2000, ISBN 0–87170–686–5, p. 18.

[21] W. Behr et al.: "Diffusion bonding of aluminium materials for the manufacturing of cooling structures to be used in ultra-high vacuum", DVS-Media, vol. 263, 2010, ISBN: 978–3–87155–589–3, pp. 127–133.

[22] T. Gietzelt, V. Toth, A. Hüll, R. Dittmeyer: "Dependency of Deformation during Diffusion Welding on Aspect Ratio using Samples made of SS 304 (1.4301)", 2016, submitted Adv. Eng. Mater.

[23] Heißisostatische Pressen HIP, ABRA Fluid AG, 2016, Available from: http://www.abrafluid.ch/heissisostatische-pressen.html [Accessed: 2016–06–23].

[24] TZM, Plansee Composite Materials GmbH, 2016, Available from: http://www.plansee.com/de/werkstoffe/molybdaen.html [Accessed: 2016–06–23].

[25] Hochvakuum-Diffusionsschweißen - Einstieg in die industrielle Serienanwendung ist in Vorbereitung, PVA TePla AG, 2016, Available from: http://www.pvatepla.com/pvatepla-service/news/b/hochvakuum-diffusionsschweissen---einstieg-in-die-industrielle-serienanwendung-ist-in-vorbereitung [Accessed: 2016–06–23].

[26] K. Hauffe: "Reaktionen in und an festen Stoffen", 2nd ed., Berlin, Heidelberg, New York, Springer, 1966, ISBN 978–3–642–88043–8, p. 463.

[27] Bergamnn, Schaefer, R. Kassing (ed.): "Lehrbuch der Experimentalphysik, Band 6", 2nd ed., Walter de Gruyter, Berlin, New York, 2005, ISBN 3–11–017485–5, p. 939.

Chapter 10

Applying Heat for Joining Textile Materials

Simona Jevšnik, Savvas Vasiliadis,
Senem Kurson Bahadir, Dragana Grujić and
Zoran Stjepanovič

Additional information is available at the end of the chapter

http://dx.doi.org/10.5772/64309/

Abstract

The middle of the last century presents the beginning of a wide use of heat technologies for joining of the textile materials. Up to now, adhesive bonding/fusing of textile materials by application of heat and pressure during the determinate time has become a wide-use technology for manufacturing numerous kinds of textile products, such as outdoor and sport garments, underwear, swimming suits, medical gowns, toys, and automotive seating fabrics. Fusing and welding technologies of textiles represent today a significant competition to traditional sewing, because the technological process is quick and energy efficient. The welding and fusing of textiles represent a great opportunity for providing a good performance as well as aesthetic appearance. New types of fusing/welding machines with high technological solutions regarding the functions, low energy consumptions, and environmental-friendly effects are placed on the market. This chapter presents fusing, hot air and hot wedge welding techniques, suitable for joining of textile materials. The theoretical background and fundamental working principles of the equipment for each technology are presented in the first part of the chapter. Special attention is given to presentation of thermoplastic adhesives, textile substrates for fusible interlinings, and welding tapes. Next, the fusing/welding methods and their parameters are described, and the methods for quality evaluation of fused/welded panels are presented. The effect of fusing/welding parameters, selected methods depending on applied fabrics, fusible interlinings and welding tapes, and used fusing/welding machines are discussed separately. Factors effecting the quality of fused and hot air/wedge welded panels are supported with latest scientific findings. The advantages and disadvantages of the presented techniques are discussed together with the applications areas of each of the presented techniques. The new application opportunities are also highlighted at the end of the chapter.

Keywords: textile materials, fusing, hot air and wedge welding techniques, fusible interlinings, welding tapes

1. Introduction

Textile materials can be joined by sewing, fusing, and welding. Sewing is the oldest and also today most popular technology for joining the textile materials, when minimum two textile components are join together to reach the new two- or three-dimensional shape of a textile product. Irrespective, if sewing is made by hand or sewing machine, the sewing thread and needle is needed to design, stitches, and seams. As the first application of permanently seamless connection of textile components can be considered the fusing of the fusible interfacing on backside of fabric to achieve more stiff part of a garment. Later, the welding technologies have become a significant competition to the sewing technologies. The first attempt of joining the textile materials in terms of seams without the sewing threads was done by hot air/wedge welding. That technology presents a milestone in manufacturing of seamless textile products because there were no yarn breakages during the sewing, no missed stitches, and seam puckering on produced textile products [1–4].

The wider use of welding technologies for textile applications can be traced from new millennium. First welding machines for commercial production of garments and technical textiles appeared in that time [5]. Since then welding has become widely used techniques for textile products, where we require special features and high functional utility, which include water permeability, resistance to pressure, wear and tear of seams, air permeability, and aesthetic appearance. In this group of textile, products are inflatable boats, protective and sports clothing, tents, truck awnings, bag filters, inflatable toys, etc. Hot air and wedge, ultrasonic, laser, and radio frequency technology can create the welded area [6, 7]. In comparison, the fusing technologies were first introduced on large scale in the 1950s and it became an integral part of clothing technology [4]. The highest development level of fusing the interning on textile materials has been detected at the end of the previous century, when a number of new types of fusible interlining and fusing machines were developed.

This chapter gives an overview of the basic principles of hot air/wedge and fusing technology as the most used seamless technologies. The theoretical background and fundamental working principles of both technologies is presented. The thermoplastic adhesives and textile substrates for fusible interlinings and welding tapes is discussed in a separate subchapter. Furthermore, the fusing/welding methods and their parameters in connection with the evaluation of end quality of fused/welded panels are presented with latest scientific findings. The advantages and disadvantages of the presented techniques using the real examples is discussed together with the certain application areas in terms of influences of fusing parameters on fused panel properties, properties of hot air welded seams, and e-textile transmission lines.

2. The principles of joining the textiles using heat

The principle of joining the textiles is based on heat applied on textile components at defined pressure and time. For all of the technological processes to join textiles components by heat, irrespective of used equipment, and application area, are common [1, 7]:

- heating elements; they can be in the form of a narrow bar, a wide rectangular platen or shaped iron, or a roller,
- pressure application method, and
- process control elements for controlling the fusing/welding procedures.

Different applications require different types of equipment; therefore, a variety of equipment can be used, such as [1, 7]:

- flat-bed press,
- stationary press with molded buck/mold/form,
- continuous fusing machines,
- rotary machines with hot air or wedge.

According to the above-mentioned backgrounds, the techniques for assembling the textiles using heat can be grouped into the following technologies: fusing, hot air/wedge welding, and ironing.

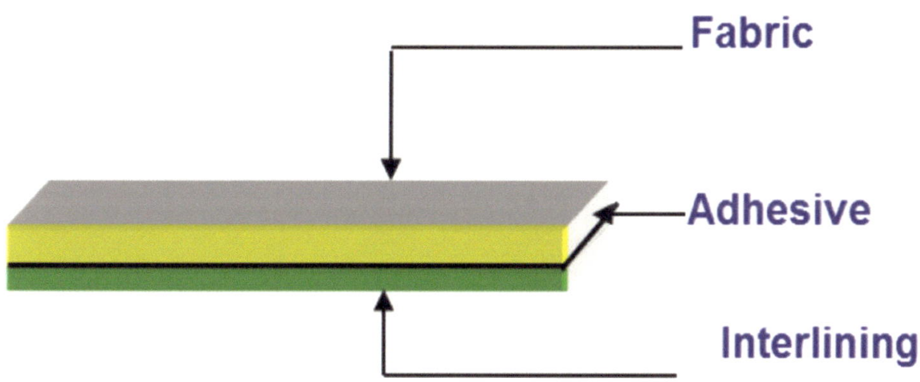

Figure 1. Schematic presentation of a fused panel.

2.1. Fusing technologies

Fusing technologies were firstly introduced bonding technologies for joining the textile materials using heat. The fused panel arises during the fusing process as a joined composite between the fabrics and fusible interlining. During the fusing process, the fabric and fusible interlining are in contact with the heating elements. The applied pressure at the end of the fusing process provides the penetration of the thermoplastic adhesive into the fabric. The right balance among fusing temperature, time, and pressure is very important for the end quality of a fused joint between the fabric and fusible interring. Moreover, it also depends on the construction and raw material of the shell fabric as well as on the type and manner of coating the thermoplastic adhesive. The properties of a fused panel have specific values with respect to the shell fabrics and fusible interlinings [8, 9]. Therefore, it is not possible to add together mechanical properties of shell fabrics and fusible interlining; every fused panel must be analyzed separately [8, 10], **Figure 1**.

2.2. Hot air/wedge welding technologies

The heat technology based on hot air and wedge for joining the textiles is called welding. It is mainly applicable to fully or partially fabrics with thermoplastic components. They can be fibers or the fabric, a thermoplastic coating or a film. Hot air welding is used to thermally bond foils and textiles, where the delivery of the heat is not in the contact with the textile material. On the other side, in hot wedge welding, a small metal wedge is in the contact with fabric during the welding process [7, 11]. Hot air welding is applied to the hot melt adhesive, which is activated at precisely controlled temperature. Under applied pressure after the cooling phase, a strong bond is formed between the tape and the seam. **Figure 2** presents the characteristic schematic presentation of hot air/wedge welded seams. Mainly those seams have water- and wind-proof properties and if they are welded on the face side of the fabric, they can contribute to the design effect of a manufactured textile product.

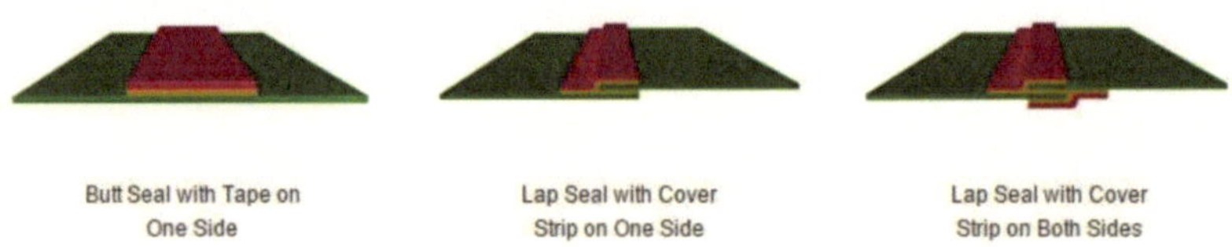

Figure 2. Schematic presentation of a hot air/wedge welded seams [12].

2.3. Ironing as welding technology

By ironing the temperature is controlled by a thermostat while pressure and time is mainly depending on operator. Wide range of ironing equipment can be used for joining the textiles by ironing. The maximum joining area should have the size of the flat of iron to get uniform joint between the textile layers [7].

3. The processing parameters for joining the textiles using heat

The joint between a minimum of two fabric layers can be made by applying the temperature and pressure at the prescribed time. Irrespective of the joining method and type of used machinery, the heat joining is controlled by four processing parameters: temperature, time, pressure, and cooling.

3.1. Temperature

The temperature presents a heat required to melt the adhesive coated on the fabric surface or synthetic fibers of used fabric. During the fusing process, the temperature of a heating plate and intermediate temperature between the fabrics to be joined together are both important. The intermediate temperature is the actual temperature, at which the adhesive begins to melt and penetrates into the structure of heat-joined fabrics. The temperature of the heating plates is for 20–30°C higher than the intermediate temperature due to heat losses, which occur at the transition temperature of the heating plates of the material [8].

The temperature depends on the time of heating, the thickness of the fabrics, thermal conductivity, and the types of thermoplastics and used method for joining the textiles using heat.

The thermal sensitive paper strips or electromechanical calorimeters can be used for measuring the actual applied temperature in the joining process or for indicating the temperature during the calibration of machine settings, **Figure 3** [13].

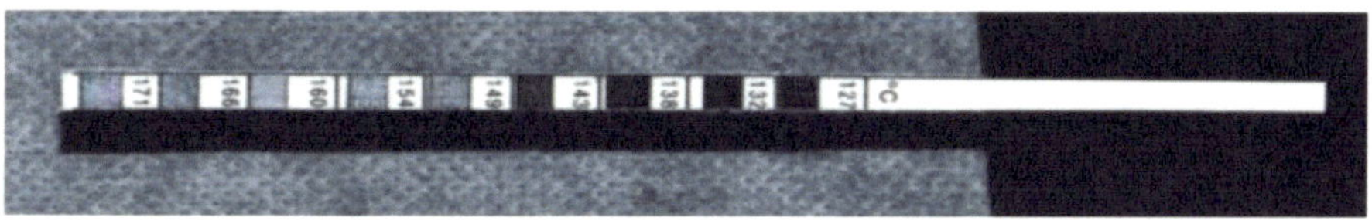

Figure 3. Thermal sensitive paper strip for temperature control [13].

On the other side, the temperature of the air during hot air welding should be set according to used materials in order to melt properly the material at the joint area. The temperature has an important effect on the viscosity of the melt materials, therefore it should be carefully controlled. The temperature is controlled using a sensor in the wedge or in the air stream. The temperature set needs to be significantly higher (100–250°C) than the material melting point. Hot air systems typically allow operation at up to 600–750°C, hot wedge at up to 500°C. It is recommended to carry out tests at a range of temperatures in order to achieve the required joint performance. The ambient temperature of the working place or effects of the sun also have the effect on the optimal welding parameters [7].

3.2. Pressure

During heat joining of textiles the following types of the pressure can be applied: planar, linear, and combination of both. The pressure accelerates the transition of the thermoplastic adhesive into the textile structure in the joining procedure and provides the uniformity of joint parts. The pressure is dependent on the type of the used fabrics and type of the adhesive. The pressure must be constant during the process of fixing, as it enables uniform transition of thermoplastic adhesive in the process [7, 11].

3.3. Time

It is important that between the fabrics a suitable intermediate temperature is reached. This can be obtained by the manipulation of the time of heating at the constant temperature settings. The time depends on the type of adhesive, type and thickness of the fabric, its thermal conductivity, and the mutual action of pressure and temperature [7, 11].

3.4. Cooling

Cooling is a very important process at the end of the technological process. In praxis, two types of cooling can be applied. Cooling presents the time needed for stabilizing the inner micro-structure of the heated textiles and adhesive resins. The most time consuming is living the assembles after heating on flat, horizontal area at least 24 h in the environmental temperature. The enforced cooling can be carried out by water-cooled plates, forced-air circulation, and vacuum. The assembly should be cooled in a horizontal plane [7, 11].

4. Machines for joining the textiles using heat

4.1. Fusing machines

Fused panel can be manufactured either on a flat-bed presses in static situation or conveyed continuously on a continuous fusing machine through a press in which they are heated, fused, and cooled without stopping.

Flat-bed press and continuous fusing machine are the basic types of fusing machines. Flat-bed press is a static method whereby the assembly is fed or positioned onto the bottom plate and the head, or top plate, is closed onto the assembly. At the end of a fusing cycle, the assemblies are cooled and then removed. In comparison, the continuous fusing machine has an endless-conveyor system for transporting the assemblies successively through the heating, pressure, and cooling stages, **Figure 4**.

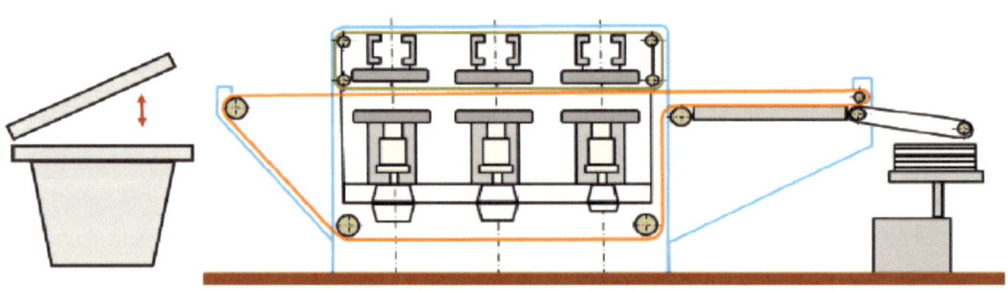

a b

Figure 4. Flat-bed press (a) and continuous fusing machine (b).

Modern high-tech interlinings often have a small temperature range for optimal bonding. Consequently, exact temperature control is very important, therefore the company VEIT has developed a new, innovative heating control system, **Figure 5**. It is intergraded in FX Diamond continuous fusing machine [14]. The new control element measures the temperature directly at the belt and therefore reacts extremely fast to any changes. The preset temperature can be maintained at a constant level and be precisely controlled.

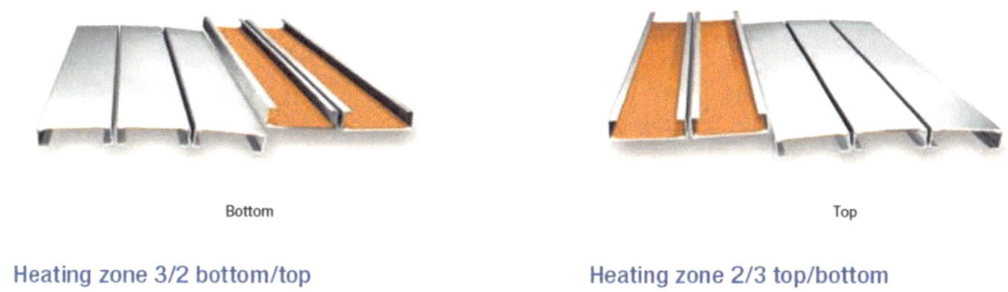

Figure 5. Heating zone integrated in FX Diamond continuous fusing machine [14].

For materials, sensitive to pressure, a combination of FLEXO and standard pressure systems were integrated into the continuous fusing machine, **Figure 6**. The air-filled FLEXO rollers offer flexibility as they adjust to the fabric thickness. Both FLEXO and standard pressure rollers operate independently from each other and allow continuous loading [14].

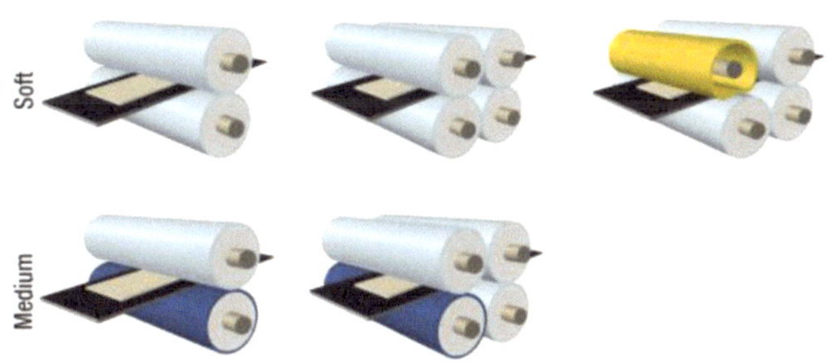

Figure 6. FLEXO Double Pressure System DXT 1400L/1600 CFC [14].

4.2. Hot air/welding machines

In order to create a welded area by hot air a heat-transfer mechanism is applied between two thermoplastic surfaces by shaped or perforated nozzle to deliver hot air to heat between fabric surfaces before they are pressed between driven rollers to apply pressure and complete the welding. Hot wedge welding uses an electrically heated wedge that contains one or more cartridge heaters. The drive rollers pull the fabric through the machine and press the heated surfaces together, **Figure 7** [1].

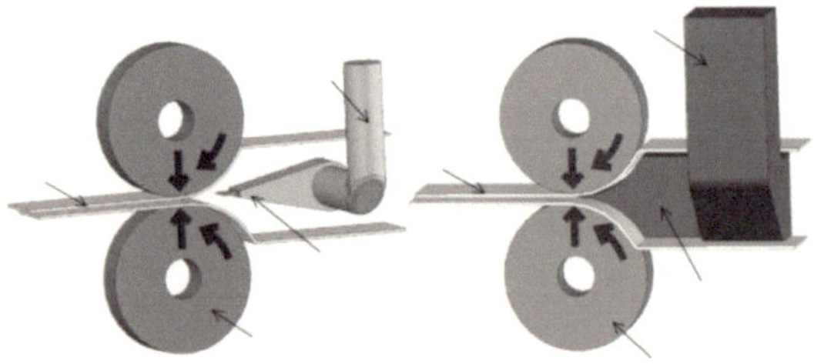

(a) (b)

Figure 7. Hot air (a) and wedge (b) welding equipment [1].

The Pfaff Industrial is a leader in manufacturing of machine for programmed seam sealing. The last model 8303i offers an increase in productivity of up to 20% compared with other taping machines, **Figure 8a** [15]. Manufacturer of welding machines and welding tapes company Framis [17] offers various welding machines for seamless joining and welding. Their trademark NoSo®, **Figure 8b**, heat bonding machines ensures depended on machine model raw cut edge for elasticated hem, raw cut edge with two sided adhesion tape for overlapped seams, folded edges in a single operation, seam sealing taping, decorative taping on open or finished garments, join of two fabrics where an adhesive tape has to have been applied between them.

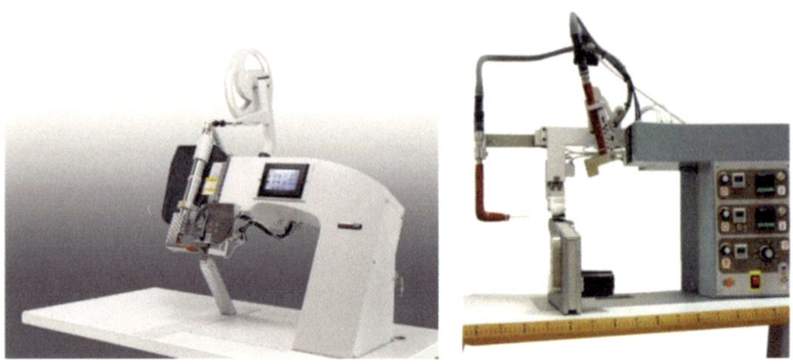

a-PFAFF 8303i [15] b- Framis MX 212 [16]

Figure 8. Hot air welding machines. (a) PFAFF 8303i [15] and (b) Framis MX 212 [16].

5. Materials, suitable for joining the textiles using heat

5.1. Fusible interlining

The term "fusible interlining" is generally defined as a base fabric having a deposit of thermoplastic adhesive resin, which can be bonded to another fabric by the application of heat and pressure. The purpose of fusible interlinings is to achieve better shape retention in the garment, particularly after laundering, increase strength and stability of the outer fabric, create a firm structure and improve the shape of the garment, add warmth, increase bulkiness, improve handling during sewing, improve the feel or crease resistance, as well as to give a better folded edge in the finished garment.

According to the structure, the support material of fusible interlining can be woven, knitted nonwoven. The thermoplastic adhesive is coating on the base fabric by the different kind of coating methods, **Figure 9** [4, 17]. The end form can be scatter powder, powder print, paste print dot, and 3P dot-process [13].

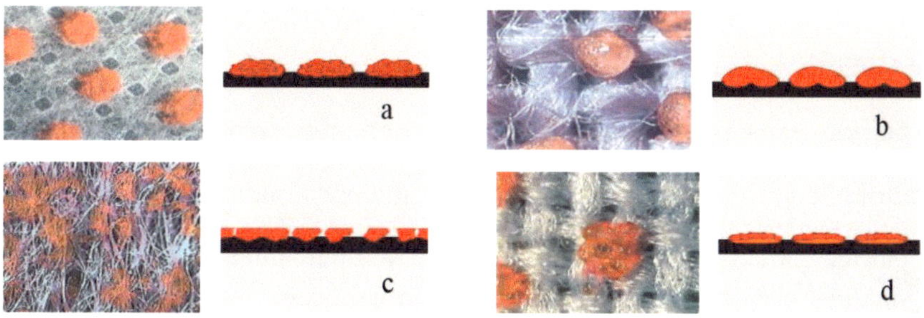

Figure 9. The types of coated adhesive on base cloth or substrate [13].

For the paste print process, the adhesive polymer powder is mixed into an aqueous dispersion. This is applied with a rotation screen printing. The paste is forced through the preformation onto the interlining fabric. Afterward, the adhesive is dried, **Figure 9a** [13]. The 3P dot is a combination of the pate printing and powdering process, **Figure 9b** [13]. First, the paste dot is deposited onto the fabric, then the powder is scattered onto it. Afterward, the adhesive is dried and sintered. The underlying dot acts as a barrier in order to prevent the adhesive from penetrating into the interlining and direct it toward the outer fabric. In this way, the bond is optimized, difficult to fuse outer fabrics. The powder point process is a gravure printing process of adhesive onto the preheated interlining, **Figure 9c** [13]. In scatter powder process, the thermoplastic powder is drawn from a powder by a spiked roller. The adhesive is then heated with infra-red heaters and bonded to the base material and flattened by pressing rollers, **Figure 9d** [13].

The fusible interlining has not only great influence on the hand value of a fused panel, but also on the aesthetic appearance, functionality, shape or model stability, as well as on the end use of the garment.

5.2. Welding tapes

Welding tape for textile purposes presents a textile material coated with an adhesive on one or two sides of a textile substrate. The adhesives can be supplied in a variety of forms, such as film, web, tape, multilayer, fabric backed and release film backed, and the width can be adjusted according to the application issues. They may also be preapplied to the fabric, i.e., substrate in the form of dots or nonwoven layer, or built into the structure of the fabric as a low melting point fiber among higher melting point material. The following adhesives are used for manufacturing of the welding tape (two/three-layered) according to used raw material of the substrate [1]:

- polyester fabrics—polyurethane, polyamide, polyolefin, and polyester-based adhesive films,
- polyamide (nylon) fabrics—some polyurethane films, polyamide, and polyester films,
- acrylic fabrics—polyamide films,
- acetate fabrics—some polyamide, polyester, and polyolefin films,
- aramid fabrics – polyurethane films,
- cotton fabrics—some polyurethane, polyamide, polyester, polyolefin, and vinyl films,
- wool fabrics—some polyurethane, copolyester, and polyamide films,
- elastic fabrics (containing lycra) —polyurethane and polyamide films have been designed that retain their elastic properties after the bonding process,
- polyurethane foams—polyurethane and copolyester films,
- PVC foams—polyurethane, polyester, and vinyl films,
- polyethylene foams—polyolefin films.

The textile substrates can be woven, knitted, nonwoven fabric, or synthetic films, i.e., 100% polyamide or 100% polyurethane. According to the number of layers, the welding tapes are divided as follows:

- one-layered (synthetic film),
- two-layered (i.e., 1. layer-synthetic film and 2. layer-synthetic film), and
- three-layered (i.e., 1. layer-synthetic film, 2. layer-synthetic film, and 3. layer-textile substrate).

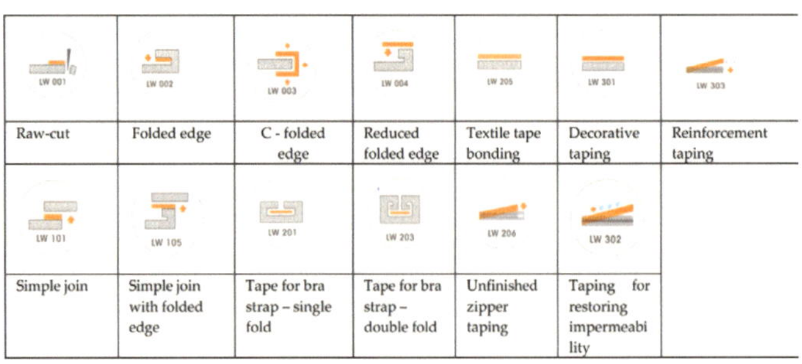

Figure 10. Welded seams manufactured by a welding tape coated on one side [16].

Mainly, the functions of the welding tapes are to cover and reinforce seams, as well as they can have also some decorative functions. Those welding tapes are extremely versatile and are used in a wide range of applications. Some of the welded seams are shown in **Figure 10**.

5.3. Selection of materials for joining the textile materials using heat

The selection of suitable materials for joining the textile materials using heat is mainly based on experiences of experts who use different nonuniform methods. The process of selection of fusible interlinings was thoroughly studied by Japanese authors [9], **Figure 11**.

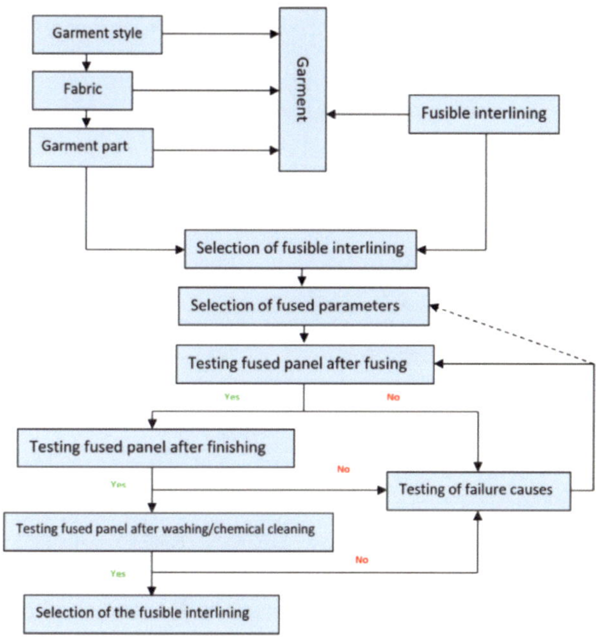

Figure 11. Systematic approach for selection of the fusible interlining.

Before any fusing process, it is necessary to perform the testing of a fused panel in order to determine if the resulting fused panel would meet the desired properties both during the manufacturing process and later during the application. The mentioned selection methodology can be also transformed for selection of the welding tapes, because the same characteristics of joins should be taken into account.

5.3.1. Selection of the fusible interlinings

To achieve the desired silhouette during the construction and planning of a garment, the kind and quality of both the shell fabric and the fusible interlining must be harmonized. The shell fabrics are chosen according to the fashion style of the garment, but the fusible interlinings are selected not only for the fashion style of the garment but also taking into account the type of the shell fabrics.

The quality of the fused panel can be determined on the basis of the bond strength, the visual appearance of the fused area, as well as the results of measurement of the mechanical and physical properties, such as tensile, bending and shear properties, formability and its dimensional stability under the process of fusing, and later after washing or chemical laundering [8]. Bond strength of the fused panel is determined by the force, required to separate the interlining and shell fabrics and is determinate by standard DIN 54310 [18]. The bond strength is the major parameter for quality estimation of the fused panel. The comparison of a bond strength of fused panel after fusing and chemical cleaning for six different fusible interlinings fused with the woolen fabric, suitable for upper garments, is shown in **Table 1**. Fusible interlinings differed regarding the raw material, weave, and type of the adhesive. Fabrics have had different surface fabric weight, weft and warp density, color, and weave.

Code of a fused panel	Bond strength/N/5 cm			
	After fusing process		After chemical cleaning	
	Minimum value	Maximum value	Minimum value	Maximum value
F-M-1	7.10	13.81	5.58	11.80
F-M-2	6.96	13.09	6.36	12.84
F-M-3	7.61	17.33	8.43	17.14
F-M-4	5.92	15.20	5.17	13.53
F-M-5	6.78	12.31	4.56	12.24
F-M-6	6.47	14.03	3.07	12.92

Table 1. Comparison of bond strength of a fused panel after fusing and chemical cleaning.

It can be seen that the fusible interlining's structure influenced the bond strength after the fusing process and chemical cleaning. The bond strength is decreased after chemical cleaning irrespective of used fusible interlining. There may be several causes. During the chemical cleaning the garment is affected by different influences simultaneously or following one after another. The chemical cleaning process is combined with the following steps: cleaning clothes in the cleaning solution, followed by drying and ironing. The chemical solvent can affect the engorgement of fibers, thus the adhesion of fibers in the fabric is reduced to the thermoplastic adhesive of a fusible interlining. This results in decreasing of the bond strength [6]. The quality of the fused panel after dry cleaning is also reduced with the formation of blisters, longitudinal lines, and with the punching of the thermoplastic adhesive on the face side of a fabric [6]. The cause for occurrence of blistering can also be in the properties of the thermoplastic adhesive; they may not be resistant to the cleaning solution, or the fused parameters could be improperly selected, i.e., they could be too high or too low. The chemical cleaning machine operation can also reduce the bond strength of a fused panel because of the mechanical forces generated during the movement of the machine and thereby appeared friction among the treated garment.

http://dx.doi.org/10.5772/64309/

The visual appearance of the garment is evaluated on the basis of visual appearance of the fused panel. The main influencing factors are breakthrough of the thermoplastic material through the surface of a shell fabric or interlining, changes in the surface structure of the shell fabric, and the moiré effect phenomena [6].

The presented facts have confirmed the importance of a suitable selection of all parameters that affect the fusing process. In general, the selection of an appropriate interlining can be defined on the basis of adjusting to:

- basic characteristics, which determine the properties of certain interlinings in the process of stabilization and the end use, and
- the mechanical properties of the fusible interlining with respect to the used shell fabric.

5.3.2. Example of selection of the fusible interlinings using machine learning

A system for automatic knowledge acquisition from a given set of examples presents an alternative way to build a knowledge base to determine the quality of a fused panel. The design of a knowledge base of an expert system is very challenging and responsible work because it must include the entire expert knowledge from the specific area [6, 19–22].

The studies have shown that machine learning from examples allows quick and systematic selection of fusible interlinings with regard to desired final mechanical properties of a fused panel [6]. For the operation of such a system, it is necessary to collect and systemize the data representing learning examples of which the rules are generated in the form of decision trees. The example of one of the constructed rules for predicting the formability of a fused panel based on 300 examples is shown in a part of the regression tree, **Figure 12**.

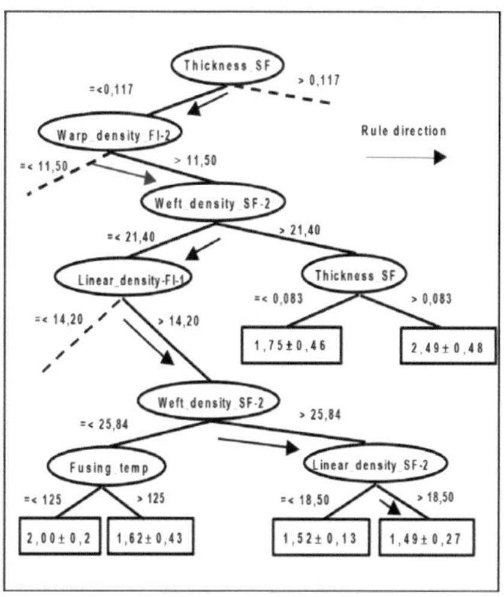

Figure 12. One constructed rule in a part of a regression tree for predicting the formability of a fused panel in a weft direction.

In the same way, it is possible to predict also mechanical and physical properties as well as bond strength of fused panel. The comparison between predicted and measured values of bond strength can be elaborated in the form of a linear coefficient of correlation. In the study [22], a very high correlation, 0.87, between the measured and predicted values has been found. Linear correlation coefficient is shown in **Figure 13**. Unfortunately, in the market there are no commercial software programs for selection of the fusible interlinings.

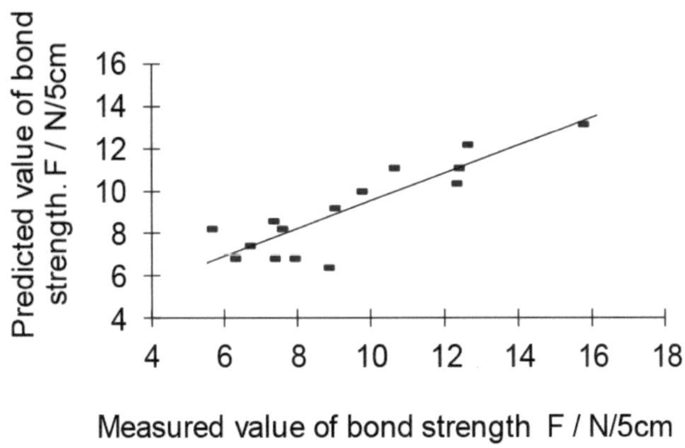

Figure 13. Correlation between the measured and predicted values of bond strength.

5.3.3. Example of selection of welding tapes

The selection of a welding tape is based on the polymer or natural fiber type, type of textile fabric, tightness of the weave or knit, the weight of the substrate, the mechanical property requirements of the joint, and the environmental conditions of use. The welding parameters, such as the temperature of the hot air, velocity, pressure, and pressure of hot air should be balanced according to selected fabric, welding tape, and machine applied. The suitable selection of a welding tape is important in defining the quality, performance, feel, stretch, and longevity of the joint made. The selection procedure can be the same as for the selection of the fusible interlining [23]. The bond strength is valid for the prior parameter when welding tapes are selected. The influence of welding parameters on bond strength of welded samples is presented in **Table 2**.

Welding tape	Temperature (°C)	Speed (m/min)	Pressure (MPa)	Bond strength (N/5 cm)
WT	280	1.5	0.65	17.55
	280	2.5	0.65	14.13
	300	1.5	0.65	16.4
	300	2.5	0.65	12.75

Table 2. Bonding strength of welded samples.

First of all, the joint between the welding tape and fabric should be adjusted from the bonding strength point of view when hot air welding parameters are planned. The results have shown that samples welded at lower temperatures and high speeds have lower bond strengths. Hence, the bonding strengths were higher when the samples were exposed to slower welding processes irrespective of applied welding temperature. The bonding strengths of the welded transmission lines, which have the value of the bond strength within the range of 12.75–17.55 N/5 cm, represent the desired bond strength for applicable purposes, such as waterproof protection or for appropriate joining of different textile components together [23].

6. Application of materials joined using heat

Hot air welding is mainly used in the healthcare and personal protective product sectors for making seams in nonwoven and coated nonwoven fabrics, also in welding of neoprene wet or dry suits. Hot wedge welding is used for joining heavy fabrics and films in outdoor applications, such as lining of swimming pools, reservoirs, and landfill sites [7]. On the other hand, fusing is mainly used for improving the shape and visual appearance of produced garments [24].

6.1. Drape properties of fused panels

The final form of a produced garment depends on the quality of the build-in material and its construction requirements. The fused panel properties can be estimated subjectively or objectively after the garment is finished. Mainly, they are evaluated on the basis of mechanical properties, bond strength, and drapability. The drapability of the fabric is one of the most significant properties, which characterize the shape of a produced garment and its adaptation to the human body.

The drape parameters depend on construction parameters of a fabric, row materials, and steps of finishing processes of fabric manufacturing, as well as on fusing technology used for production of a garment. The interlining can change the fabric properties, such as stiffness and extension properties, hand, visual appearance of incorporated pattern of a garment regarding to the desired design requirements.

It is a ratio of a projected pleating fold area formed by a piece of fabric after draping under its own weight to the original area of this piece of fabric without draping. The higher the fabric drape coefficient, the lower the fabric drapability [25]. It is the percentage of the ring, between radius R_1 of the fabric and radius R_2 of the disc holding the fabric, which is covered by the projected shadow (**Figure 14**), and it can be determined by Cusick [25]:

$$CD = \frac{S_p - \pi R_1^2}{\pi R_2^2 - \pi R_1^2} \quad (1)$$

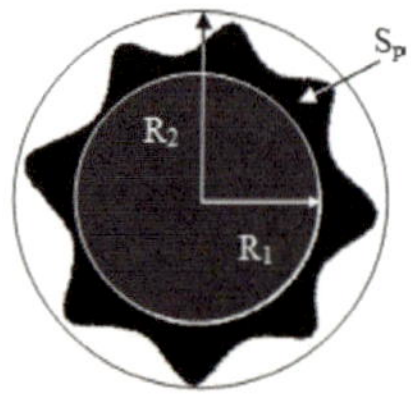

Figure 14. Determination of a drape coefficient.

where:

CD—drape coefficient,

S_P—projection area of draped specimen, mm^2,

R_1 - radius of horizontal disk, mm,

R_2—radius of nondeformed specimen, mm.

The interpretation of a drape coefficient value is connected with the number, form, amplitude and distribution of folds, and their positions according to weft and warp direction. The high value of drape coefficient means that the fabric is stiff and therefore it could be difficult to reform. Alternatively, low value of drape coefficient means easier reform and at the same time, better adaptation of fabric to the shape of cloth. The shape and number of folds depend on fullness and fabric stiffness. A fabric with higher stiffness has larger and wider folds and less stiff fabrics have narrower folds. **Table 3** shows the drapability of some investigated fused panels [26].

Sample code	F1	F1–FI1	F1–FI2	F2	F2–FI1	F2–FI2
Drape coefficient (%)	0.33	0.65	0.70	0.40	0.58	0.71
Number of folds	8	7	6	8	7	6
Minimum amplitude (mm)	92.9	106.0	120.2	98.2	117.8	122.5
Maximum amplitude (mm)	138.2	142.3	143.7	137.4	143.7	145.9
Figures of draped samples		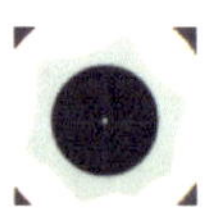	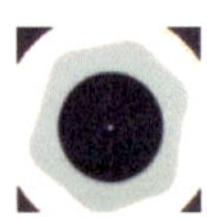			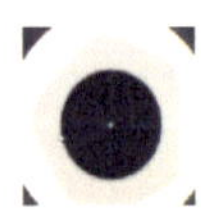

Table 3. The drapability of fused panels.

The presented results show that after the fusing process all the fused panels have higher value of drape coefficient in comparison with the shell fabric. The properties of a thermoplastic resin (type of adhesive, amount of adhesive) are the main reason for this effect, because the adhesive blocks up the moving of fabric threads in both warp and weft direction. The fusible interlining with stiffer properties have less nodes and the values of a drape coefficient are higher [26].

6.2. Seam properties of hot air welded multilayer materials

The seam properties of hot air welded multilayer materials will be discussed on the basis of the application of this technology in the shoe manufacturing process.

For tracking sport shoes inner part of shoes, i.e., inner sock is usually made from the multilayer, waterproof, and breathable materials. The seams of the inner sock should also provide enough smoothness to avoid the friction between the feet and shoes during the wearing. The multilayer material with the integrated waterproof layer such as SympaTex® or GoreTex® are the most frequently used materials. The traditional joining technique used for joining the multilayer waterproof material for the inner socks of the shoes was carried out in two steps in order to keep the waterproofness [27]. The multilayered fabrics were sewn by zig-zag stitches with the predefined specifications. For sewing, a needle size 100 N m with the rounded point shape and waterproof sewing thread with fineness of 30 tex/3 S were used. The length and width of stitches were both 5 mm. In the second step, the seam was covered with a waterproof tape using the hot air welding technique. The waterproof tape has to be welded exactly on the middle (±1 mm) of the seam and during welding puckering of the membrane should be avoided. Next, the hot air welding was carried out at the following processing conditions: temperature of hot air was 365–380°C, the speed of welding was 2.5 m/min., the pressure of welding was 0.7 bar, the pressure between welding wheels was 2.5 bar, and the pressure of hot air blowing was 0.6 bar [27]. **Table 4** presents the characteristics and visual appearance of a joining seam of a multilayered material for shoes.

Position of seamed layers	Joining method	Sewing machine	Visual appearance of seams	
			Face side	Back side
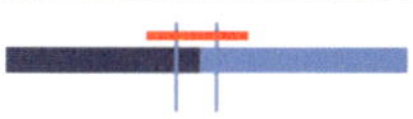 LSp Stitch type: zig-zag	Sewing + Hot air welding	Pfaff KI491-755/13		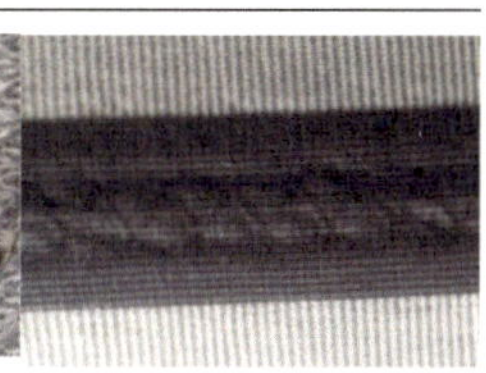

Table 4. The characteristics and visual appearance of a joining seam of a multilayer material for shoes.

From the point of shoe manufacturer view, all produced seams would reach the desired minimum of bond strength, which is between 10 in 20 N/5 cm. But the traditionally joined materials can reach even the strength of 562.5 N/5 cm [27].

The hot air welding parameters, particularly welding temperature and pressure, had influence on the seam thickness. Thus, the ideal seam thickness of the welded or joint area should be the same as the thickness of the multilayered fabric considering the comfort during the use of the shoes. In comparison, the traditional seams applied by the shoe manufacturer are thicker than the multilayered fabrics. Moreover, the seam stiffness is also higher than the multilayered fabric, therefore the used seam construction is not enough functional during the shoe wearing because the friction problem can appear. To avoid the above-mentioned problems, the ultrasonic welding as the alternative technological process were proposed by Jevšnik et al. [27–29]. The first results have shown that that the stiffness of the traditional seams applied by the shoe manufacturer was higher than those of ultrasonic welded seams. However, the results have shown that ultrasonic welding damaged the water proof membrane and low thickness and high stiffness in seam areas of multilayered fabrics also appeared, which can be stated as disadvantages of ultrasonic welding for manufacturing the joints [27].

6.3. E-textile transmission lines made using the hot air welding

Transmission lines built on electrical circuit to interconnect electrical elements, such as sensors, actuators, transistors, power sources for gathering sensitive information, monitor vital functions, and for sending information through the textile structure for further processing [1, 2, 23], can be obtained by different conductive yarns. Conductive yarns are either pure metal yarns or composites of metals and nonconductive textile materials, single or multiple strands, and mono- or multifilament. Nowadays, different integration methods for manufacturing the textile transmission lines by conductive yarns, such as woven, knitted, sewn, couched, e-broidery, printed, and welded can be used. Use of the hot air welding presents new, very promising technology for manufacturing the e-textile transmission lines [23, 30].

As mentioned above, the transmission lines can connect the electronic components integrated into textile materials. E-textile transmission line manufacture by hot air welding technologies are composed of a base fabric and endless welding tape with integrated conductive yarn. The conductive yarn is positioned on the base fabric and therefore hidden between the fabric and welding tape, **Figure 15**.

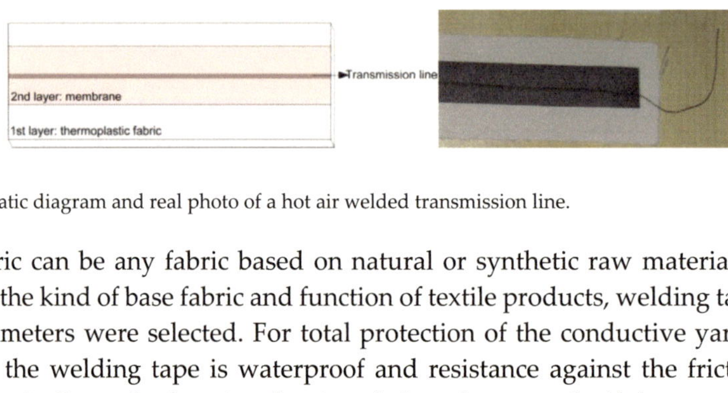

Figure 15. Schematic diagram and real photo of a hot air welded transmission line.

The base fabric can be any fabric based on natural or synthetic raw material composition. According to the kind of base fabric and function of textile products, welding tape and hot air welding parameters were selected. For total protection of the conductive yarn, it is recommended that the welding tape is waterproof and resistance against the friction and other mechanical loads. From the functional point of view, for example, if the smart garment will have protection or sports function, it is also recommended that breathability and windproof properties are considered.

Providing the suitable quality of hot air welded transmission lines, the optimization process of selection of the suitable welding tape and conductive yarn, as well as welding parameters should be very careful planned. The optimization process can be divided into three steps. In the first step, the linear electrical resistance and signal transmission noise of the conductive yarn after welding parameters should be tested. The hot air welding parameters have unpredictable influence on the linear electrical resistance and signal transmission noise [23], therefore very precise tests under standard environmental condition should be carried out to avoid the problems with the electrical components later. In the second step, the bond strength between the welding tape and base fabric should necessary be evaluated. It is recommended that the bond strength is higher than 10 N/5 cm when the tape and fabric can be smoothly divided [23]. In the literature, it has been reported that bonding problem can appear when Teflon and silicon finishes are used [3]. It is also mentioned that since some dyestuffs react differently to heat, therefore they may affect bond strength of the welded fabrics, above all when darker colors are used [3]. If all above-mentioned parameters reach satisfactory results, in the thread step we should evaluated the mechanical properties and visual appearance of welded transmission lines. Kurson Bahadir et al. [23] were studied the influence welding parameters on properties of e-textile transmission lines manufactured with seven different conductive yarns (four stainless steel and three silver-coated PA) under following hot air welding condition, **Table 5**.

	Temperature (T/°C)	**Velocity (v/m min^{-1})**	**Pressure (p/bar)**
Weld set 1	350	2.5	6.5
Weld set 2	450	2.5	6.5
Weld set 3	350	1.5	6.5

Table 5. Welding parameters for manufacturing e-textile transmission lines.

The researchers tested the influence of hot air welding parameters on textile transmission lines from different aspects, as will be presented in the next sections. The presented studies have confirmed that hot air welding techniques can be suitable for constructing reliable and durable transmission lines. The investigation tests have shown that beside the suitable selected welding tape, conductive yarns, and base fabrics also the suitable welding parameters taking into account the applied materials need to be carefully controlled.

6.3.1. Influence of the welding parameters on conductivity of e-textile transmission lines

The linear resistances of the welded conductive yarns according to the defined welding sets (**Table 5**) are presented in **Table 6**.

Table 6 illustrates the changes in the linear resistances of conductive yarns when subjected to welding processes. Mainly, if the temperature increases, the linear resistances of the conductive yarns decrease for both stainless steel and silver-coated polyamide yarns in comparison with the original linear resistance of stainless steel and silver coated yarns. The decrease of conductivity is smaller for stainless steel yarns then for silver-coated polyamide yarns. As the temperature increases the conductivities of the welded conductive yarns, those having conductivity values of less than 70 Ω/m, remain almost at the same values. However, the linear resistance of the yarn silver-coated polyamide (yarn no. 6) dramatically increases from 420

Ω/m (see **Table 1**) to 524 Ω/m when the temperature increased. In case of increasing the temperature when using the silver-coated polyamide yarn (yarn no. 7), the transmission line could not be formed because the welding parameters damaged the yarn and the conductivity was interrupted. The reason for this is melting of the coated polyamide fibers at the interface, which led to failure of the transmission line [23].

Yarn no.	Reference	Weld set 1	Weld set 2	Weld set 3
1	<12	12.33	12.00	12.33
2	<25	25.00	25.00	25.67
3	<35	30.67	33.33	34.00
4	<70	69.00	69.67	70.67
5	<50	50.33	52.00	52.40
6	<420	507.33	524.33	592.00
7	<2000	3003.33	*	3046.67

*Indicates that no seam was formed; therefore, no result was obtained.

Table 6. Linear resistances of welded conductive yarns (Ω/m).

6.3.2. Influence of the welding parameters on signal transference capabilities of e-textile transmission lines

It was found out that the signal amplitudes of the welded conductive yarns were slightly higher than their reference values after hot welding process, **Figure 16.** It can be clearly seen that the samples within the weld set 3 (T = 350 °C; v = 1.5 m/min; p = 6.5 bar), obtained higher SNR values [23].

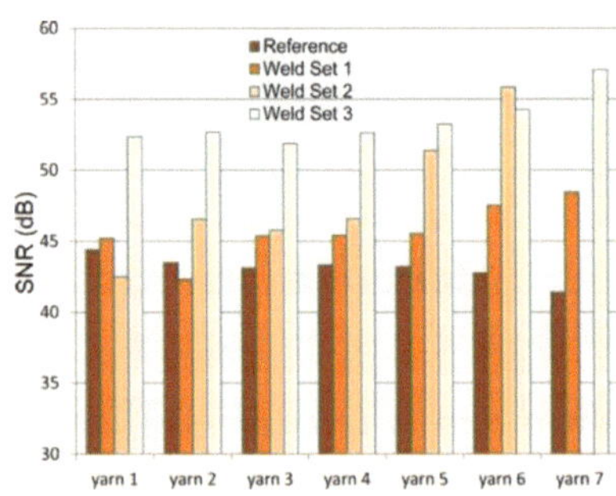

Figure 16. Comparison of SNR values of welded samples (weld set 1: T = 350°C, v = 2.5 m/min, p = 6.5 bar; weld set 2: T = 450°C, v = 2.5 m/min, p = 6.5 bar; weld set 3: T = 350°C, v = 1.5 m/min, p = 6.5 bar).

Moreover, it is also evident that after the welding processes the SNR values of the silver-coated polyamide yarns (yarn no. 5, yarn no. 6, and yarn no. 7) were higher than in stainless steel yarns (yarn no. 1, yarn no. 2, yarn no.3, and yarn no. 4). In other words, after the welding processes those transmission lines made of silver-coated polyamide conductive yarns showed better signal transference capabilities compared with those transmission lines made of stainless steel conductive yarns.

6.3.3. Influence of the welding parameters on visual appearance of e-textile transmission lines

The visual appearance of welded e-textile transmission lines plays an important role when the transmission line should be integrated on the face side of the product. From the functionality point of view, it is recommended that the transmission lines are integrated between the layers of garment in order to be protected against the mechanical damages. The investigations have shown that the visual appearance in terms of visibility of conductive yarns trough out of welding tape, and puckering of the welding tape around the conductive yarns after welding are the most frequently appearing phenomena. Those parameters are important for the final quality of the manufactured product [22, 30].

The visibility of conductive yarns after hot air welding depend on used thickness and twisting of conductive yarns, as well as on thickness of the welding tape. **Figure 15** shows the visibility of conductive yarns when three-layer welding tape and two different conductive yarns are used, i.e., stainless steel and silver-coated PA. Due to three-layer construction of the welding tape, it thoroughly overlaps the structure of the used conductive yarns in terms of visibility after the welding processes irrespective of the selected welding parameters. Thus, stainless steel conductive yarn (yarn no. 1, **Table 4**) despite of diameter 315 μm is slightly visible, while the silver-coated PA conductive yarns (**Table 4**) are after welding processes completely invisible because they are thinner and with less twists.

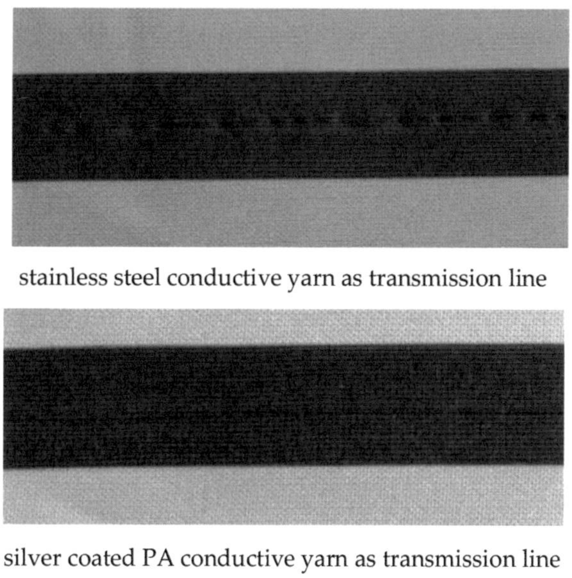

Figure 17. Estimation of a visual appearance of welded samples.

The puckering phenomena were observed after all welding processes with stainless steel conductive yarns where thinner welding tapes are used, **Figure 17** [30]. The stainless steel yarn is stiffer with higher twisting ratio and weight in comparison of PA-coated silver yarns. To sum up, the visual appearance of the welded transmission lines mainly depends on the selected conductive yarn properties and textile materials for layers rather than the selected welding parameters.

7. Conclusions

Today, the textiles materials are involved almost in all areas of products. We require from the textile products special features and high functional utility, which include water permeability, resistance to pressure, wear and tear or seams, air permeability, and aesthetic appearance. For more demanding products, such as inflatable boats, protective and sports clothing, etc., the traditional sewing methods no longer meet the technical requirements for joining the garment parts nor the aesthetic appearance of the products. The textiles materials, methods, machinery, and adhesives necessary for the heat joining such as fusing and hot air welding apparel seams are today more and more involved in the commercial market.

In comparison with the hot air welding, the fusing technology has much longer tradition. In fact, the main purpose of fusing is to improve the aesthetic appearance of a garment, as well as to simplify the garment manufacturing process. Further, hot air welding represents today a significant competition to the traditional sewing and becomes one of the most important technologies in the production of various functional and nonconventional textiles. It will be even more important in the future. The hot air welding provides faster production, low waste material, and it is energy-efficient, therefore it is expected that the merged parts in textile end-products should completely meet the users' expectations.

Besides fusing and hot air welding for seamlessly joining the textile materials, we can also use other welding techniques, such as laser, ultrasound, warm air, and high-frequency welding. The selection of a welding technique depends on the function of the final product, type of material used, and the desired quality of welded joints. To summarize, welding of textile materials has a significant potential for manufacturing of textile products of the future.

Acknowledgements

The study, described in this chapter, relates to the project, which has received funding from the European Union's Horizon 2020 research and innovation program under the Marie Sklodowska-Curie grant agreement no. 644268, project title: Welding of E-Textiles for Interactive Clothing-E-TexWeld.

Author details

Simona Jevšnik[1*], Savvas Vasiliadis[2], Senem Kurson Bahadir[3], Dragana Grujić[4] and Zoran Stjepanovič[5]

*Address all correspondence to: simonajevsnik@gmail.com

1 Inlas d.o.o., Slovenske Konjice, Slovenia

2 Piraeus University of Applied Sciences (TEI Piraeus), Department of Textile Engineering, Athens, Greece

3 Faculty of Textile Technologies and Design, İstanbul, Turkey

4 University of Banja Luka, Faculty of Technology, Republic of Srpska, Bosnia and Herzegovina

5 University of Maribor, Faculty of Mechanical Engineering, Maribor, Slovenia

References

[1] Jones I: The use of heat sealing, hot air, and hot wedge to join textile materials. Joining textiles. Stylious G (Ed.), Principle and Application. 1st ed. Woodhead Publishing, Cambridge, 2013, pp. 355–373, DOI: 10.1533/9780857093967.3.355.

[2] Jevšnik S, Stjepanovič Z, Rudolf A, Eryürük, Kalaoglu F, Kraner Zrim P, Rijvec T: The characteristics of welded seams using high quality textile materials. 8th International Conference Textile Science, TEXSCI'13, Technical University of Liberec, 2013, pp. 1–7.

[3] Shishoo R: The Global Textile and Clothing Industry, 1st ed. Padstow, Cornwall, 2012, 204 p., DOI: 10.1533/9780857095626.

[4] Cooklin G: Fusing Technology, 1st ed. Textile Institute, Manchester, 1990, 94 p.

[5] Jevšnik S, Vuherer T, Gotlih K: Imb 2006 - The future lies in innovation - sewing without thread and sewing needle. Tekstilec, 49, 7/9: pp. 127–132.

[6] Jevšnik S: The effect of the temperature on the bond strength and durability of fused panel. Symposium of clothing engineering '98, Maribor, 1998, p. 75-82.

[7] Jones I, Stylios GK: Joining Textile, Principle and Application. 1st ed. Philadelphia: Woodhead Publishing Series in Textiles, 2013, 594 p. DOI: 10.1016/b978-1-84569-627-6.50022-6.

[8] Jevšnik S: Selection of fusible interlining and prediction of properties of fused part of clothing with system for automatic knowledge. Master thesis, University of Maribor, Maribor, 1999.

[9] Wang Y, Chen DS, Chen H, Wang WZ: Study on fusible interlining shrinkage rates of garment materials, Materials Research Innovations, 2014, 19, sup 6, pp. S6-85–S6-88.

[10] Jevšnik S: Predicting mechanical properties of fused panel. Fibres & Textiles in Eastern Europe, 2000, 8, 2, pp. 54–56.

[11] Prabir J: Assembling technologies for functional garments—an overview. Indian Journal of Fibre & Textile Research, 2011, 36: pp. 380–387.

[12] Jiang B: Hot air inflatable boats welding machine, [Internet] Available from: http://www.sewingtechnology.blogspot.com [Accessed: 2016-11-2].

[13] Vilene Fusing Guide, Freudenberg, Weinheim, 1999, 57 p.

[14] [Internet]. Available from: http://www.veit.de/fileadmin/user_upload/pdfs/pdf/produkte/prospekte/englisch/Overview_Fusing_Machine_FX_e2.pdf [Accessed: 2016-11-2].

[15] Pfaff Industrial, [Internet]. Available from: http://www.pfaff-industrial.com/en/applications/schweissen?set_language=en [Accessed: 2016-11-2].

[16] Framis Italia, [Internet]. Available from: http://www.framis.it/termosaldatrici-per-produzioni-industriali/eng// [Accessed: 2016-11-2].

[17] Laković, M, Soljačić, I: Thermoplastic resins of fusible interlining for front fusing. Tekstil, 1985, 34, 6, pp. 395–406.

[18] DIN 54310: Testing of textiles; delamination of fusible interlinings from upper fabrics, mechanical delamination test. German Institute for Standardization e.V., Berlin, Köln, 1980.

[19] Stjepanovič Z, Strah H: Selection of suitable sewing needle using machine learning techniques. International Journal of Clothing Science and Technology, 1998, 10, 3/4: pp. 209–218.

[20] Stjepanovič Z, Jezernik A, Nikolić M, Karalič A: A knowledge based system for prediction of quality properties of polyester/cotton yarns. Fibres & Textiles in Eastern Europe, 1998, 1, 1: pp. 39–41.

[21] Stjepanovič Z, Jezernik A, Nikolić M: Machine learning from examples for effective yarn engineering. Applied Informatics: Proceedings of the Fourteenth IASTED International Conference, held in Insbruck, Austria, 1996. Anaheim, 1996, pp. 74–76.

[22] Jevšnik S: Using inductive learning to predict the mechanical properties of the fused panel. Annals of DAAAM for 1999 & Proceedings of the 10th International DAAAM Symposium, 1999, Vienna, pp. 233–234.

[23] Kursun Bahadir S, Kalaoglu F, Jevšnik S: The use of hot air welding technologies for manufacturing e-textile transmission lines. Fibers and Polymers, 2015; 16, 6: pp. 1384–1394, DOI: 10.1007/s12221-015-1384-z.

[24] Xin L, Gui LW: Influence of fusible interlining on draping modeling ability of fabrics, Applied Mechanics and Materials, 2014, 685: pp. 72–75.

[25] Cusick GE: The dependence of fabric drape on bending and shear stiffness. Journal of Textile Institute, 1965 56, 11: pp. 596–606.

[26] Jevšnik S, Rudolf A, Kreševič Vraz S, Stjepanović Z, Saricam C, KalaoğluF: Drapability —parameter for aesthetic appearance of garment. Second scientific-professional meeting with the international participation. "Development tendency in textile industry – Design, Technology and Management, 2010, Beograd, pp. 76–80.

[27] Jevšnik S, Eryürük SH, Kalaoğlu F, Karagüzel Kayaoğlu B, Komárková P, Golombikova V, Stjepanović Z: Seam properties of ultrasonic welded multilayered textile materials. Journal of Industrial Textiles, 2015, pp. 1–19.

[28] Violleau E: Ultrasonic welding for composite materials. Composites Magazine 2014; 87: pp. 92–95.

[29] Eryürük SH, Karagüzel Kayaoğlu B, Kalaoğlu F, Jevšnik S, Stjepanović Z, Rudolf A, Rijavec T: Ultrasonic welding of technical textile components used in cold weather shoes. 4th ITMC International Conference, Proceedings Book. Roubaix: Ensait, 2013, pp. 109–114.

[30] Jevšnik S, Kalaoğlu F, Kursun Bahadir S: Quality assessment of hot air welded e-textile transmission lines. 5th International Istanbul Textile Congress 2015: Innovative Technologies "Inspire to Innovate", 2015, Istanbul: Istanbul Technical University, Faculty of Textile Technologies and Design, 2015.

Chapter 11

Magnetic Pulse Welding: An Innovative Joining Technology for Similar and Dissimilar Metal Pairs

T. Sapanathan, R. N. Raoelison, N. Buiron and
M. Rachik

Additional information is available at the end of the chapter

http://dx.doi.org/10.5772/63525

Abstract

Once it was widely thought to be an exceptional innovative welding solution, the magnetic pulse welding, dragged the related manufacturing industries and particularly automobile companies for its complex assembly solutions in early 2000s. Although this technique has been implemented by some giant manufacturers for various joining tasks, the process still has not been well adopted by industries. However, in recent years, many researchers turned their attention to the potential applications and insight investigations of this process due to the existence of bottlenecks and the prime novelty of this technique. This chapter clearly highlights the process, applications, requirements, interfacial kinematics of the welding, numerical predictions of interfacial behaviours and multi-physics simulations. This chapter recommends that the overall outlook of the process is promising while it requires extra attention in the individual welding cases and its material combinations.

Keywords: magnetic pulse welding, interface, numerical modelling, weldability, plastic deformation

1. Introduction

Recently, multi metallic hybrid materials have been produced by joining dissimilar metals at their solid state [1–5]. Those types of materials become attractive in modern engineering applications as they can provide multiple attributes in a single solution [6–9]. The magnetic pulse welding (MPW) is an innovative joining technology that has been mainly considered for this purpose [10]. This technology involves a significantly high speed collision and high strain rate plastic deformation. In general, the workpiece experiences a strain rate of up to 10^2–10^4 s^{-1} [11] while this could reach an ultimate value of 10^6–10^7 s^{-1} at an interface [12]. The synergetic effect between electromagnetic phenomena with metal plasticity contributes towards the advantages of this process in comparison with conventional and other high speed welding techniques. However, this process has not been widely implemented until now, even though it is known since late 1960s [13, 14] and its booming advantages have been paid attention by automotive manufacturers in early 2000s. That is, large scale implementation of this technology has always been challenged by the existence of unique complex realities of this process. But, fortunately, recent technological advancements allow for thorough investigations to understand the physical phenomena of this multi-physics process and facilitate an effective utilization of the technology into modern engineering applications. Emerging scientific technology provides more room to explore such high speed manufacturing processes using sophisticated engineering tools such as high speed measurements, experimental observations, microscopic analyses and advance computing techniques.

MPW is believed to bring innovative solutions in joining technology and the merits of the process are covered in this context. The objectives of this chapter are divided into five main sections including position of the MPW today and its potentials, description of the process, weld features and variance, identification of the weld nature by simulating the interface behaviour during the collision, and computation of the in-flight dynamics using coupled multi-physics numerical simulations.

2. Position of the MPW process today and its influences

Joining processes technically and financially offer strong potentials and represent a significant global market. An overview of the added value brought by the joining activities is briefly presented to address the growing interest in this engineering field. This part of the chapter explains the technical-economic status of the welding technology including recent developments in 2005–2007 about the added value with its increasing trend for the global market of the welding. Current status of the technical evolution marks a major technological transition with the emergence of the solid state welding method for the MPW. A brief review of the technical solutions developed by MPW is addressed. It also provides various representations of successful configurations, suitable combinations of materials and major advantages offered by the MPW process.

2.1. Socio-economic influences of the joining and welding processes

A recent survey focused on the European zone provides data related to the technical and financial consequences of the joining technology and the added value brought by the joining manufacturing activities [15]. In the case of Germany, the data gives a representative indication since this country is the leading manufacturer of joining machines which represent about one-third of the EU's production [15]. Available data inform that, for this country alone, the added value generated by the assembly industry is increased more than 22 billion euros by 2000s without any significant decrease [15], which corresponds to a proportion of 26% increase at European level. But contributions of other countries are also important; this includes Italy, France, Poland and United Kingdom which have respectively provided 18%, 10%, 9%, and 8% [15]. This assessment does not include either the worldwide data or the recent data, but represents an indication of the socio-economic influence of the joining technology.

Welding activities generally cover a substantial part of the assembly industry. For reference purposes, they returned a total market turnover of 19.3 billion euros for Germany in 2003, providing 6% of the jobs linked to this industry, which represents 1.7% of increase in employment opportunities including all sectors [16]. A comparative study carried out between 2001 and 2005 has shown an added value increased by 18% of job creation including 5% directly linked to the welding activities [16]. Such expansion highlights the socio-economic benefits brought by the welding technology. Furthermore, note that the welding represents a non-negligible investment in several industrial branches including the most advanced sectors in transportations, energy and medicine.

In the specific case of metal joining, welding methods bring some useful flexibilities. It does not require intermediate joining component (bolt, rivet, adhesive layer, brazing material etc.) allowing thereby possibilities to produce structures with the benefits of cost and weight reduction. A weld can confer as a permanent joint which is suitable for many mechanical performances. In addition, the welding methods can be applied at varied length scales, from micrometric (micro welding) to several hundreds of millimetres. Furthermore, welding practices include several techniques and processes, making them robust, widely used and intrinsic to technological advances and innovations.

2.2. Innovative nature of the electromagnetic pulse technology (EMPT)

Conventional welding processes show difficulties in joining new metal combinations. The current innovations increasingly introduce dissimilar assemblies that enable to meet new challenges such as light weight requirement, structural reinforcement, and other functional specifications. In this respect, innovative solutions have led to the consideration of complex functional material combinations including metallic assemblies with different melting temperatures, where the fusion welding processes fail when producing such joints at the interface. The discrepancy between the melting points of two dissimilar metals prevents a successful joint formation by solidification of a molten pool as usually achieved during a fusion welding process. The exploration of new methods have led to various welding principles among which high velocity impact welding (HVIW) methods enable bonding dissimilar metallic combinations. High pressure, short duration and low temperature bonding form the main particular characteristic of these methods [17]. The welding involves a strong interfacial collision in various high velocity impact methods using the explosive detonation (explosive welding), the laser shock impulse (laser spot welding), the magnetic impulse (MPW), or the vaporizing foil actuation (vaporizing foil actuation welding).

The use of electromagnetic impulse to provide a significant Lorentz force makes the MPW as an attractive method with respect to other high speed collision welding processes. The EMPT is particularly different in terms of cost, reliability, ease of use, flexibility, rate of work, no

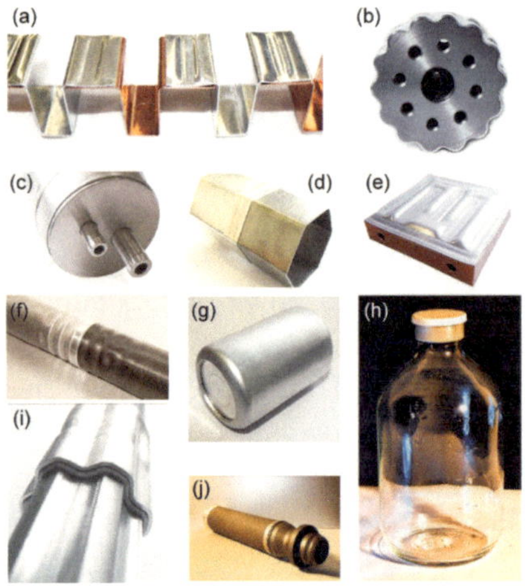

Figure 1. EMPT for industrial applications implemented by "PSTproducts" (a) Al/Cu electric bus bar [www.pstproducts.com], (b) EMPT crimped gear box part [www.pstproducts.com] (c) EMPT welded Al pressure vessel for air conditioning system [28], (d) EMPT welded Al/steel crash box [www.pstproducts.com], (e) EMPT welded Al/Cu cooling plate [www.pstproducts.com] (f) EMPT crimped Al/steel tube instrumental panel beam [28], (g) EMPT for hemming of a Al pressure vessel [28], (h) EMPT crimped Al lid on a pharmaceutical glass bottle [29], (i) EMPT crimped drive shaft [28] and (j) EMPT crimped air suspension [28].

requirement of being consumable and eco-efficiency [18]. This method simply uses a standard electrical source intermittently and a magnetic coil. The welding test does not require either a surface treatment or a long experimental preparation, and is performed in a very short period of time, i.e. it only takes less than a few hundred microseconds to produce a joint. This is a precise joining method that has been successfully applied on several similar and dissimilar metallic combinations for different configurations such as overlap, half lap, cross lap, end lap etc. This joining method is also suitable for various geometrical components including tubular assembly, plates or any specific shape. It is possible to generate a complex distribution of a magnetic pulse force due to the strong flexibility of electromagnetic welding tool design [19]. Current potentialities of the EMPT are depicted by Kapil et al. [18]. The authors addressed a comprehensive review of successful applications, where some are being industrialized, and the growing interest given to the process in several industrial sectors such as in automobile, aerospace, nuclear, electrical and microelectromechanical systems (MEMS), ordinance and packaging [18]. Although the pragmatic results are numerous, concisely its applications are well suited to any tubular assembly, regular or irregular shapes, as well as to any flat shape connections (**Figure 1**). EMPT is successfully implemented to perform various manufacturing tasks using semi and fully automated lines by "PSTproducts GmbH" that also offers engineering and industrial solutions including a robotic arms to effectively handle the portability of the unit in industrial welding cases (**Figure 2**) [20]. In addition, the process covers a broad range of material combinations including Cu/Zr-based metallic glass [21], Al/metallic glass [22], Cu/Manganin [23], flexible circuit boards [24], Cu/Brass, Cu/steel, Cu/Al, Al/steel, Al/Mg, Al/Ni, Al/Fe, Al/Ti and Ti/Ni [25–27]. With all these aforementioned benefits, the EMPT is continuously explored and progressively optimized to bring new potential advancements for effective industrial implementation.

http://dx.doi.org/10.5772/63525

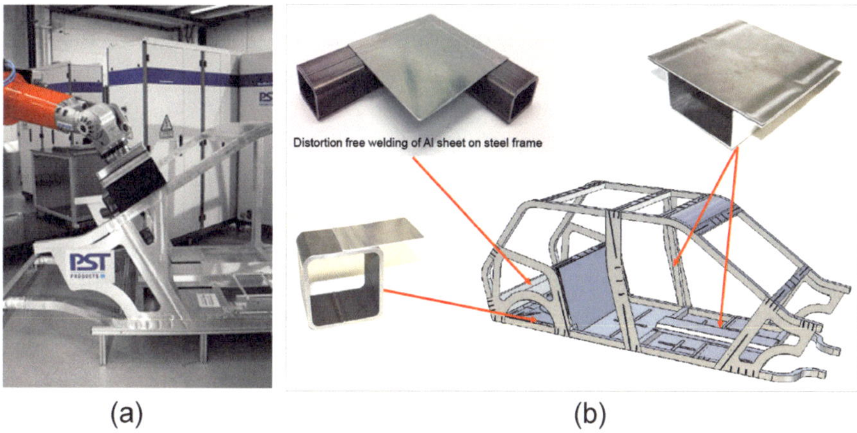

Figure 2. (a) Automated robotic arm used to implement EMPT during a Body in White (BIW) construction and (b) various welded components produced using the robotic arm by "PSTproducts" [20].

3. Description of the MPW process

In general, the MPW process is a user-friendly joining method. The working principle of the process is simple and the welding procedure is fast, easy, and viable. This section briefly explains the general principle of the process including the architecture of the welding machine and the welding parameters. Interactions between process and welding parameters are provided including the specifications of their controllable and measurable natures. This gives a holistic understanding of the process principle with different variables involved in the selection of the welding parameters.

3.1. Magnetic pulse welding architecture

Figure 3 shows typical magnetic pulse welding architecture with overlap configuration used to weld a core clad combination. The MPW is sufficiently flexible to weld various shapes of components for different joint configurations such as half lap, overlap, cross lap and end lap (Section 2.2). Basically, a MPW setup consists of a pulse generator, a coil and an optional field shaper. The generator contains a transformer which transforms a low-voltage power supply into a high voltage charge in the range of kilo-Volts stored in a capacitor bank. This generator set, connected to an inductive coil through a control switch, delivers a high discharge current in the range of a few hundred kilo-Ampere. The electric discharge flowing through the coil generates a magnetic field which creates significantly large Lorentz force within the external tube (the flyer) in the case of a tubular assembly. Thus, the flyer tube undergoes a high strain rate plastic deformation and collides onto the fixed inner rod to produce a high velocity collision. The discharge pulse frequency depends on the parameters of the electromagnetic circuit (Equation 1) and which lies in between 10–200 kHz, but usual operational frequencies are in between 10–20 kHz during the applications. The inductive multi turn coil can be used with a field shaper that concentrates the magnetic field in the working area while increasing the magnetic field intensity. Moreover, one can also control the electromagnetic field and consequent deformation of the material by utilizing various geometries of the field shaper for the same coil [30]. The process can also be performed without the field shaper depending on the required process parameters, or the coil itself can be manufactured which includes the field shaper geometry, especially suitable for single turn coils. However, using a multi turn coil with a single turn field shaper is the notable practice in the current manufacturing that reduces the replacement cost in case of damage of either tool. **Figure 3** shows a MPW configuration including a coil with a separate field shaper and also indicates the direction of the electromagnetic forces on the external tube where the compressive Lorentz force facilitates the deformation of the flyer tube.

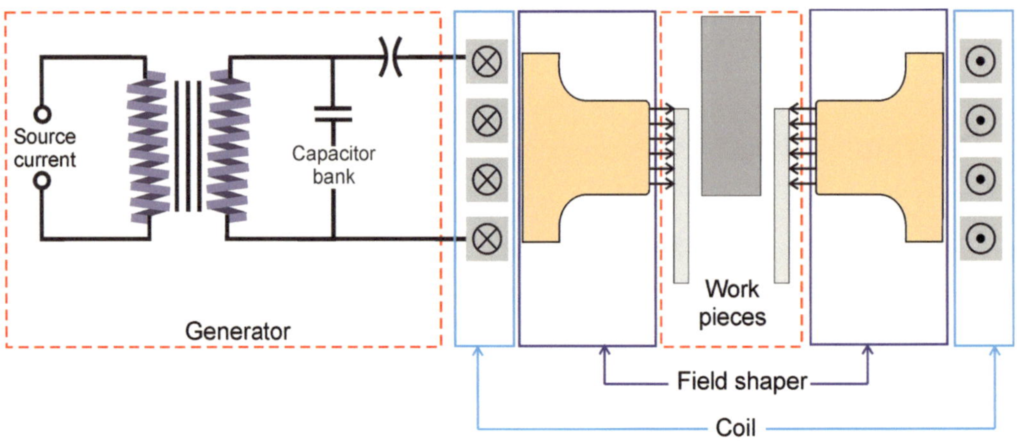

Figure 3. A typical architecture of the MPW and crimping of tubular assemblies used in overlap configuration.

The circuit frequency f can be determined by other parameters using Equation (1).

$$f = \frac{1}{2\pi}\sqrt{\frac{1}{LC}} \tag{1}$$

Where L is the circuit total inductance (H) and C the capacitance of the generator (F).

3.2. Description of the influencing parameters and working conditions

Basically, a MPW test consists of a simple procedure. The workpieces are placed inside the working section for a tubular assembly. Diameters of both flyer and inner rod determine the air gap, which lies in between the inner surface of the flyer and the outer surface of the rod. The formation of the joint in the overlap configuration decides the weld length depending on the inserted lengths of both the rod and tube inside the working area in an assembly. The MPW user prepares this installation where the initial air gap is one of the main geometrical parameter affecting the collision condition, and thus the weld formation. A command console allows for setting up the discharge voltage that is the main adjustable parameter controlling the welding test. The discharge voltage value is indicated as it is or in terms of discharge energy (*E*) given by Equation (2).

$$E = \frac{1}{2} C U^2 \tag{2}$$

Where *E* is the discharge energy (J), *C* the used total capacity of the bank capacitor (F) and *U* the discharge voltage (V). Generally, the discharge voltage and the initial air gap provide a set of controllable parameters denoted as (*U*, *g*). They are indicated for each corresponding welding test. However, the discharge pulse frequency also becomes a crucial process parameter since it decides the penetration of the magnetic field from the coil through the thickness of the flyer. This creates an eddy current in the external region of the flyer that penetrates in accordance with the skin depth defined by Eq. (3). The interaction of the eddy current with the magnetic field from the coil produces the Lorentz force that accelerates the flyer till its onset of collision with the inner rod. Hence, without the skin depth effect, the motion of the flyer is impossible. Therefore, for an effective collision, the discharge pulse frequency should be selected to generate a skin depth lower than the wall thickness of flyer. Eq. (3) also provides an indication of the influence of skin depth on the discharge pulse frequency. More details of the skin depth could be found elsewhere [31].

$$\delta = \sqrt{\frac{\rho}{\pi f \mu_o}} \tag{3}$$

Where δ, ρ, f and μ_o respectively denote the thickness of the zone affected by the skin depth effect (m), the electrical resistivity of the flyer (Ωm), discharge current frequency (Hz) and magnetic permeability in free space.

High frequency current is generally recommended for high resistive metals. One should also carefully consider the discharge pulse frequency of a setup requirement for a generator that also modifies the capacity of a welding machine. In addition, increase in discharge frequency reduces the impulse duration that subsequently increases the strain rate of the flyer and the consequent dynamic response of the material and collision conditions. Those interrelated effects can make difficult in the selection of a suitable discharge frequency that is conducive to provide expected results. Generally, the discharge voltage and the air gap are the most adjustable parameters for a MPW test. The discharge pulse frequency value is indicated by an experimental measurement using a Rogowski probe.

4. Interface behaviour and weld variances

In a conventional welding process, a weld is basically defined by three distinct zones: the solidified molten zone at the joint, the surrounding heat affected zone, and the base metals whose properties remain the same. The weld is generally produced at large scale to ensure an efficient joint at the interface during a conventional welding process. But, in MPW, since the bonding principle is completely different, the notion of weld nature is particularly varied in term of size and morphology. High velocity impact welds are typically confined at the interface within a few micrometre thick zones. A permanent bond occurs immediately between the two components but the behaviour of the interface represents different weld variances. Investigations of welded specimens from Al6060-T6/Al6060-T6 joints have allowed identifying the typical variances for a similar metal pair [32–35]. The major observations of such investigations are reported in the following sections, which also cover the effect of joining dissimilar combinations.

4.1. Onset of weld without apparent interfacial deformation

The interface experiences a progressive kinematic phenomenon that governs the generation of various interfacial morphologies when subjected to the high speed collision. In MPW, the weld natures are generally identifiable at the microscopic level, and the first case is an apparent bonding showing a metal continuity across a straight bonded interface (**Figure 4**). This corresponds to the onset of weld produced by a predominant high compressive stress which is a hydrostatic stress since the grains adjacent to the interface remain undeformed (**Figure 4**). The interfacial zone exhibits an equiaxed grain structure without any noticeable deformation supporting the bond formation due to hydrostatic stresses. Generally, the contact pressure is expected to be in the range of 1–20 GPa according to a simple assessment based on the expression of collision pressure $P = \rho_1\rho_2C_1C_2V_i / (\rho_1C_1 + \rho_2C_2)$, where ρ_1 and ρ_2 are the material densities, C_1 and C_2 are the speed of longitudinal waves in those materials and V_i is the impact velocity. The welded joint can extensively be straight as long as the interface remains stable during the complete collision (**Figure 4b**). The hydrostatic bonding is uniform along the interface and the collision is essentially governed by a normal stress. However, due to the oblique collision, the tangential component of the impact velocity can be high enough to cause

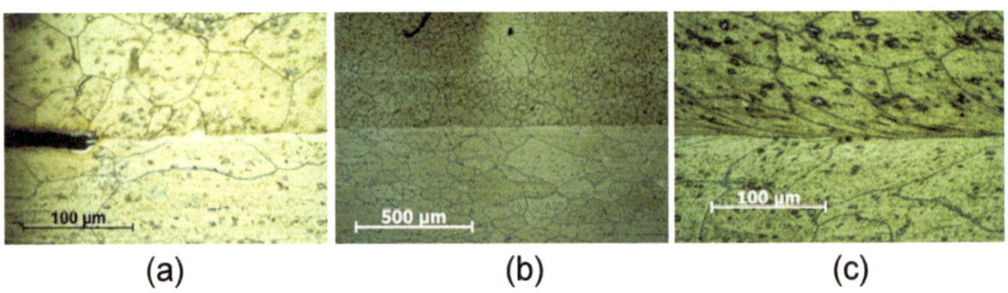

Figure 4. Typical interfacial features of an onset of bonding [32, 33]: (a) magnified view of a straight bonded interface, (b) typical large bonded zone and (c) typical bonded interface with onset of interfacial shearing.

a shearing action at the interface. Therefore, as a result of the confined shearing, the nearby grains at the vicinity of the interface become flattened and elongated as shown in **Figure 4c**.

4.2. Interfacial deformation and wavy nature of the bonded interfaces

The interfacial shearing causes a series of kinematic instabilities that changes the morphology of the interface whose behaviour is therefore similar to the Kelvin-Helmholtz instability that occurs at fluidic interfaces subjected to an interfacial shearing. The straight interface becomes wavy at the onset of instability. The development of wave at the interface is governed by two main phenomena: interferences of compressive shock waves due to the dynamic collision and a jetting phenomenon governed by an alternate inversion. As suggested by Ben-Artzy et al., the first phenomenon is a mechanism for the creation of periodic humps with regular shapes [36]. The wave height and periodicity depend on the size of the structure, the compressive stress intensity, and the interference of shock waves along the interface. Under these conditions, the initiation and development of hump can be considered as a consequence of deformation of the bonded and sheared interface by the reflection of mechanical waves. All the humps regardless of their amplitude reveal shear strain along the interface that would be regularly shaped the wavy interface (**Figure 5**).

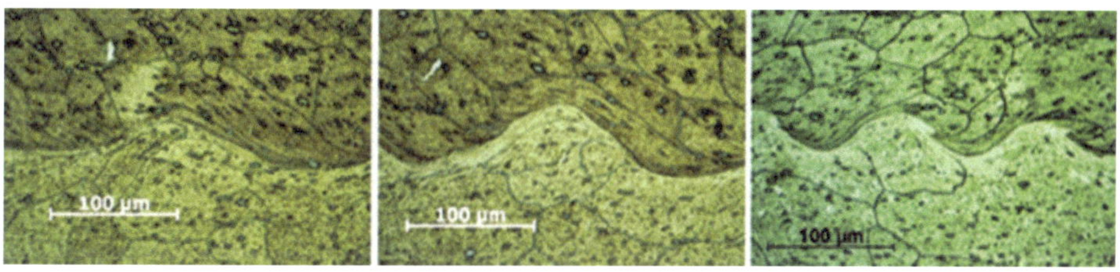

Figure 5. Typical magnetic pulse welded interfaces with wavy morphologies showing grains shearing along the wavy pattern [32, 33].

4.3. Jetting phenomenon and interfaces with irregular wavy shapes

The interfacial shearing can reach an instability threshold due to a high strain rate and severe shearing. In the literature of impact welding, a notion of shear instability is used to demonstrate this circumstance. The confined shearing remains at the interface, but becomes abruptly excessive so that a jetting phenomenon occurs. A tangential jet forms ahead of the collision point and the jet kinematics is controlled by the normal force component along the interface that develops from both shock waves and impact velocity. The jet may evolve following a series of upward and downward jetting to form a sequence of inverted curves along the interface. In that way, the interface develops by jetting phenomenon which is interpreted as both a weld indicator and a mechanism of interfacial humps formation. However, the kinematics of the sheared interface is complex so that it is difficult to clearly state which mechanism really forms the wavy interface: the interference of compressive shock waves or the jetting phenomenon. These two factors can be concomitant or asynchronous or due to their consequences. Particularly, for regular shape humps, a simultaneous prominence may prevail as observed in **Figure 6**. The deformation of the interface creates a hump with nearly symmetric shape (regular shape) while the grains adjacent to the bonded interface are strongly sheared with an upward kinematics at the left front of the hump and a downward kinematics at the right front (**Figure 6a**). If the jetting phenomenon becomes predominant, irregular shapes arise from the progression of humps. A jetting aspect and its orientation can be clearly observed that depend on the governing stresses. On the typical case evidenced in **Figure 6b**, a downward jet occurs. This could initially be an upward jet that is inverted by an interference of shock waves at the interface, or a pure downward jet guided by the local stress evolution, or kinematic instability governed by an opposed shear alone similar to that of Kelvin-Helmholtz instability occurs at fluidic interfaces subjected to a shearing. Hence, the mechanism of wavy interface formation is to be fully explored despite the experimental identification of particular morphologies.

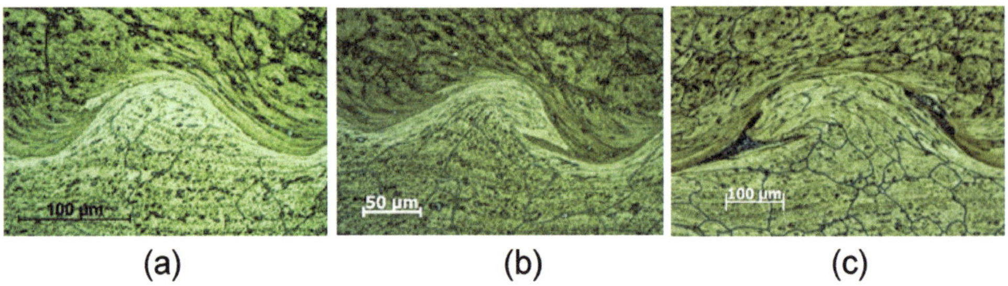

Figure 6. Wavy morphology with strong confined shearing, jetting, irregularly shaped interface and defect onset [32, 33]. (a) Wave formation with upward kinematics, (b) downward kinematics and (c) wavy interface with onset of defects.

4.4. Vortex development and formation of defective welds

Generally, the wavy shape is identified as a particular feature of a high speed welded joint and it has also been suggested as a weld indicator. Nevertheless, distinction between regular and irregular development of wavy interfaces could be useful. Formations of irregular humps due to the predominance of the jetting phenomenon indicate a severe shearing that can promote a confined heating. The plastic work can also be high enough to cause a strong thermomechanical softening, even a melting, and thus the strong shear stress would be unfavourable in the weld formation. Potential matters from ejection also lead to the formation of cavities within the welded interface. The interfacial observation in **Figure 7** represents a typical onset of defects appearing within the irregular wavy interface produced by the jetting phenomenon. Formation of holes is evidenced at the jet tip and at the vicinity of the strongly sheared interfaces (**Figure 7**). Moreover the defects are confined at the bonded joint. The defective weld can be also caused by the kinematic progression of the interface. The jet flow may evolve towards complex kinematics such as a swirling flow. Zones of jetting occurrence become potential sites of vortex development similar to the Kelvin-Helmholtz instability in fluids. Experimentally, it was evident that the jetting affected zone (JAZ) containing prominent voids with circular morphology supports the suggestion of vortex development (**Figure 7**). These voids increase in size at the vortex affected zone (VAZ), and their enlargement depends on the local kinematics and stress distribution. Note that the size of voids can reach up to several hundred μm, in the same range of the height of the humps, i.e. approximately equal to the size of the weld width. The mechanism of void formation could be due to an ejection molten fluid, a solidification shrinkage or a local fragmentation combined with particulates jetting governed by shear stresses [32]. In any event, the welded interface becomes defective due to the formation of discontinuous voids (**Figure 7**).

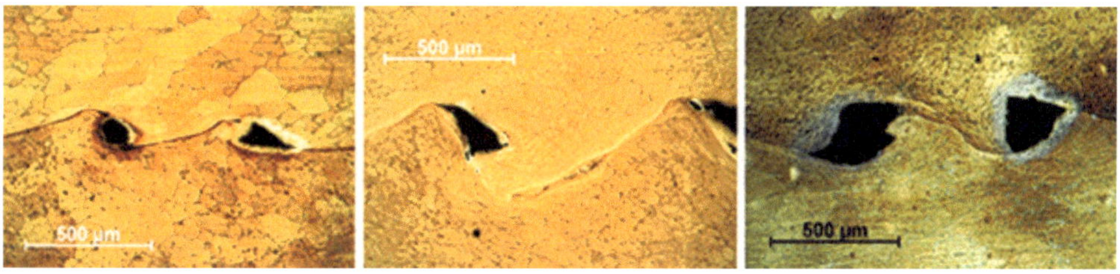

Figure 7. Typical magnetic pulse welds with defects at the VAZ [33, 34].

The MPW of similar metal pairs generally reveal the weld natures among those aforementioned variances. This large depiction is useful to identify the weldability. It is clear that the single case of wavy morphology is not sufficient to substantiate the weld formation as commonly suggested. Although this weld indicator is valuable, the bonding is able to occur independently of the interface shape thus the development of humps can evolve a defective joint. In addition, the relevance of wavy morphology is discussed for dissimilar metal combinations. However, this situation is identified as no longer a reliable consideration of a bond formation. Albeit the literature of impact welding includes several cases of regular or irregular wavy shaped dissimilar joints, there exist some cases where the formation of humps is conducive to brittle joints. This scenario was found when replacing the previous Al6060-T6/Al6060-T6 pair by a dissimilar Al6060-T6/Cu combination, the copper being the inner rod in order to ensure the same in-flight behaviour for the flyer prior to the collision. Further investigations for in-flight behaviour of similar and dissimilar cases are also explored using multi-physics simulations in Section 6.

4.5. Weld with an interfacial mixing or intermetallic compounds

In term of material dissimilarity, the copper is softer, having a higher melting temperature and dissipating the interfacial heating more quickly than those of the aluminium alloy. The combination of these properties produces a different response at the Al/Cu interface during the collision. In contrast to the previous case (Al/Al), the major noticeable changes are associated with the interface nature and particularly prominent with the vortex development. When the vortex instability develops, the copper forms a solid spiral due to its higher malleability and higher melting temperature than those of the flyer part. The VAZ consists of a roll-up so that the copper and the aluminium are locally intermixed (**Figure 8a**). The kinematics of the vortex contributes to the interface bonding by an interlocking mechanism governed by the intermixing phenomena. Note that the vortex instability evolves at high strain rate and excessive deformation so that the plastic work heating enables to melt the aluminium during the swirling phenomenon. The molten phase is quickly solidified due to confinement of the heating and the rapid heat dissipation is facilitated by the good thermal conductivity of both copper and aluminium. These two factors promote a high cooling rate in the range of 10^{4-6} K/s that prevents the atomic structural changes such as crystallisation during slow thermal kinetic. The hyperquenching freezes the random allocation of atoms within the initial molten phase and produces an intermetallic phase which was proven to be amorphous [37]. Hence, the dissimilar Al/Cu combination is conducive to intermetallic formation, within the VAZ and along the interface that alters the physical nature of the weld. The intermetallic phases may appear within discontinuous pockets (**Figure 8b**) or as a continuous layer with a non-uniform thickness (**Figure 8c**). In any case, the presence of the intermediate intermetallic media introduces a new weld variance that is particularly suggested for dissimilar metal joints produced in MPW.

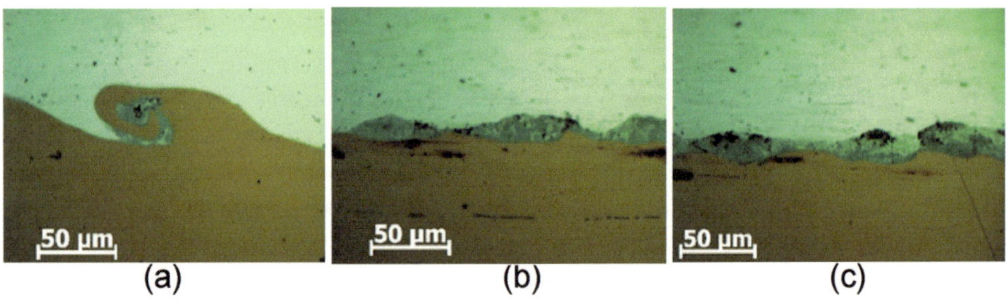

Figure 8. Typical dissimilar magnetic pulse welds with (a) vortex development, (b) intermetallic pockets and (c) continuous intermetallic layer at the interface [33].

4.6. Fracture within intermetallic phases and detrimental welds

The formation of a permanent bonding becomes difficult with the accumulation of intermediate intermetallic compounds (IMCs). The fast shrinkage during the solidification stage involves cold cracking phenomenon governed by the heterogeneous heat conduction combined with the incompatibilities of the thermal expansion coefficient. For the case of thin IMC layer, light microscope observations at low magnification reveal transverse cracks across the thickness without further propagations outside the intermetallic zone (**Figure 9a**). These microcracks randomly coalesce and formarbitrary multidirectional crack propagation. The intermetallic media facilitates the cracks to fragment which lead to a catastrophic failure of the joint (**Figure 9b**). It was evidenced that the thickening of the intermetallic phase is favourable to form numerous cracks so that its occurrence can strongly impair the weld integrity. Eventually, the interface can completely break due to a propagation of a macro crack within the intermetallic zone along the interface. According to experimental analyses, this major fracture is attributable to the development of predominant shrinkage stresses during the intermetallic solidification or to a detrimental shearing stress arises from the liquid or solid ejection of intermetallic phases from the interface [33–35]. Finally, the interface debonding could be also resulted from a separation of molten IMCs from the aluminium solid wall while opening stresses could act during the confined melting. Such hypothesis was supported by experimental observations revealing typical features of a liquid that freely spreads prior to the solidification [33–35]. Indeed, the top surface of an identified intermetallic region exhibits a smooth appearance without fracture pattern due to debonding, and the heterogeneous nature of this surface evidences a free fluid flow [33]. These entire phenomena associated with intermetallic formation adversely affect the welding efficiency of dissimilar metal combinations. The sole alternative is to minimize the thickness of inter metallic layer using an identified lowest critical impact energy. Hence, from similar to dissimilar metal combination, the notion of bonding and weldability is phenomenologically different so that their identification should be accurately established. The next section is numerical simulation to reproduce the interfacial morphologies, which could be used to identify the potential weld variances including the weldability window for similar and dissimilar assemblies.

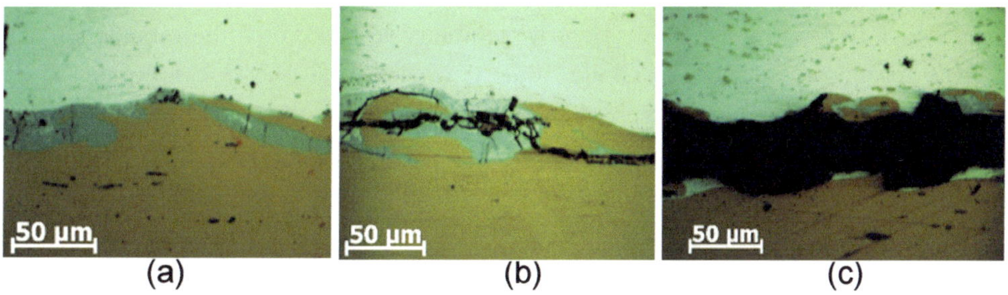

Figure 9. Typical defective dissimilar magnetic pulse welds with (a) onset of cracks within the intermetallic phases, (b) fragmentation, and (c) catastrophic failure [33].

5. Numerical simulation of the interface behaviour

This section begins with a literature survey regarding the numerical simulation of the interface behaviour during impact welding, prior to the suggestion of a suitable method, namely Eulerian simulation to compute the collision and weld generation. This method is applied for both similar and dissimilar metal combinations (Al/Al and Al/Cu) to show the convincing predictions of typical interfacial features including the wavy morphology and formation of defects.

5.1. A brief literature review of impact welding simulations

The numerical simulations of the weld generation during impact welding processes can be classified into five distinct methods known as Lagrangian, Adaptive Lagrangian-Eulerian (ALE), Eulerian, smooth particle hydrodynamic (SPH) and molecular dynamics (MD). Generally, Lagrangian computation fails during the development of excessive interfacial shearing. Due to the large strain produced by the kinematic instability of the interface, the mesh experiences strong flattening and distortion that eventually aborted the computation [38–48]. A wavy interface is difficult to perform even the onset of jet kinematics can be detected using this method. Alternatively, ALE method is suggested to improve the mesh quality by utilizing a node relocation algorithm during the computation [39–41, 46, 49], but its real capability to produce persuasive wavy patterns has to be demonstrated. ALE method also suffers from the bad mesh quality due to the interfacial shearing and jetting. To overcome mesh issues, particle based methods were investigated among which the MD computations allow for the accurate simulation of complex interfacial morphologies but they meet a scale limitation [50–52]. MD method is rather appropriate for small scale, in the range of 10–100 nm unlike the SPH method that enables the computation of the interfacial jetting and wavy morphology at a large scale [42, 43, 53, 54]. However, the accuracy of the SPH method is discussed regarding the consideration of the dissipative terms [55]. The method becomes unsuitable if such physical phenomenon prevails.

At present, Eulerian computation offers a possibility to reproduce the kinematics of the interface during the collision. Generally, Eulerian method is used in computational fluid dynamics but can be applied to a solid to simulate a material flow, with acceptable results using a Johnson-Cook constitutive model for describing the material [44, 45, 56, 57]. In the literature, some simulations of weakly shaped interfaces were presented [42, 43, 50–53]. The method merits to be further explored to compute the full development of the wavy morphology as well as the defects' formation. Sections 5.2–5.4 include a description of the Eulerian procedure and convincing simulation results encouraging the enactment of such method in MPW.

5.2. Eulerian method

Eulerian method is mainly used in computational fluid dynamics that solves the conservation equation of mass, momentum and energy. The history of state variables is computed at any point (M) of a domain. The method uses a fixed computational grid and the time dependant variation of these variables on each grid describes the fluid flow. This is called pure Eulerian computation applied to Newtonian and non-Newtonian fluids. The major difference between fluids and solids relies on the description and treatment of mechanical behaviour. For the peculiar case of solid dynamic, the high strain rate dependency of the stress governs the mechanical behaviour and a constitutive law, the Johnson-Cook law, is generally used. For a suitable numerical treatment of this law, a split method is suggested to solve the governing equations of Eulerian method whose differential forms are expressed as follows:

$$\frac{\partial \psi}{\partial t} + \vec{u}.\overrightarrow{grad(\psi)} = -\psi\, div(\vec{u}) \tag{4}$$

$$\frac{\partial \vec{u}}{\partial t} + \vec{u}.\underline{\underline{grad(\vec{u})}} = \frac{1}{\psi}\overrightarrow{div\left(\underline{\underline{\sigma}}\right)} + \frac{1}{\psi}\overrightarrow{F_v} \tag{5}$$

$$\frac{\partial e}{\partial t} + \vec{u}.\overrightarrow{grad(e)} = \underline{\underline{\sigma}} : \underline{\underline{D}} \tag{6}$$

where, $\psi, \vec{u}, \underline{\underline{\sigma}}, \overrightarrow{F_v}, \underline{\underline{D}}, e$ respectively denotes density, velocity vector, Cauchy stress, internal body force vector, strain rate tensor and specific internal energy which is assumed to be an enthalpy to compute the heating due to the plastic work $\underline{\underline{\sigma}} : \underline{\underline{D}}$.

The split strategy consists of an operation that decomposes each conservation equations into two parts in order to separately compute advection variables. In a one dimensional form, the generic conservation equation is given by (Equation 7) and the splitting gives the pair of equations (Equations 8 and 9).

$$\frac{\partial \varnothing}{\partial t} + u_x \frac{\partial \varnothing}{\partial x} = F \tag{7}$$

$$\frac{\partial \varnothing}{\partial t} = F \tag{8}$$

$$\frac{\partial \varnothing}{\partial t} + u_x \frac{\partial \varnothing}{\partial x} = 0 \tag{9}$$

Equation (8) represents the usual dynamic equation of solid which is solved by the usual Lagrangian method, i.e. the mechanical computation of spatio-temporal motion given by the mesh deformation depending on the material behaviour. This is called the Lagrangian step. Afterwards, the converged state variables given by the Lagrangian step are advected using Equation (9) on spatial fixed mesh. This is called the Eulerian step during which the advection computes the material flow. The material interface is calculated by the volume of fluid (VOF) front tracking method. The sequential Lagrangian/Eulerian computation eliminates the problems of mesh distortion and flattening, thus enables to reproduce the complex kinematics of the interface during the collision progression.

5.3. Virtual testing of the wavy interface generation

Virtual tests are performed under the same conditions as for the experimental results provided in Section 4. **Figure 10** shows the results obtained from a 2D simulation. First computation considers a similar Al/Al combination with a flyer thickness of 1.5 mm and an air gap of 2 mm. A Johnson-Cook law is used to describe the material behaviour with the assumptions of isotropic hardening, a von Mises yield surface and a geometric non-linearity.

The Eulerian finite element computation can provide accurate predictions with respect to the weld generation and the singular feature found in impact welding including the development of interfacial wavy morphology during the collision progression. In **Figure 10**, the collision propagates from the right to the left. The bonding begins with a straight interface as previously described in the Section 4.1. Although, this situation can be computed successfully by a usual Lagrangian method based on a mesh deformation, it fails to produce the wavy interface. In contrast, Eulerian computation enables the simulation of the weak or big waves' formation along the interface as clearly shown in **Figure 10**. The complex kinematics of jetting is reproduced with a fine description of the material interface. **Figure 11** shows three types of typical jetting stages ahead of collision points. Developments of successive inverted jets [downward (**Figure 11a**) and upward (**Figure 11b**, **c**)] are evidenced. The material flow governed by this jetting phenomenon forms the progressive wavy morphology. Ejection of particles is also evidenced in the simulation. In the numerical point of view, this result arises from the advection procedure that calculates the temporal evolution of each state variable over the computational grid. Material flows occur where they are expected to appear. Note that the ejection is a physically realistic phenomenon in MPW. It was evidenced that solid fragments or aggregates are ejected from the interface during the collision [32, 33].

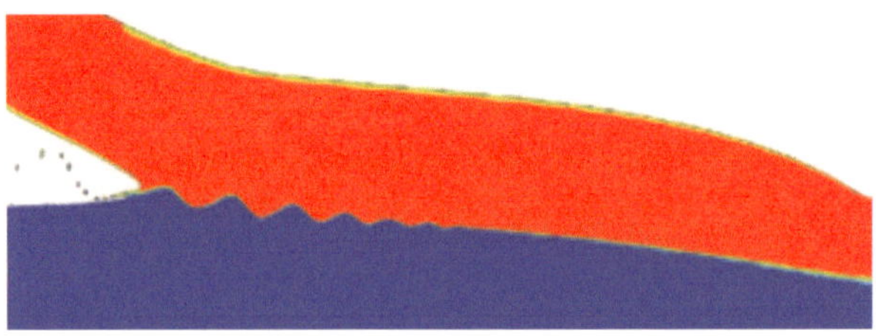

Figure 10. Eulerian simulation showing the development of wavy pattern at the welded interface.

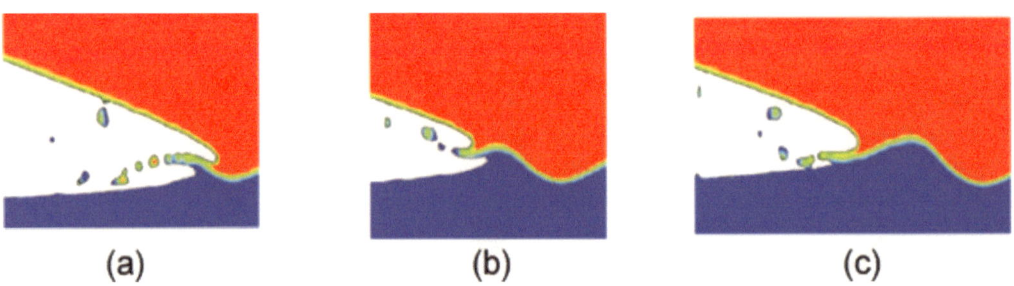

Figure 11. Eulerian simulation of jetting kinematics during the collision propagation with downward jet in (a) and upward jets in (b) and (c).

5.4. Computation of the thermomechanical phenomena

The plastic work due to the interfacial shearing plays a significant role in the interfacial behaviour. It produces localised heating phenomenon at the vicinity of the excessive sheared zone where a temperature change arises. With the accurate computation of the interfacial deformation, Eulerian simulation captures the interfacial heating while providing accurate predictions in terms of heating location and shape of the heat affected zone (HAZ). **Figure 12** presents a comparison between a computed HAZ and experimental observations. The simulation results exhibit particular HAZ where the highest temperature indicates a potential site of defect's formation. This zone undergoes a thermomechanical softening which is conducive to a development of failure when detrimental conditions (critical strain, stress, damage parameters etc.) occur. Experimentally, along the wavy pattern, formation of voids is observed within the numerically predicted extremely heated sites. Experimental observation of the voids' shapes also concur the shapes of the high temperature distribution during the numerical simulations for the Al/Al joint (**Figure 12a,b**). In the case of Al/Cu combination, Eulerian simulation also reproduces an interfacial heating that corroborates the corresponding experimental observations (**Figure 12c–e**). The computed heating clearly indicates the development of a confined heated layer for Al/Cu joint. The temperature distribution of the layer reveals a convincing shape and size that leads the possibility to predict the formation of intermediate intermetallic phases at the interface. Furthermore, the ejection of significant quantity of material due to the strong interfacial shearing combined with the relatively softer material of the copper than that of aluminium (**Figure 12c–e**) causes a potential detrimental phenomenon that can explain the large experimental cracks (**Figure 12d**). These overall predictions demonstrate the capability of Eulerian simulation to investigate the weldability conditions for MPW by a direct computation of the interfacial behaviour during the collision. Such approach will significantly facilitate the accurate depiction of welding conditions and contribute thereby to the process proficiency.

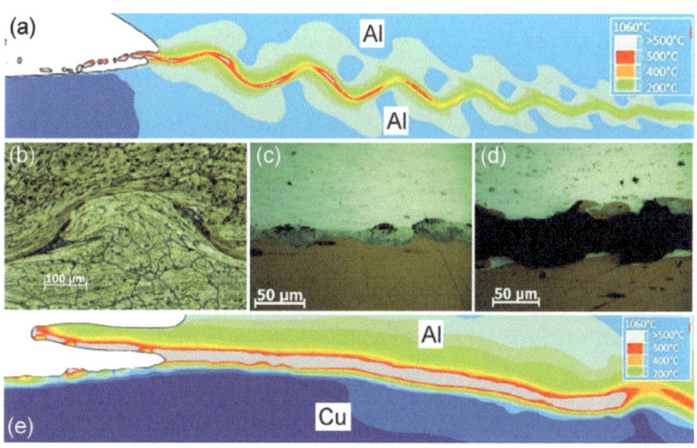

Figure 12. Typical simulation results of the weld natures for (a) similar and (e) dissimilar materials, predicting wavy interface and localised thermal effects. Experimental observations of wave formation in (b) and detrimental phenomena in b–d.

6. Multi-physics computation of the flyer in-flight behaviour

In impact welding, the in-flight behaviour of the flyer determines the collision conditions. Generally, the flyer velocity prior to the impact governs the interfacial phenomena. This is the characteristic parameter that should be known depending on the process and adjustable parameters. Experimental measurements using laser velocimetry methods provide an accurate assessment of the flyer velocity but numerical computation offers a better description of the flyer velocity in terms of spatial and temporal distribution. This section presents a multi-physics computation of the MPW process behaviour. It covers the electromagnetic discharge through the coil and the coupled electromagnetic-mechanical computation of the flyer behaviour. A 3D model is described, including the physical interactions of the process, the governing equations, the resolution procedure, and both boundary and initial conditions. It is employed to show the capability of the model to compute the process behaviour and particularly, the flyer kinematics and macroscopic deformation. Illustrations of spatially distributed impact velocity simulation are presented.

6.1. Governing physics and multi-physics interaction during MPW

Figure 13 describes the multi-physics phenomena involved in the MPW process. The in-flight behaviour of the flyer is mainly governed by the electromagnetic induction and the mechanical response of the material via the Lorentz force while the structural deformation modifies the distribution of magnetic field which in turn affects the electromagnetic interaction between the coil and flyer. This process produces the macroscopic deformation of the structure. Note that the skin depth effect and associated current confinement causes a Joule effect which heats the external part of flyer where the Lorentz force occurs. Generally, the eddy current intensity is high enough, up to several hundred of kA, to generate a strong heating that diffuses within the flyer. A conductive material is expected to involve a good heat transfer and vice versa for a material with low thermal conductivity. Metals such as steel suffer from strong heating due to this phenomenon whereas aluminium or copper seems to limit such heating effect. The consequence of which will make the variance in electromagnetic properties with the temperature that can change the Lorentz force. Therefore, a full physical description of the phenomenon governing the flyer in-flight behaviour should include this electromagnetic-thermal-mechanical interaction. However, under suitable conditions, an electromagnetic-mechanical coupling provides an accurate computation of the flyer kinematics.

The interfacial collision rather involves microscopic phenomenon that can be separately treated using the flyer kinematics given by the electromagnetic-mechanical macroscopic computation. The time-dependent flyer velocity distribution becomes the initial condition for the computation of the impact as previously described in section (Section 5.3). The structural changes involved by both interfacial dynamics and thermal kinetics imply the consideration of specific metallurgical phenomena that decide properties of the joint. However, the multi-physics simulation of the interface can be limited to the mechanical and thermal aspects to reproduce the morphological features of the interface. In the interface simulations (Section 5.3), the mechanical computation describes the interfacial kinematics with the plastic work heating while the heat transfer take into account of the mechanical property changes due to the thermal effect.

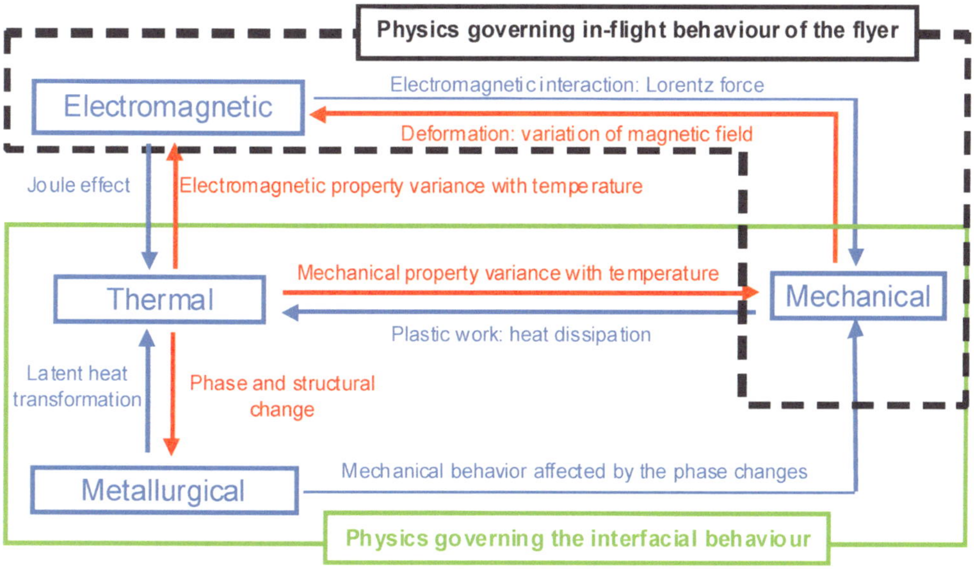

Figure 13. Synoptic depiction of the multi-physics interactions involved in MPW process.

Moreover, solution time steps are important input parameters that govern the convergence of a simulation. In an electromagnetic-mechanical coupling, that requires for both electromagnetic and mechanical time steps during a simulation. In general the electromagnetic time step $\Delta T \leq p^2/2D$, where p and D are characteristic mesh size and characteristic diffusion time. The characteristic diffusion time D is determined by, $D = 1/\mu\sigma$. The mechanical time step (Δt) is always smaller than the electromagnetic time step, $\Delta t \ll \Delta T$.

6.2. 3D coupled electromagnetic-mechanical model

The electromagnetic problem is governed by the Maxwell equations (Eqs. 10–13) and the electrical and magnetic constitutive relations (Equations 14 and 15). The calculation of magnetic field and eddy current can be performed using these equations.

$$\vec{\nabla} \times \vec{\mathrm{E}} = -\frac{\partial \vec{\mathrm{B}}}{\partial \mathrm{t}} \tag{10}$$

$$\vec{\nabla} \times \vec{\mathrm{H}} = \vec{\mathrm{j}} + \varepsilon \frac{\partial \vec{\mathrm{E}}}{\partial \mathrm{t}} \tag{11}$$

$$\vec{\nabla}\cdot\vec{B}=0 \tag{12}$$

$$\vec{\nabla}\cdot\vec{E}=\frac{\rho}{\varepsilon} \tag{13}$$

$$\vec{J}=\sigma\vec{E}+\vec{J_s} \tag{14}$$

$$\vec{B}=\mu\vec{H} \tag{15}$$

In these equations σ, μ and ε respectively represent electrical conductivity, magnetic permeability and electrical permittivity. $\vec{E}$, $\vec{B}$, $\vec{H}$, ϱ, j and $\vec{J_s}$ denote the electric field, magnetic flux density, magnetic field intensity, total charge density, total current density, and source current density respectively. In magnetic pulse forming and welding processes, there involves no charge accumulation and eddy current approximation follow a divergence free current density, those implies $\varrho = 0$ and $\varepsilon\frac{\partial \vec{E}}{\partial t}=0$.

Due to the divergence condition of Eq. (10) and Eq. (12), they should satisfy the following correlations written in Eq. (16) and Eq. (17) respectively.

$$\vec{E}=-\vec{\nabla}\Phi-\frac{\partial\vec{A}}{\partial t} \tag{16}$$

$$\vec{B}=\vec{\nabla}\times\vec{A} \tag{17}$$

where Φ and $\vec{A}$ are respectively the electric scalar potential the magnetic vector potential. Since the mathematical degree of freedom satisfy the magnetic vector potential $\vec{A}$, a gauge equation is applicable. Using the aforementioned co-relations with the generalized Coulomb gauge condition, $\nabla(\sigma\vec{A})=0$, one could separate the vector and scalar potentials as shown in Equation (18) and Equation (19) respectively.

$$\nabla\left(\sigma\vec{\nabla}\Phi\right)=0 \tag{18}$$

$$\sigma\frac{\partial\vec{A}}{\partial\mu}+\vec{\nabla}\times\left(\frac{1}{}\vec{\nabla}\times\vec{A}\right)+\sigma\vec{\nabla}\Phi=\vec{j_s} \tag{19}$$

http://dx.doi.org/10.5772/63525

Therefore, by solving these Equations 18–19, the two unknowns $\vec{A}$ and Φ in an electromagnetic system can be solved. Finally, based on these solutions procedure and magnetic pressure calculations, Lorentz force is estimated.

In this study, the electromagnetic coupled numerical simulations were carried out using LS-DYNA® package with the solver version R8. The resolution scheme in the electromagnetic-mechanical solver uses both finite element method(FEM) and boundary element method (BEM) [58]. BEM is used toe valuate the surface current and electromagnetic field thus the magnetic field in the air is not required in LS-DYNA® simulations. FEM is used during the computation of eddy current and Lorentz force in the workpieces. At each electromagnetic time step, the electromagnetic and mechanical computations are coupled.

A typical situation of one turn coil with a separate field shaper model is considered as an illustrative 3D simulation case (**Figure 14**). The model consists of solid8 node elements for both workpieces and tools to handle the electromagnetic algorithm in LS-DYNA. The largest element size was selected based on the skin depth and ensured that the element size is sufficient to accurately capture the electromagnetic skin effect (Eq. 3).

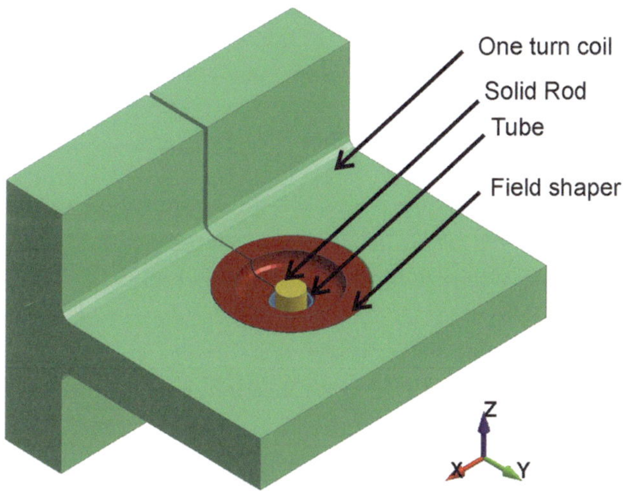

Figure 14. 3D geometrical model.

Material model was described using a simplified Johnson-Cook model (Eq. 20) in the numerical simulations to capture the high strain rate deformation behaviour of the work pieces.

$$\bar{\sigma} = \left(A + B\bar{\varepsilon}^{n}\right)\left[1 + Cln\left(\frac{\dot{\varepsilon}}{\dot{\varepsilon}_0}\right)\right] \tag{20}$$

where, $\bar{\sigma}$ and $\bar{\sigma}$ are the von Mises equivalent stress and strain respectively, $\dot{\varepsilon}$ is the strain rate. Here $\dot{\varepsilon}_0$ is the quasi-static threshold strain rate, treated as equal to *1/s*. *A*, *B*, *C* and *n* are constants, obtained from literatures, are listed in **Table 1**. Other mechanical, electromagnetic and thermal quantities used in the model are listed in **Table 2**.

Johnson-Cook parameters	*A* (MPa)	*B* (MPa)	*C*	*n*
Aluminium alloy	352	440	0.0083	0.42
Commercially pure copper	90	292	0.025	0.31

Table 1. Johnson-Cook parameters used to prescribe the constitutive behavior of workpieces.

Material	Part	Density (kg/m^3)	Young's modulus(GPa)	Poisson's ratio	Electrical conductivity (S/m)
Aluminium alloy 2024	Tubeand Rod	2700	73	0.33	1.74×10^7
Commercially pure copper	Rod	8900	124	0.34	3.48×10^7
Copper alloy	Field shaper	7900	210	0.29	2.66×10^7
Steel	Coil	--------------------Rigid-------------------			4.06×10^6

Table 2. Physical properties of the materials and theirs corresponding parts.

6.3. Boundary and initial conditions and other specifications

This section addresses a general procedure for the specification of boundary and initial conditions for an electromagnetic-mechanical model. In the current example, the bottom side of the tube and the top side of the rod are fixed during the simulation. Coil was considered as rigid and fixed during the simulations. The field shaper geometry was placed inside the coil and left free without any boundary conditions to well represent the experimental condition. Automatic surface to surface contact was prescribed between the rod and the tube to capture the contact behaviour during the collision. Electrical conductivity (σ) is defined by electromagnetic card and the relative magnetic permeability (μ_r) was considered as one for all the materials, which indicates that the ferromagnetic effects of materials are neglected during the simulation, whose μ_r values are generally greater than one. The electric and magnetic fields are not only depending on the geometry, assembly configurations and input parameters but also they depend on the input and output surface of the current definitions in a coupled simulation. Therefore, in order to accurately represent the experimental conditions, the exact same connecter point areas are used in the numerical simulations.

In general, an electromagnetic simulation requires to be defined with at least one electrical circuit. In order to define an electrical circuit, a voltage or current curve is applied across the input and output surfaces. Alternatively, the circuit could be specified with R, L, C (resistance, load, and capacitance respectively) parameters of the exterior parts in a circuit without including these parameters of the coil and other components used in a particular simulation.

http://dx.doi.org/10.5772/63525

The specification of the test case and the input current used in the numerical simulations are shown in **Figure 15**.

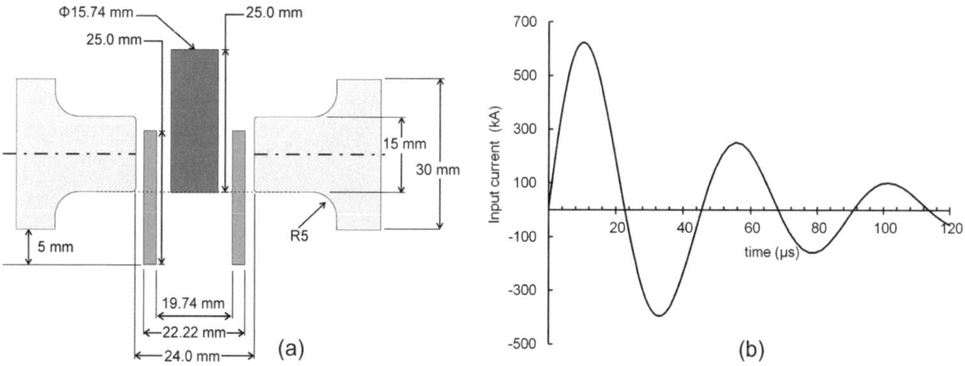

Figure 15. (a) Schematic illustration showing the main working area of the model (field shaper and workpieces) except the coiland (b) the source current used in the model.

6.4. Simulation results of the in-flight behaviour of the flyer

In these numerical simulations, a similar Al/Al and Al/Cu combinations were investigated for their in-flight behaviour in terms of impact velocity and the collision angle at the onset of impact. The properties used in these simulations are corresponding to aluminium alloy 2024 and a commercially pure copper respectively for Al and Cu. The collision angles were calculated estimated on the angle between the radial and longitudinal velocity components (respectively V_r and V_z, see **Figure 17** for detail) from the simulations. The impact velocity was calculated inside of the tube along the longitudinal direction. Sudden change in the resultant velocity was used to determine the onset of the impact and subsequent conditions. That is, immediately at the onset of the impact, the resultant velocity of the tube rapidly reduces. Based on the prediction of the corresponding onset time, resultant velocity and the angle of attack [$tan^1 (Vz/Vr)$] were calculated.

Impact velocity and impact angle against the longitudinal distance from the top edge of the tube are provided in **Figure 16**. The highlighted regions in **Figure 16a** and **b** are able to come in contact during the simulation, which is in agreement with the ~9 mm contact distance observed in experimentally welded samples (**Figure 16c**) obtained under same assembly configuration.

These results suggest that the high speed dynamics for the particular welding case as shown in **Figure 15**. That is, for the particular configuration of the simulation (**Figure 15a**), the top edge of the tube is located slightly above the horizontal mid-plane of the field shaper. This condition causes the highest velocity and the first occurrence of impact slightly below the top edge of the tube. The angle measurements were followed the sign convention given in **Figure 17a**. A closer look of the in-flight collision dynamics shown in **Figure 17b** illustrates the potential collision condition for the particular assembly configuration used in this numerical study.

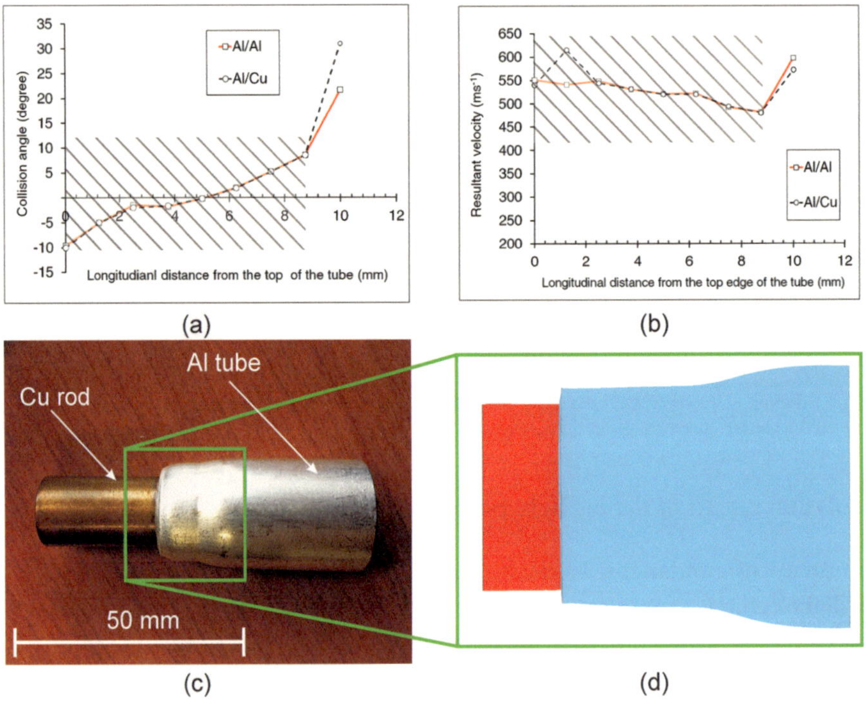

Figure 16. Impact angles along the longitudinal distance from the top edge of the tube for the simulation in (a) and instantaneous resultant velocity at those corresponding points during the onset time in (b). The highlighted regions in (a) and (b) well represent the onset of impact. (c) MPW of Al/Cu sample and (d) final shape of the workpieces from a numerical simulation.

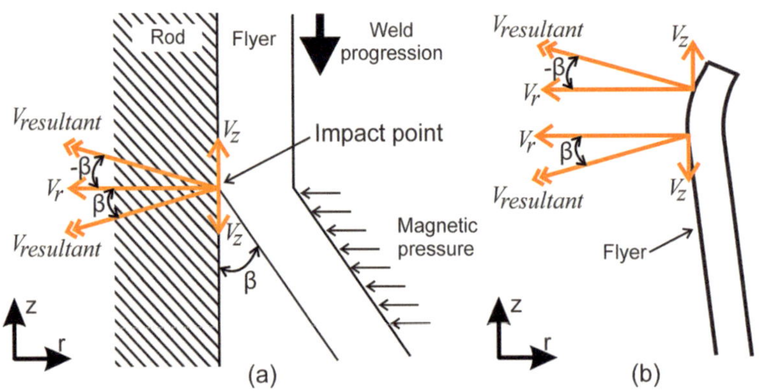

Figure 17. (a) Convention of the angle measurement and (b) a closer look of the in-flight dynamics of the flyer velocity and the impact angle variation.

Although these investigations show that the variance in impact angle and impact velocity, they are not apparent for different inner rods, the influence is suggested as highly depended on the current frequency. That is, one could neglect the difference of the inner rod when predicting the in-flight behaviour at higher current frequencies than that of a critical frequency that can be estimated by equating the first collision time with the full diffusion time of magnetic field through thickness from the exterior to the inner surface of the flyer. In contrast, the difference in the impact condition is not negligible for various inner rods at lower frequencies than that of the critical one. At those lower frequencies, the impact can be influenced by the conductivity of the inner rod [59].

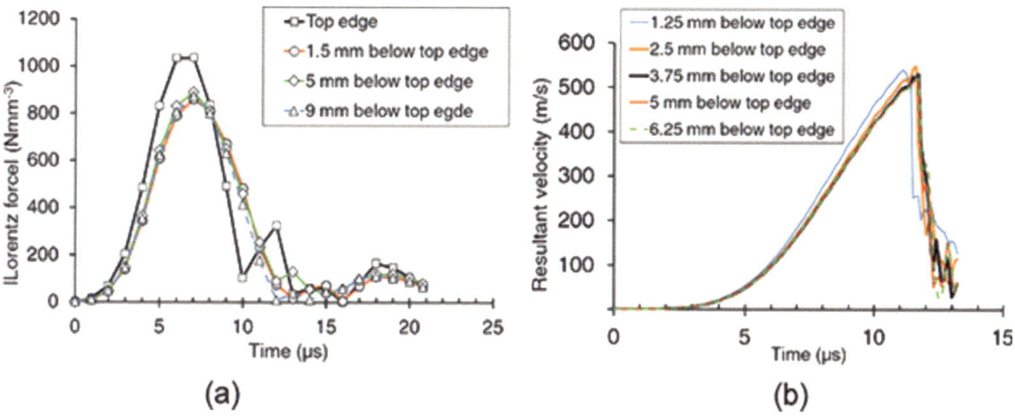

Figure 18. Time dependent radial component of the (a) Lorentz force and (b) velocity obtained from various points on the flyer.

The multi-physics coupled simulations particularly have the benefits of predicting the time dependent the Lorentz body force (**Figure 18a**) and velocity (**Figure 18b**) during the process. Those macroscopic data determines the collision conditions which in turn decides the weld generation. They are crucial for the computation of the interface behaviour during the welding and thus serve as required input condition. Generally, the velocity distribution is appropriate for the interface behaviour simulation that can reproduce physically realistic results (Section 5.4).

7. Conclusions

The magnetic pulse welding is identified as a promising alternative to produce multi material assemblies, while it provides the attractive benefits in terms of cost, reliability, ease of use, flexibility, rate of work, no requirement of consumable and environmental friendliness. The operational conditions were highlighted including the input voltage and initial gap between the flyer and rod were identified as important assembly parameters for a particular overlap configuration of the MPW process. Moreover, the MPW process is also highly influenced by the current frequency and the electromagnetic skin depth effect. After that interfacial natures and weld variances were investigated for both similar and dissimilar material combinations, where it was identified that the weld variances develop under onset of bonding, wavy interface formation, irregular interfaces, jetting, occurrence of vortices, interfaces with defects and intermetallic formation. The vortex formation is highly apparent in the dissimilar assembly while this also forms the intermetallic phases at the interface as a clear distinct feature from the similar metal assembly. After that, numerical simulations were utilised to capture the interfacial features, this could serve as a potential method to identify the influencing parameters during the weld formation. These simulations well capture the interfacial features including jetting and ejection phenomena. Moreover these simulations reveal the interfacial heating which closely resemble with the defects in terms of side occurrence and shape in those assemblies that indicates these simulations could be utilised to predict the optimum weldability window for various combinations. Finally an investigation of coupled electromagnetic-mechanical simulations provides insight understanding and in-flight kinematics of the flyer that predicts the collision conditions during the MPW process. In summary, all these results indicate the potential outlook and innovative nature of the process among other existing welding technologies. The multi-physics nurture with high speed dynamics and associated high strain deformation particularly makes the process more complex that requires extra attention while manipulating the process. However, promising benefits and the evidence of potentially permanent weld formation always pave the way and keep attracting the manufacturing industries.

Acknowledgements

Authors acknowledge the funding for the MSIM project from "Région Picardie" and for COILTIM project financial support from "Région Picardie" and "Le fonds européen de développement économique régional (FEDER)". Authors also thank "PlateformeInnovaltech" for its collaboration. Moreover, authors greatly appreciate and acknowledge the permission from "PSTproducts GmbH" to reuse their images in this chapter.

Author details

T. Sapanathan[1], R. N. Raoelison[2*], N. Buiron[1] and M. Rachik[1]

*Address all correspondence to: rija-nirina.raoelison@utbm.fr

1 Sorbonne universités, Université de Technologie de Compiègne, Compiègne cedex, France

2 Université de Bourgogne Franche Comté, Université de Technologie de Belfort Montbéliard, IRTES EA7274, Belfort, France

References

[1] R. Lapovok, H. P. Ng, D. Tomus, and Y. Estrin, "Bimetallic copper–aluminium tube by severe plastic deformation," *Scripta Materialia*, vol. 66, pp. 1081-1084 2012.

[2] M. Zebardast and A. K. Taheri, "The cold welding of copper to aluminum using equal channel angular extrusion (ECAE) process," *Journal of Materials Processing Technology*, vol. 211, pp. 1034-1043, 2011.

[3] T. Sapanathan, S. Khoddam, S. H. Zahiri, A. Zarei-Hanzaki, and R. Ibrahim, "Hybrid metallic composite materials fabricated by sheathed powder compaction," *Journal of Materials Science*, vol. 51, pp. 3118-3124, 2016.

[4] T. Sapanathan, S. Khoddam, and S. H. Zahiri, "Spiral extrusion of aluminum/copper composite for future manufacturing of hybrid rods: astudy of bond strength and interfacial characteristics," *Journal of Alloys and Compounds*, vol. 571, pp. 85-92, 2013.

[5] M. Knezevic, M. Jahedi, Y. P. Korkolis, and I. J. Beyerlein, "Material-based design of the extrusion of bimetallic tubes," *Computational Materials Science*, vol. 95, pp. 63-73, 2014.

[6] O. Bouaziz, H. S. Kim, and Y. Estrin, "Architecturing of metal-based composites with concurrent nanostructuring: a new paradigm of materials design," *Advanced Engineering Materials*, vol. 15, pp. 336-340, 2013.

[7] D. Embury and O. Bouaziz, "Steel-based composites: driving forces and classifications," *Annual Review of Materials Research*, vol. 40, pp. 213-241, 2010.

[8] T. Sapanathan, S. Khoddam, S. H. Zahiri, and A. Zarei-Hanzaki, "Strength changes and bonded interface investigations in a spiral extruded aluminum/copper composite," *Materials and Design*, vol. 57, pp. 306-314, 2014.

[9] T. Sapanathan, "Fabrication of axi-symmetric hybrid materials using combination of shear and pressure," PhD Thesis (Doctorate), Faculty of Engineering. Department of Mechanical and Aerospace Engineering, Monash University, Melbourne, 2015.

[10] T. Aizawa and M. Kashani, "Magnetic pulse welding (MPW) method for dissimilar sheet metal joints," 57th Annual Assembly of the International Institute of Welding (IIW) 2004, Osaka, Japan.

[11] J. R. Johnson, G. Taber, A. Vivek, Y. Zhang, S. Golowin, K. Banik, *et al.*, "Coupling experiment and simulation in electromagnetic forming using photon doppler velocimetry," *Steel Research International*, vol. 80, pp. 359-365, 2009.

[12] Y. Zhang, S. Babu, and G. Daehn, "Interfacial ultrafine-grained structures on aluminum alloy 6061 joint and copper alloy 110 joint fabricated by magnetic pulse welding," *Journal of Materials Science*, vol. 45, pp. 4645-4651, 2010/09/01 2010.

[13] A. Turner, P. H. Zhang, V. Vohnout, and G. S. Daehn, "Spot impact welding of sheet aluminum," *Materials Science Forum*, vol. 396, pp. 1573-1578, 2002.

[14] T. Aizawa, M. Kashani, and K. Okagawa, "Welding and forming of sheet metals by using magnetic pulse welding (MPW) technique," presented at the 4th International conference on High Speed Forming, Columbus, 2010.

[15] I. K. Middeldorf, "The economic importance of welding and joining in Europe production values, values added and employees," *Düsseldorf, Germany: German Welding Society*, 2009.

[16] K. Middeldorf, "Trends in joining: value added by welding," *Conference Magnetic Pulse Welding and Forming*, Munich, 2008.

[17] V. Lysak and S. Kuzmin, "Lower boundary in metal explosive welding. Evolution of ideas," Journal of Materials Processing Technology, vol. 212, pp. 150-156, 2012.

[18] A. Kapil and A. Sharma, "Magnetic pulse welding: an efficient and environmentally friendly multi-material joining technique," *Journal of Cleaner Production*, vol. 100, pp. 35-58, 2015.

[19] V. Psyk, D. Risch, B. L. Kinsey, A. E. Tekkaya, and M. Kleiner, "Electromagnetic forming —a review," *Journal of Materials Processing Technology*, vol. 211, pp. 787-829, 2011.

[20] P. Pasquale and R. Schäfer, "Robot automated EMPT sheet welding," in *5th International Conference on High Speed Forming*, TU Dortmund, Germany, 2012.

[21] N. Hutchinson, Y. Zhang, G. Daehn, and K. Flores, "Solid state joining of Zr-based bulk metallic glass," *TMS, San Francisco, CA*, 2009.

[22] M. Watanabe, S. Kumai, G. Hagimoto, Q. Zhang, and K. Nakayama, "Interfacial microstructure of aluminum/metallic glass lap joints fabricated by magnetic pulse welding," *Materials Transactions*, vol. 50, pp. 1279-1285, 2009.

[23] M. Kashani, T. Aizawa, K. Okagawa, and Y. Sugiyama, "Welding of manganin and copper sheets by using magnetic pulse welding (MPW) technique," *IEICE Technical Report EMD*, vol. 109, pp. 29-31, 2009.

[24] T. Aizawa, K. Okagawa, and M. Kashani, "Application of magnetic pulse welding technique for flexible printed circuit boards (FPCB) lap joints," *Journal of Materials Processing Technology*, vol. 213, pp. 1095-1102, 2013.

[25] V. Shribman, "Magnetic pulse welding for dissimilar and similar materials," in *3rd International Conference on High Speed Forming*, Dortmund, 2008, pp. 13-22.

[26] V. Shribman, Y. Livshitz, and O. Gafri, "Magnetic pulse welding and joining–a new tool for the automotive," *SAE Technical Paper*, pp. 3401-3408, 2001.

[27] A. K. Jassim, "Magnetic pulse welding technology," in *Energy, Power and Control (EPC-IQ), 2010 1st International Conference on*, 2010, pp. 363-373.

[28] S. Kallee, R. Schäfer, and P. Pasquale, "Automotive applications of electromagnetic pulse technology (EMPT)," *Publication, PSTproducts GmbH*, 2010.

[29] R. Schäfer, P. Pasquale, and S. Kallee, "The electromagnetic pulse technology (EMPT): forming, welding, crimping and cutting," Institute of Welding bulletin, Poland, vol. 58, pp. 50-57, 2014.

[30] Z. Fan, H. Yu, and C. Li, "Plastic deformation behavior of bi-metal tubes during magnetic pulse cladding: FE analysis and experiments," *Journal of Materials Processing Technology*, vol. 229, pp. 230-243, 2016.

[31] W. H. Hayt and J. A. Buck, *Engineering electromagnetics* vol. 7: McGraw-Hill, New York, 2001.

[32] R. Raoelison, N. Buiron, M. Rachik, D. Haye, G. Franz, and M. Habak, "Study of the elaboration of a practical weldability window in magnetic pulse welding," *Journal of Materials Processing Technology*, vol. 213, pp. 1348-1354, 2013.

[33] R. N. Raoelison, T. Sapanathan, N. Buiron, and M. Rachik, "Magnetic pulse welding of Al/Al and Al/Cu metal pairs: consequences of the dissimilar combination on the interfacial behavior during the welding process," *Journal of Manufacturing Processes*, vol. 20, Part 1, pp. 112-127, 2015.

[34] R. Raoelison, N. Buiron, M. Rachik, D. Haye, and G. Franz, "Efficient welding conditions in magnetic pulse welding process," *Journal of Manufacturing Processes*, vol. 14, pp. 372-377, 2012.

[35] R. Raoelison, M. Rachik, N. Buiron, D. Haye, M. Morel, B. Dos Sanstos, *et al.*, "Assessment of gap and charging voltage influence on mechanical behaviour of joints obtained by magnetic pulse welding," in *5th International Conference on High Speed Forming*, Dortmund, 2012, pp. 207-216.

[36] A. Ben-Artzy, A. Stern, N. Frage, and V. Shribman, "Interface phenomena in aluminium–magnesium magnetic pulse welding," *Science and technology of welding and joining*, vol. 13, pp. 402-408, 2008.

[37] R. Raoelison, D. Racine, Z. Zhang, N. Buiron, D. Marceau, and M. Rachik, "Magnetic pulse welding: interface of Al/Cu joint and investigation of intermetallic formation effect on the weld features," *Journal of Manufacturing Processes*, vol. 16, pp. 427-434, 2014.

[38] E. Uhlmann and A. Ziefle, "Modelling pulse magnetic welding processes–an empirical approach," in *4th International Conference on High Speed Forming, Columbus, Ohio, USA*, 2010, pp. 108-116.

[39] A. Nassiri, G. P. Chini, and B. L. Kinsey, "arbitrary lagrangian eulerian fea method to predict wavy pattern and weldability window during magnetic pulsed welding," in *ASME 2015 International Manufacturing Science and Engineering Conference*, Charlotte, North Carolina, USA, 2015.

[40] A. A. Mousavi, S. Burley, and S. Al-Hassani, "Simulation of explosive welding using the Williamsburg equation of state to model low detonation velocity explosives," *International Journal of Impact Engineering*, vol. 31, pp. 719-734, 2005.

[41] S. A. Mousavi and S. Al-Hassani, "Finite element simulation of explosively-driven plate impact with application to explosive welding," *Materials and Design*, vol. 29, pp. 1-19, 2008.

[42] X. Wang, Y. Gu, T. Qiu, Y. Ma, D. Zhang, and H. Liu, "An experimental and numerical study of laser impact spot welding," *Materials and Design*, vol. 65, pp. 1143-1152, 2015.

[43] X. Wang, Y. Zheng, H. Liu, Z. Shen, Y. Hu, W. Li, *et al.*, "Numerical study of the mechanism of explosive/impact welding using smoothed particle hydrodynamics method," *Materials and Design*, vol. 35, pp. 210-219, 2012.

[44] S. A. A. A. Mousavi, "Numerical studies of explosive welding of three-layer cylinder composites-part 2," *Materials Science Forum*, Vol. 580, 2008, pp. 327-330.

[45] A. A. Mousavi, and S. Al-Hassani, "Numerical and experimental studies of the mechanism of the wavy interface formations in explosive/impact welding," *Journal of the Mechanics and Physics of Solids*, vol. 53, pp. 2501-2528, 2005.

[46] A. Nassiri, "Investigation of wavy interfacial morphology in magnetic pulse welding: mathematical modelling, numerical simulation and experimental tests," PhD Thesis (Doctorate), Univseristy of New Hampshire, Durham, NH, 2005.

[47] A. Oberg, J. Schweitz, and H. Olfsson, "Computer modeling of the explosive welding process," in *Proceedings of the 8th International Conference on High Energy Rate Fabrication*, San Antonio, Texas, 1984, pp. 75-84.

[48] A. Nassiri, G. Chini, A. Vivek, G. Daehn, and B. Kinsey, "Arbitrary Lagrangian–Eulerian finite element simulation and experimental investigation of wavy interfacial morphology during high velocity impact welding," *Materials and Design*, vol. 88, pp. 345-358, 2015.

[49] A. Nassiri, G. Chini, and B. Kinsey, "Spatial stability analysis of emergent wavy interfacial patterns in magnetic pulsed welding," *CIRP Annals-Manufacturing Technology*, vol. 63, pp. 245-248, 2014.

[50] S. Kiselev and V. Mali, "Numerical and experimental modeling of jet formation during a high-velocity oblique impact of metal plates," *Combustion, Explosion, and Shock Waves*, vol. 48, pp. 214-225, 2012.

[51] S. Kiselev, "Numerical simulation of wave formation in an oblique impact of plates by the method of molecular dynamics," *Journal of Applied Mechanics and Technical Physics*, vol. 53, pp. 907-917, 2012.

[52] O. Saresoja, A. Kuronen, and K. Nordlund, "Atomistic simulation of the explosion welding process," *Advanced Engineering Materials*, vol. 14, pp. 265-268, 2012.

[53] M. Liu, D. Feng, and Z. Guo, "A modified SPH method for modeling explosion and impact problems," in *APCOM & ISCM*, Singapore, 2013.

[54] S. Kakizaki, M. Watanabe, and S. Kumai, "Simulation and experimental analysis of metal jet emission and weld interface morphology in impact welding," *Materials Transactions*, vol. 52, pp. 1003-1008, 2011.

[55] D. J. Price, "Smoothed particle hydrodynamics: things I wish my mother taught me," in *Advances in Computational Astrophysics: Methods, Tools, and Outcome*, Cefalù, Italy, 2011.

[56] F. Grignon, D. Benson, K. Vecchio, and M. Meyers, "Explosive welding of aluminum to aluminum: analysis, computations and experiments," *International Journal of Impact Engineering*, vol. 30, pp. 1333-1351, 2004.

[57] W . Xu and X. Sun, "Numerical investigation of electromagnetic pulse welded interfaces between dissimilar metals," *Science and Technology of Welding and Joining*, DOI 10.1179/1362171815Y.0000000092, 2015.

[58] I. Çaldichoury and P. L'Eplattenier, EM Theory Manual, Livermore Software Technology Corporation, California, USA, 2012.

[59] T. Sapanathan, K. Yang, R. Raoelison, N. Buiron, D. Jouaffre, and M. Rachik, "Effect of conductivity of the inner rod on the collision conditions during a magnetic pulse welding process, " in 7th International Conference on High Speed Forming, Dortmund, DOI 10.17877/DE290R-16981, 2016.